7th International Conference on Acoustic and Radio EeV Neutrino Detection Activities (ARENA 2016)

EPJ Web of Conferences Volume 135 (2017)

Groningen, The Netherlands
7 - 10 June 2016

Editors:

S. Buitink
J.R. Horandel
S. de Jong
R. Lahmann
R. Nahnhauer
O. Scholten

ISBN: 978-1-5108-3783-6

Printed from e-media with permission by:

Curran Associates, Inc.
57 Morehouse Lane
Red Hook, NY 12571

Some format issues inherent in the e-media version may also appear in this print version.

Printed by Curran Associates, Inc. (2017)

For additional information, please contact EDP Sciences – Web of Conferences
at the address below.

EDP Sciences – Web of Conferences
17, Avenue du Hoggar
Parc d'Activité de Courtabœuf
BP 112
F-91944 Les Ulis Cedex A
France

Phone: +33 (0) 1 69 18 75 75
Fax: +33 (0) 1 69 28 84 91

contact-edps@webofconferences.org

Additional copies of this publication are available from:

Curran Associates, Inc.
57 Morehouse Lane
Red Hook, NY 12571 USA
Phone: 845-758-0400
Fax: 845-758-2633
Email: curran@proceedings.com
Web: www.proceedings.com

TABLE OF CONTENTS

List of participants

Name		Affiliation
Jaime	Alvarez-Muñiz	Universidad de Santiago de Compostela
Antonio	Bonardi	IMAPP - Radboud University Nijmegen
Justin	Bray	University of Manchester
Florian	Briechle	RWTH Aachen University
Nikolay	Budnev	Irkutsk State University
Stijn	Buitink	Vrije Universiteit Brussel (VUB)
David	Butler	Karlsruhe Institute of Technology (KIT)
Fabrizia	Canfora	Radboud University Nijmegen/Nikhef
Antonio	Capone	University "La Sapienza" and INFN, Roma
Chih-Ching	Chen	XAO, CAS / NTU
Arthur	Corstanje	IMAPP - Radboud University Nijmegen
Richard	Dallier	SUBATECH, Nantes
Sijbrand	De Jong	Radboud University Nijmegen / Nikhef
Krijn	de Vries	IIHE, Vrije Universiteit Brussel (VUB)
Cosmin	Deaconu	UChicago/KICP
Qader	Dorosti	KVI-CART, Univ. of Groningen
Florian	Gaté	SUBATECH, Nantes
Christian	Glaser	RWTH Aachen University
Kael	Hanson	University of Wisconsin - Madison
Dirk	Heinen	RWTH Aachen University, III. Physikalisches Institut
Freddie	Hendriks	KVI-CART, Univ. of Groningen
Jörg	Hörandel	Radboud University Nijmegen / Nikhef
Tim	Huege	Karlsruhe Institute of Technology (KIT)
Clancy	James	University of Erlangen-Nuremberg
Olga	Kambeitz	Karlsruhe Institute of Technology (KIT)
Dominik	Kießling	Universität Erlangen Nürnberg
Dmitriy	Kostunin	Karlsruhe Institute of Technology (KIT)
Robert	Lahmann	Friedrich-Alexander-Univ. Erlangen
Ming-Yuan	Lu	University of Wisconsin - Madison
Olivier	Martineau	LPNHE, CNRS-IN2P3, Paris
Keiichi	Mase	Chiba University
Pragati	Mitra	Vrije Universiteit Brussel (VUB)
Katharine	Mulrey	Vrije Universiteit Brussel (VUB)
Rolf	Nahnhauer	DESY, retiree
Anna	Nelles	University of California Irvine
Carl Gilbert	Pfendner	Ohio State University
Julian	Rautenberg	Bergische Universität Wuppertal
Laura	Rossetto	IMAPP - Radboud University Nijmegen
Danny	Sardian	KVI-CART, Univ. of Groningen
Olaf	Scholten	KVI-CART, Univ. of Groningen

Frank	Schröder	Karlsruhe Institute of Technology (KIT)
Francesco	Simeone	INFN, Roma 1
Sander	ter Veen	ASTRON
Rick	Timmer	KVI-CART, Univ. of Groningen
Ad	van den Berg	KVI-CART, Univ. of Groningen
Mike	van der Schaar	Universitat Politècnica de Catalunya/LAB
Lucas	van Sloten	KVI-CART, Univ. of Groningen
Arjen	van Vliet	Radboud University Nijmegen
Eli	Waxman	Weizmann Inst. of Science
Tobias	Winchen	Vrije Universiteit Brussel (VUB)
Stephanie	Wissel	Cal Poly
Enrique	Zas	IGFAE / Departamento de Física de Particulas
Sybrand	Zeinstra	KVI-CART, Univ. of Groningen
Anne	Zilles	Karlsruhe Institute of Technology (KIT)

edp sciences

EPJ Web of Conferences

Statement of Peer review

In submitting conference proceedings to EPJ Web of Conferences, I certify to the Publisher that I adhere to the **Policy on Publishing Integrity** of the journal in order to safeguard good scientific practice in publishing.

1. All articles have been subjected to peer review administered by the proceedings editors.
2. Reviews have been conducted by expert referees, who have been requested to provide unbiased and constructive comments aimed, whenever possible, at improving the work.
3. Proceedings editors have taken all reasonable steps to ensure the quality of the materials they publish and their decision to accept or reject a paper for publication has been based only on the merits of the work and the relevance to the journal.

Title, date and place of the conference

7th International Conference on Acoustic and Radio EeV Neutrino Detection Activities
ARENA-2016 , Groningen, June 7-10, 2016

Proceedings editor(s):

Stijn Buitink, Jörg R. Hörandel, Sijbrand de Jong, Robert Lahmann, Rolf Nahnhauer, and Olaf Scholten

Date and editor's signature

Stijn Buitink, 9 Feb 2017

Editorial

The seventh edition of the biennial international workshop on "Acoustic and Radio EeV Neutrino Detection Activities", ARENA 2016, was held at the Hampshire Plaza Hotel in Groningen from June 7th to 10th, 2016. The workshop encompassed the fields of acoustic and radio detection techniques for particle cascades initiated by high-energy cosmic-neutrino and cosmic-ray interactions. The latest research work was presented by about 50 speakers from the USA, Europe and Asia. There was a large attendance of PhD candidates and young postdocs which made for an enthusiastic atmosphere with lively and fruitful discussions. At the meeting some spin-offs also were presented of acoustic and radio detection methods outside the field of cosmic-particle research.

The workshop included an excursion to the LOFAR core near Exloo, the memorial center of the historic transit camp of Westerbork, the Westerbork Radio Synthesis Array, the historic Dwingeloo radio observatory and the ASTRON headquarters, providing a nice mix of cultural and more scientific activities.

 DOI: 10.1051/epjconf/201713501001

Status of air-shower measurements with sparse radio arrays

Frank G. Schröder[1,*]

[1] *Institut für Kernphysik, Karlsruhe Institute of Technology (KIT), Germany*

Abstract. This proceeding gives a summary of the current status and open questions of the radio technique for cosmic-ray air showers, assuming that the reader is already familiar with the principles. It includes recent results of selected experiments not present at this conference, e.g., LOPES and TREND. Current radio arrays like AERA or Tunka-Rex have demonstrated that areas of several km^2 can be instrumented for reasonable costs with antenna spacings of the order of 200 m. For the energy of the primary particle such sparse antenna arrays can already compete in absolute accuracy with other precise techniques, like the detection of air-fluorescence or air-Cherenkov light. With further improvements in the antenna calibration, the radio detection might become even more accurate. For the atmospheric depth of the shower maximum, X_{max}, currently only the dense array LOFAR features a precision similar to the fluorescence technique, but analysis methods for the radio measurement of X_{max} are still under development. Moreover, the combination of radio and muon measurements is expected to increase the accuracy of the mass composition, and this around-the-clock recording is not limited to clear nights as are the light-detection methods. Consequently, radio antennas will be a valuable add-on for any air shower array targeting the energy range above 100 PeV.

1 Introduction

Immense progress has been made in the last years regarding the understanding of the radio emission by air showers and regarding the development of experimental techniques. Because of this progress current radio arrays can measure air-shower observables such as the arrival direction, the energy, and the depth of the shower maximum, X_{max}, with accuracies similar to those of the established and leading optical techniques. While these classical techniques of air-Cherenkov and fluorescence light detection are limited to clear nights, radio detection is possible around the clock. Only a few percent of the total time is not usable for cosmic-ray physics, since thunderstorm clouds alter the radio signal [1, 2]. The advantages and the recent success of the radio technique are outlined in detail in longer review articles [3, 4], which also contain plenty of introduction to the field. This proceeding gives a short summary and points out some open tasks for the future development of the radio technique for air showers. On the one hand, technical improvements are necessary for using antennas arrays as stand-alone detectors, e.g., the demonstration of an efficient and pure self-trigger. On the other hand, there are aspects in which the radio technique brings intrinsic advantages over other techniques, e.g., for the detection of inclined air showers ($\theta \approx 70°$) [5, 6], or for determination of the cosmic-ray energy scale [7].

*e-mail: frank.schroeder@kit.edu

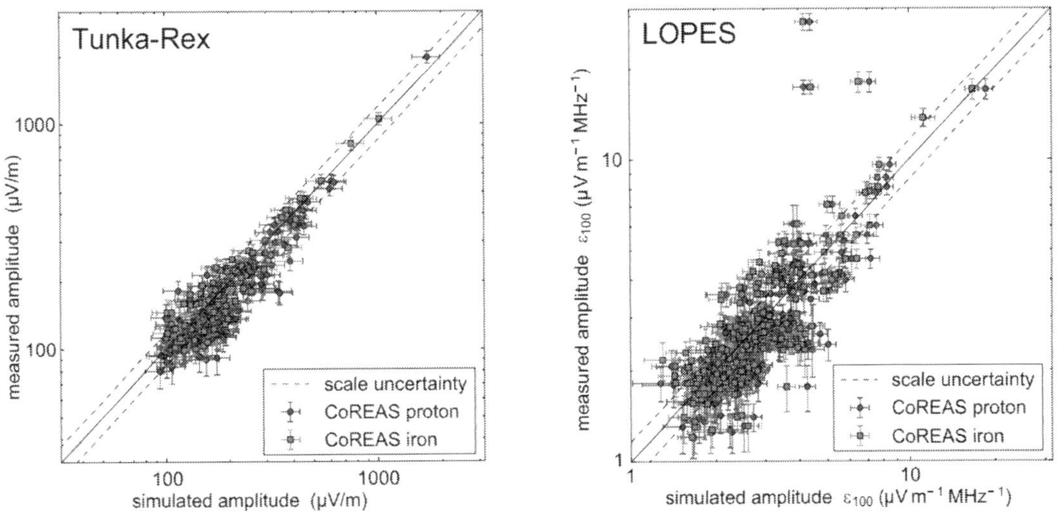

Figure 1. Comparison of radio amplitudes measured by Tunka-Rex and LOPES to CoREAS simulations of air-showers initiated by protons and iron nuclei. The energy and shower geometry was set to the values reconstructed by the host experiments Tunka-133 and KASCADE-Grande (from Ref. [18]).

2 Understanding of the radio emission

There is consensus among theorists working in the field, that the radio emission of air-showers is mostly due to the geomagnetic deflection of electrons and positrons [8–10], and to a smaller extent by the Askaryan effect due to the time-variation of the negative charge excess [11–14]. The strength of the Askaryan relative to the geomagnetic effect for a given geomagnetic angle α depends on the distance to the shower axis and on the zenith angle θ (see Refs. [4, 15] for a compilation of various measurements). There likely is a small phase shift between both emission mechanisms, and first experimental indications by LOFAR and SLAC T-510 have been discussed at this conference. Moreover, there is agreement and experimental evidence [16, 17] that the refractive index of the air affects the coherence conditions. Therefore, the emission is enhanced at the Cherenkov angle which is around 1° for air-showers leading to a Cherenkov ring of about 200 m diameter at ground for vertical showers. While at lower frequencies $\lesssim$ 100 MHz the signal fills the complete area inside the Cherenkov ring and extends beyond, at higher frequencies of several 100 MHz up to a few GHz, the emission is detectable almost exclusively at the Cherenkov angle.

This experimentally confirmed knowledge of the radio emission is implemented implicitly [19, 20] or explicitly [21, 22] in a variety of codes calculating the radio emission of simulated air showers. Until now only CoREAS has been extensively tested against measurements: absolute amplitude measurements by LOPES and Tunka-Rex are consistent inside of a 20 % scale uncertainty with CoREAS (see figure 1). The apparent disagreement of LOPES and CoREAS published earlier [23], has recently been solved by a reevaluation of the absolute calibration of LOPES [24]. Moreover, Tunka-Rex [25] and LOFAR have been [26] calibrated with exactly the same reference source as LOPES: LOPES and Tunka-Rex measurements agree within the calibration uncertainty [18], and a comparison with LOFAR measurements has still to be done. AERA features an independent absolute calibration [27], and the measured radiation energy is at least on average compatible with CoREAS and ZHAireS [7]. Finally, measurements at SLAC under controlled laboratory conditions agree with simulations, but still feature a larger systematic uncertainty of 40 % [28].

Table 1. Selected antenna arrays dedicated to air-shower detection, and the local strength of the geomagnetic field according to the international geomagnetic reference model IGRF [30]. Some experiments have been operated in additional configurations, and some are missing in the table, e.g., those located at Antarctica.

Name of experiment	Operation period	Latitude	Longitude	B_{geo} in μT	Number of antennas	Area in km^2	Band in MHz
Yakutsk [31]	since 1972	61°42' N	129°24' E	59.7	6	0.1	32
LOPES [32]	2003–2013	49°06' N	8°26' E	48.4	30	0.04	40 – 80
CODALEMA [13]	since 2003	47°23' N	2°12' E	47.6	60	1	2 – 200
TREND [33]	2009–2014	42°56' N	86°41' E	56.3	50	1.2	50 – 100
AERA [34]	since 2010	35°06' S	69°30' W	24.0	153	17	30 – 80
LOFAR [35]	since 2011	52°55' N	6°52' E	49.3	several 1000*	huge*	10 – 240
Tunka-Rex [25]	since 2012	51°49' N	103°04' E	60.4	63	1	30 – 80
SKA-low [36]	planned	26°41' S	116°38' E	55.5	60,000	1	50 – 350

* Air showers are measured by several 100 antennas within a few 100 m from the center of LOFAR.

In summary, the physics of the radio emission seems to be understood to a level of at least 20 %, and a better test is hampered by the calibration uncertainty of current experiments. The same calibration uncertainty also limits the scale accuracy of the shower energy determined from radio measurements. Currently the scale accuracy is between 10 % [29] and 20 % [24] for different experiments, where the difference primarily originates from the scale accuracies of the calibration sources claimed by the manufacturers. The calibration uncertainty also limits the scale accuracy of the shower energy determined from radio measurements. Consequently, one of the most important tasks is to improve the absolute calibrations of antennas. Then, the radio technique has a chance to provide an even more accurate measure of the absolute shower energy than the currently leading fluorescence technique, since the strength of the radio signal amplitude depends less on atmospheric conditions.

3 Experiments

Already in the 1960s a variety of analog radio experiments measured cosmic-ray air showers [8]. In the 2000s digital radio experiments started to successfully measure air-showers, and several second-generation digital arrays do so now. Apart from antenna arrays dedicated to this purpose (see table 1), also radio experiments aiming mainly at neutrino searches, such as ANITA [37] and ARIANNA [38], have detected cosmic-ray air showers. Articles on most of these experiments can be found in this issue, with a few exceptions: **LOPES** was the radio extension of the KASCADE-Grande air-shower array at the Karlsruhe Institute of Technology. Several methods have been developed by LOPES, e.g., for amplitude and time calibration [39, 40], or for the measurement of the energy [41], and position of the shower maximum [41–43]. These methods are now applied to newer experiments in more radio-quiet environments. While LOPES was stopped in 2013, its data analysis still continues, and the data are planned to be released to the public as part of the KCDC project [44]. **TREND** was a prototype experiment in Tianshan, China. It has shown that self-triggering on the radio signal enables the detection of air showers in this radio-quiet environment [33]. Its successor, GRANDproto, will examine the self-triggering efficiency and purity in more detail, especially for inclined showers, since near-horizontal showers are an interesting target for neutrino detection. **TAROGE** is an antenna tower on top of a mountain in Taiwan [45]. The principle is similar to ANITA except that the antennas are not hanging on a balloon [37]. Observing the ocean, TAROGE aims mainly at the detection of near-horizontal air showers before and after reflection on the water.

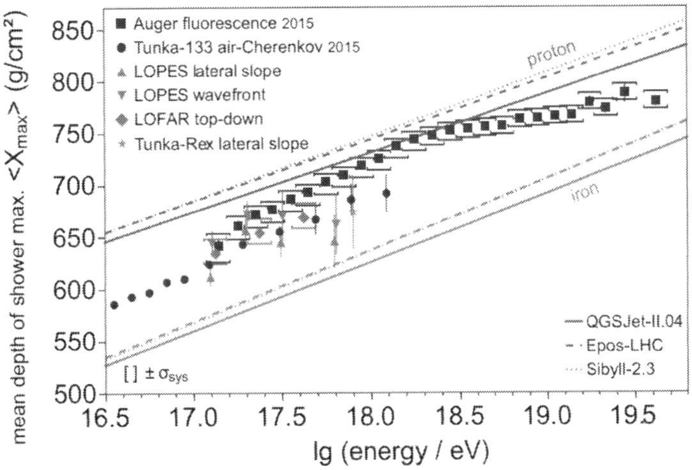

Figure 2. Mean $X_{\max}$ values measured by radio arrays using different analysis methods, and for reference measurements by the Auger fluorescence detectors and the Tunka-133 air-Cherenkov detectors, and predictions by CORSIKA simulations using various interaction models (from Ref.[4], values from Refs. [15, 32, 35, 46–48]).

4 Applications of the radio technique for air showers

Ultra-dense arrays, such as LOFAR and the future SKA [36], feature antenna spacings which partly are of the order of a wavelength and can measure the radio signal in great detail. Thus, their resolution on the main shower observables (direction, energy, $X_{\max}$) is not limited by background, but by various systematic uncertainties. Including these systematic uncertainties LOFAR already now has a resolution similar to the leading air-fluorescence technique [35], but with less exposure than the Pierre Auger Observatory [47, 49] or the Telescope Array [50] in the energy range between 10^{17} and 10^{18} eV (see figure 2). Although the SKA will be able to compete also in exposure, the real advantage of ultra-dense arrays is that their precision can be even better than that achieved by the optical techniques. Nevertheless, methods exploiting mass-sensitive radio observables going beyond a simple analysis of $X_{\max}$ distributions still have to be developed.
Radio extensions of existing air-shower arrays, such as AERA or Tunka-Rex, can enhance the total accuracy by providing an independent measurement of the electromagnetic shower component for relatively little additional costs. This use-case of radio extension is the one most advanced, with dedicated software publicly available [51]. The co-located particle detector array can provide a fully efficient trigger, which reduces systematic uncertainties due to the dependencies of the radio detection efficiency on the arrival direction and on the time-varying background level. It has been shown that even with detector spacings of about 200 m, radio measurements of the absolute shower energy are precise and accurate to about 15 – 20 % [52, 53], which is similar to other detection techniques. The $X_{\max}$ precision currently achieved with AERA and Tunka-Rex is of the order of 40 g/cm^2 [53, 54], i.e., twice worse than that of air-fluorescence measurements. However, reconstruction methods of $X_{\max}$ are still under development, and the precision likely can be improved by combining complementary radio observables [54], e.g., the footprint [55], the wavefront [43], or the frequency spectrum [56]. Furthermore, a combined analysis of the radio signal with muon measurements ought to bring additional mass-sensitivity for all zenith angles [57]. This is important since it is not yet clear whether the $X_{\max}$ precision will be sufficient for inclined showers.
Huge autonomous arrays larger than the Pierre Auger Observatory, such as the proposed project

GRAND [58], might provide a way for enhancing the world aperture on the highest energy cosmic rays and additionally for searching neutrinos above 10^{19} eV. Since the radio footprint extends over several km^2 for inclined air showers [6], detector spacings of 1 – 2 km would allow for reasonable total costs, i.e., below the costs of typical space experiments. CODALEMA [59], TREND [33], AERA [60], and ARIANNA [38] have shown that autonomous radio detection is feasible, but there remain some open questions: it has to be shown that the trigger can be pure and efficient enough for inclined showers, and there are technical challenges regarding infrastructure, data communication, and reliability of long-term and large-scale remote operation. Alternative ways to large apertures might be provided by observing air-showers with antennas in space [61], or by using the moon as a target [62]. However, these methods require successful proof-of-principle demonstrations, and it is not yet clear how they will compare in terms of measurement accuracy to ground-based antenna arrays.

5 Conclusion

The original dream of completely replacing air-fluorescence by radio detection might not become true: not because the radio technique would not be accurate enough, but simply because the dense antenna spacing required for full-sky coverage makes it more expensive than originally thought. Nevertheless, the radio technique likely is a suitable replacement for inclined showers, has a variety of other use cases, and will potentially become even more accurate than the optical techniques.

Acknowledgements

Thanks go to the conference organizers, to my colleagues at KIT, to the LOPES, Pierre Auger and Tunka-Rex Collaborations for fruitful discussions, and to DFG for grant Schr 1480/1-1.

References

[1] W.D. Apel, et al. - LOPES Coll., Advances in Space Research **48**, 1295 (2011)
[2] P. Schellart, et al. - LOFAR Coll., Physical Review Letters **114**, 165001 (2015)
[3] T. Huege, Physics Reports **620**, 1 (2016)
[4] F.G. Schröder, Prog. Part. Nucl. Phys. **93**, 1 (2017), arXiv:1607.08781
[5] T. Gousset et al., Astropart. Phys. **22**, 103 (2004)
[6] O. Kambeitz for the Pierre Auger Coll., EPJ Web of Conferences **this issue** (2016)
[7] Pierre Auger Coll., Physical Review Letters **116**, 241101 (2016)
[8] H. Allan, Progress in Elementary Particle and Cosmic Ray Physics **10**, 171 (1971)
[9] H. Falcke, et al. - LOPES Coll., Nature **435**, 313 (2005)
[10] D. Ardouin, et al. - CODALEMA Coll., Astropart. Phys. **31**, 192 (2009)
[11] Pierre Auger Coll., Physical Review D **89**, 052002 (2014)
[12] P. Schellart, et al. - LOFAR Coll., JCAP **10**, 014 (2014)
[13] A. Bellétoile, et al. - CODALEMA Coll., Astropart. Phys. **69**, 50 (2015)
[14] D. Kostunin et al., Astropart. Phys. **74**, 79 (2016)
[15] D. Kostunin, PhD Thesis, Karlsruhe Inst. Tech., Germany (2015), doi:10.5445/IR/1000053435
[16] R. Smida, et al. - CROME Coll., Physical Review Letters **113**, 221101 (2013)
[17] A. Nelles et al. -LOFAR Coll., Astropart. Phys. **65**, 11 (2014)
[18] W.D. Apel, et al. - LOPES and Tunka-Rex Colls., Physics Letters B **763**, 179 (2016)

[19] T. Huege, M. Ludwig, C. James, AIP Conference Proceedings **1535**, 128 (2013)
[20] J. Alvarez-Muñiz et al., Astropart. Phys. **35**, 325 (2012)
[21] V. Marin, AIP Conference Proceedings **1535**, 148 (2013)
[22] K. Werner et al., Astropart. Phys. **37**, 5 (2012)
[23] W.D. Apel, et al. - LOPES Coll., Astropart. Phys. **50-52**, 76 (2013)
[24] W.D. Apel, et al. - LOPES Coll., Astropart. Phys. **75**, 72 (2016)
[25] P.A. Bezyazeekov, et al. - Tunka-Rex Coll., Nucl. Inst. Meth. A **802**, 89 (2015)
[26] A. Nelles et al., JINST **10**, P11005 (2015)
[27] Pierre Auger Coll. et al., JINST **7**, P10011 (2012)
[28] K. Belov, et al. - T-510 Collaboration, Physical Review Letters **116**, 141103 (2016)
[29] F. Briechle for the Pierre Auger Coll., EPJ Web of Conferences **this issue** (2016)
[30] Helmholtz Centre Potsdam, http://www.gfz-potsdam.de/deklinationsrechner/ (2016)
[31] S. Knurenko et al., Proceedings of Science **ICRC2015**, 255 (2015)
[32] F.G. Schröder, et al. - LOPES Coll., J. Phys.: Conf. Ser. **632**, 012102 (2015)
[33] D. Ardouin, et al. - TREND Coll., Astropart. Phys. **34**, 717 (2011)
[34] Pierre Auger Coll. et al., JINST **11**, P01018 (2016)
[35] S. Buitink, et al. - LOFAR Coll., Nature **531**, 70 (2016)
[36] T. Huege et al., Proceedings of Science **ICRC2015**, 309 (2015)
[37] S. Hoover, et al. - ANITA Coll., Physical Review Letters **105**, 151101 (2010)
[38] A. Nelles, et al. - ARIANNA Coll., EPJ Web of Conferences **this issue** (2016)
[39] S. Nehls et al., Nucl. Inst. Meth. A **589**, 350 (2008)
[40] F.G. Schröder et al., Nucl. Inst. Meth. A **615**, 277 (2010)
[41] W.D. Apel, et al. - LOPES Coll., Physical Review D **90**, 062001 (2014)
[42] W.D. Apel, et al. - LOPES Coll., Physical Review D **85**, 071101(R) (2012)
[43] W.D. Apel, et al. - LOPES Coll., JCAP **09**, 025 (2014)
[44] A. Haungs, et al. - KASCADE-Grande Coll., J. Phys.: Conf. Ser. **632**, 012011 (2015)
[45] C.-C. Chen, et al. - TAROGE Coll., Proceedings of Science **ICRC2015**, 663 (2015)
[46] V.V. Prosin, et al. - Tunka-133 Coll., EPJ Web of Conf. **99**, 04002 (2015)
[47] A. Porcelli for the Pierre Auger Coll., Proceedings of Science **ICRC2015**, 420 (2015)
[48] F. Riehn, PhD Thesis, Karlsruhe Inst. Tech., Germany (2015), doi:10.5445/IR/1000052699
[49] Pierre Auger Coll., Nucl. Inst. Meth. A **798**, 172 (2015)
[50] Z. Zundel, et al. - Telescope Array Coll., Proceedings of Science **ICRC2015**, 445 (2015)
[51] Pierre Auger Coll., Nucl. Inst. Meth. A **635**, 92 (2011)
[52] Pierre Auger Coll., Physical Review D **93**, 122005 (2016)
[53] P.A. Bezyazeekov, et al. - Tunka-Rex Coll., JCAP **01**, 052 (2016)
[54] F. Gate for the Pierre Auger Coll., EPJ Web of Conferences **this issue** (2016)
[55] S. Buitink et al. -LOFAR Coll., Phys. Rev. D **90**, 082003 (2014)
[56] S. Grebe et al. -Pierre Auger Coll., AIP Conference Proceedings **1535**, 73 (2013)
[57] E. Holt for the Pierre Auger Coll., J. Phys.: Conf. Ser. **718**, 052019 (2016)
[58] O. Martineau-Huynh, et al. - GRAND, Proceedings of Science **ICRC2015**, 1143 (2015)
[59] P. Lautridou, et al. - CODALEMA Coll., Nucl. Inst. Meth. A **692**, 65 (2012)
[60] Pierre Auger Coll., et al., JINST **7**, P11023 (2012)
[61] P. Motloch et al., Astropart. Phys. **54**, 40 (2014)
[62] J.D. Bray, Astropart. Phys. **77**, 1 (2016)

EPJ Web of Conferences **135**, 01002 (2017) DOI: 10.1051/epjconf/201713501002
ARENA 2016

Recent results from CODALEMA and the Nançay radio facilities related to cosmic-ray measurements

Richard Dallier[1,2,⋆], *Didier* Charrier[1,2], *Laurent* Denis[2], *Antony* Escudie[1], *Florian* Gaté[1], *Alain* Lecacheux[3], *Lilian* Martin[1,2], and *Benoît* Revenu[1,2]

[1] *SUBATECH - Ecole des Mines de Nantes - CNRS/IN2P3 - Université de Nantes, 4 rue Alfred Kastler, 44307 Nantes cedex 3, France*
[2] *U.S. Nançay - Observatoire de Paris, CNRS/INSU, 18330 Nançay, France*
[3] *LESIA - Observatoire de Paris, CNRS/INSU, 5 place Jules Janssen, 92190 Meudon, France*

Abstract. Since 2003, the Nançay Radio Observatory hosts the CODALEMA experiment, dedicated to radio detection of cosmic ray induced extensive air showers. CODALEMA also features the R&D EXTASIS project, aiming at detecting the low-frequency signal ([2-6] MHz) produced at the sudden disappearance of the air shower particles hitting the ground. The 3 current antenna arrays present different antenna density and extent, and can be operated in a joint mode to record simultaneously the radio signal coming from air showers. Therefore, the Nançay facilities may offer a complete description of the air shower induced electric field at small, medium and large scale, and over an unique and very wide frequency band (from ~ 2 to 200 MHz).

1 Radio detection: performances and limitations

The radio technique is now used in a flurry of experiments throughout the world, among which AERA [1] and LOFAR [2] are the best known. These experiments present different bandwidth and detector densities. AERA measures showers in [30–80] MHz; LOFAR uses either a low frequency ([30–80] MHz) or a high frequency window ([110–190] MHz) (not simultaneously). On AERA, the detection technique is tested over ~ 17 km^2, with a sparse array of autonomous radio detector stations, similar in principle to the ones hereafter described for CODALEMA-3. Contrariwise, LOFAR features a very dense core of radio antennas, externally triggered by a particle detector, over an area of ~ 0.2 km^2 allowing one to study the electric field profile of individual air showers at small distances. But either triggered by ground particle arrays or in standalone mode, radio detectors achieving EAS detection still exhibit strong internal or external limitations (parasitic radio transients, limitations on detection of highly inclined EAS for externally triggered arrays, low detection range even in the [30–80] MHz band, obvious contradiction between the high antenna density required for an optimal X_{max} reconstruction and the need to cover large areas...). Overcoming those limitations would require new instrumental developments, that are difficult to implement on currently running experiments without strong modifications. Therefore the CODALEMA-3/EXTASIS facilities at the Nançay radio observatory aim to combine a very wide bandwidth (from 2 MHz up to 200 MHz), variable antenna density

⋆e-mail: richard.dallier@subatech.in2p3.fr

and smart triggering algorithms to explore some possible solutions to each of the above problems. In this paper, we will shortly describe the Nançay experimental site facilities devoted to radio detection of UHECR, and we present some performances recently reached with these multi-scale and multi-frequency facilities.

2 The Nançay facilities and the CODALEMA-3/EXTASIS experiments

Nowadays, the CODALEMA-3/EXTASIS experiment is composed of 5 arrays of detectors. The most ancient, installed for CODALEMA-2, is the particle detector array, while CODALEMA-3 was developed later with standalone radio detectors and two new cabled arrays of antennas (the so-called "compact array" and the composite trigger cluster) aiming at building a purely radio trigger. At last, a sparse array of 7 low-frequency antennas has been installed for the EXTASIS project. Figure 1 shows the map of the current setups.

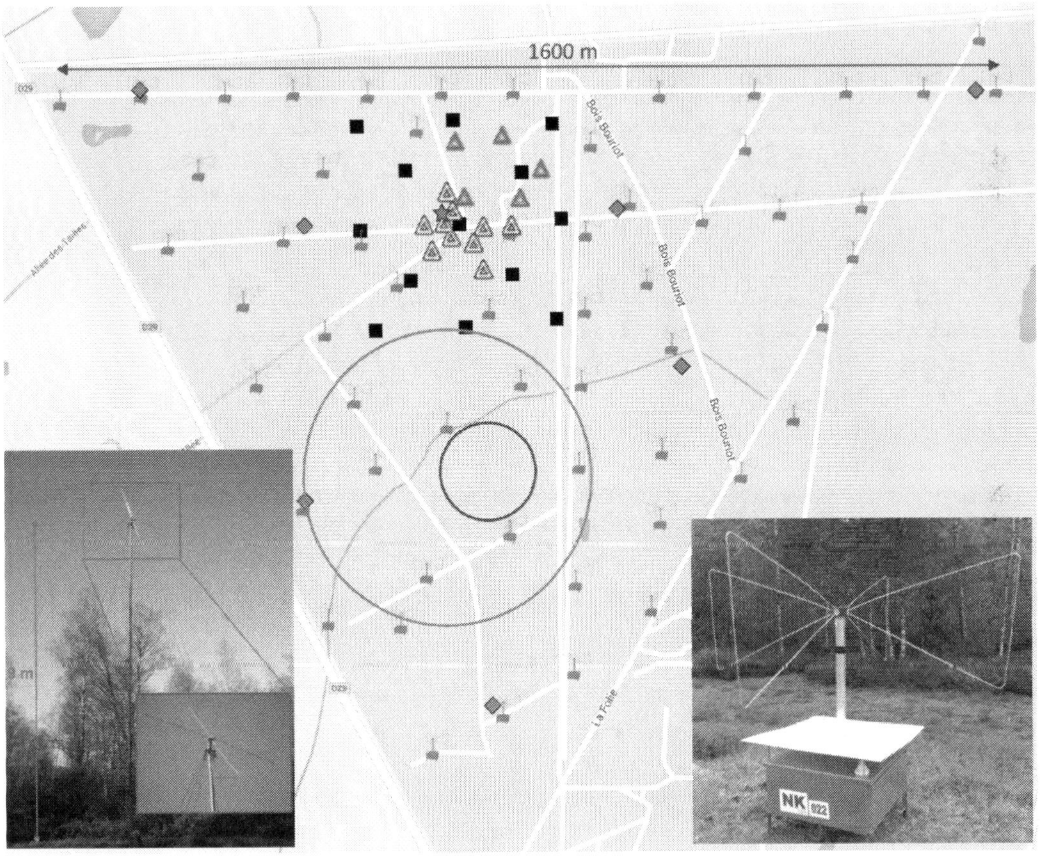

Figure 1. Map of the Nançay observatory (North on top). Red symbols: 57 CODALEMA autonomous radio detection stations; black squares: 13 scintillators; white and green triangles: 10 antennas of the compact array and 5 antenna cluster of the composite trigger; red star: "tripole" antenna; blue diamonds: 7 low-frequency antennas. Small purple circle: location of the FR606 LOFAR station; blue circle: extension of the NenuFAR radiotelescope. Bottom-right: a standalone detection station with the Butterfly antenna. The metallic box contains the electronics board crate in a shielded inner compartment. Bottom-left: a low frequency antennas on its post at 9 m height.

2.1 The Scintillator Detector Array (SD)

The goal of the particle detector (notably described in [3]) is to provide the arrival direction, the size, the core location of the atmospheric shower and an estimation of the energy of the incident cosmic ray. It also provides a master trigger, which can be broadcasted to any other system throughout the observatory, among others the compact array of antennas and a dedicated station holding the same GPS module as the radio stations described later, for time-stamping. The particle detector array is made of 13 scintillators located on a grid with a variable spacing (approximately 120 m on the diagonal and 170 m on the NS and EW directions). It covers a 340×340 m^2 area whose center roughly corresponds to the former cabled radio array center of CODALEMA-2. The particle detector geometry has been designed such that its energy threshold is located at the knee energy, and the array reaches its maximum efficiency for an energy of 10^{16} eV, with a sensitivity level of 1 particle per m^2.

2.2 The standalone antenna array

Since 2013, 57 standalone radio detection stations are installed over 1 km^2, with an average spacing of about 150 m between stations. The latter feature 2 crossed-polarization "Butterfly" antennas (also used on AERA [4]), in the North-South (NS) and East-West (EW) directions (Fig. 1). Each of the EW and NS input channels is split in two equal parts by a 3 dB splitter. One part is amplified and filtered by a wide band analog, passive filter (in the 35-80 MHz and the 150-200 MHz bands). The filtered signal is compared to a voltage threshold, remotely adjustable. If the threshold is passed on one or both of the input channels, the level 1 (T1) triggering condition is fulfilled and the event is time-stamped by a GPS board. The daily T1 yield on a single station is about 10^6. The other part of the signal (full band) continuously feeds a digitizer board, whose circular memory is frozen only when a T1 arises. A second triggering level (T2) is applied depending on the rise time of the signal [5], analyzed on-line by a local PC, and reducing the initial trigger rate by a factor of 1000, therefore getting rid of the most prominent noise sources. A third triggering level (T3) can be applied at the array scale, selecting only interesting events having hit several stations in a given time window and constituting the air shower candidates. On average about 2 events per day are finally selected as cosmic ray events in time coincidence with the particle detector ones, above a threshold of 10^{17} eV. A very large bandwidth is still available, from 20 MHz to 85 MHz and from 110 MHz to 200 MHz, thanks to the response of the Butterfly antenna and the fast digitizing system (1 GS/s). CODALEMA-3 is currently the only radio experiment able to access such a wide band at once.

The high frequencies could be of paramount interest to sample the Cerenkov ring. As shown in figure 2 from [6], close to the axis the expected Cerenkov signal in 120-200 MHz is much higher than in the band 30-80 MHz (with which it competes, giving a "double-peak" structure), and furthermore its variation with the distance to the axis is faster. Unless a high antenna density is used (LOFAR case), it is difficult to correctly sample the electric field profile in 30-80 MHz, while using the information above 120 MHz could greatly help determining the latter with only few antennas spaced by a few tens of meters (CODALEMA-3 case). This is currently studied on CODALEMA-3 events simulated with the SELFAS code [7].

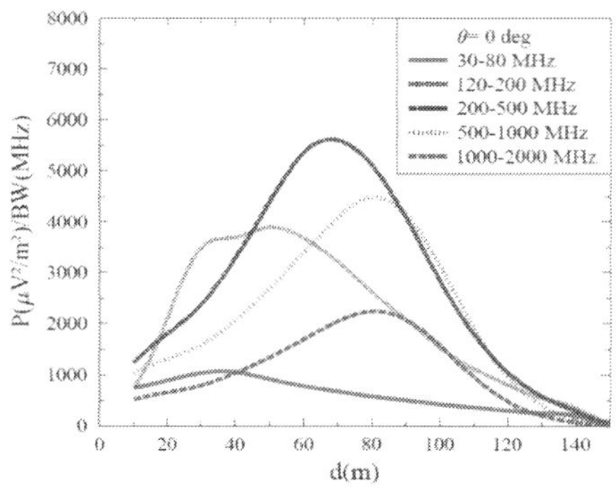

Figure 2. Electric field distribution vs distance to the shower axis for various frequency bands. Vertical proton at 10^{17} eV [6].

2.3 A vertical component measurement

A three-fold antenna system has been installed nearby the center of the particle detector to measure the complete electric field of air showers, *i.e* along 3 polarizations EW, NS and vertical. Indeed on all currently operating radio detection arrays, only the horizontal polarizations are measured, either directly along NS and EW directions or later projected on these axes. This allows reconstructing the vertical electric field component provided that the far-field assumption is valid, but although strong hints based on the theories of air shower radio emission tend to support this hypothesis (the main component should be the transverse current due to the separation of charges in the geomagnetic field), the latter has never been experimentally verified. Several attempts for measuring directly the vertical component of the E-field have been made but no convincing results have been shown, probably due to the low performances of the antennas used by now. We chose to reconsider the problem and to use 3 regular Butterfly antennas equipped with the LONAMOS LNA, rotated twice around the NS and vertical axes (Fig. 3). Triggered by the particle detector, the 3 signals are recorded with the same type of electronics as for the standalone stations. We have now to properly deconvolve it together with the regular autonomous station signals recorded at the same time, to derive conclusions about the initial hypothesis.

Figure 3. The tripole antenna.

2.4 A Compact Array as a composite triggering system

Two majors problems occur with a sparse array of antennas or standalone stations in the band 30-80 MHz: the detection range is quite small (few hundreds of meters) and the trigger selectivity is low in standalone mode. To study possible solutions to both these problems, a cluster of 10 antennas in dual linear horizontal polarisation has been installed, arranged in star-shape of 150 m width (24 to 146 m spacing). This cluster can operate in two observing modes:

- SD triggered, 20 channels recorded, 6 μs snapshots, 400 MHz ADC, bandwidth [10-200] MHz;
- self detecting mode (development in progress), with a continuous sampling of (up to) 8 channels, in circular polarisation (linear output added in quadrature), with an ADC at 100 MHz, a bandwidth of 50 MHz (2nd Nyquist band) and real-time software (using NVIDIA K20 GPU at 5 Tflops).

The SD triggered mode is used to test, off-line, the reconstruction algorithms intended to be implemented on-line in self-detecting mode. About 10^3 radio events from 6.10^5 SD triggers have been reconstructed so far, allowing one to finely tune the reconstruction algorithms. Notably, the beam-forming method helps avoiding triggering on natural and man made transient noise (*i.e.* lightnings, electric/electronic devices...) and moreover increases the sensitivity of antennas by a factor of $\sqrt{N}$, N being the number of antennas in the cluster. We are currently developing an on-line detection scheme based on continuous sampling (F_S=100 MHz, bandwidth 50 MHz) of selected antenna output, stored in a ring buffer. A pre-conditioning is made by Fourier transforming each channel, synthesizing a circularly polarized channel, then computing the signal envelop for each successive time-window corresponding to the array time aperture (500 ns). It generates ~2000 beams on the sky (~ 2° apart) via beamforming, in which a search for beam(s) containing signal above some intensity threshold is conducted. Additional criteria (*e.g.* rise time, known RFI direction, polarization) may be used to increase the selectivity of the online trigger. This should allow deriving the size and the spacing between the antenna clusters needed to cover a very wide area in the case of a further giant observatory based on radio detection.

2.5 EXTASIS and the low-frequency components

The coherent emission due to the variation of the transverse current in the shower, also known as "geomagnetic effect", is expected to be the dominant effect in the global emission of the charged particles of the shower during its development in the atmosphere. A secondary process, related to the time-varying charge excess, has also been observed (see for example [8]). All the simulation codes show that the maximum of emission due to these mechanisms peaks at a frequency of ~10 MHz, whatever the energy and the geometry of the shower are. Unfortunately, man-made broadcasting at low and medium frequencies (5 to 25 MHz) generate strong AM lines, that are visible all around the Earth. For that reason, all the recent experiments of radio detection of UHECR use a rather "clean" band located between AM and FM lines, *i.e* ~[30-80] MHz. Although several detections at low and very low frequencies have been made up to the 90's (see [9] and references therein) and have shown that the signal has a higher amplitude at low frequencies (below 10 MHz) than in the range [20-80] MHz, the various mechanisms proposed to explain this were not satisfactory. We therefore chose to reinvestigate low frequency measurements, by adding dedicated antennas in Nançay and by using the SELFAS simulation code to figure out what could be the emission mechanism. Especially, we will focus on the signal predicted by SELFAS. This new mechanism proposed in [9] is the coherent emission of the secondary electrons in excess in the shower front when reaching the ground level. These particles suffer a sudden deceleration when they disappear below the ground level; we proposed to call this the "sudden death mechanism", which creates the sudden death pulse (SDP). Detecting it is the goal of the EXTASIS project. 7 antennas on 9 m high poles have recently been installed within CODALEMA-3 (see figure 1). Their acquisition is triggered by the particle detector, and is made thanks to digital oscilloscopes (500 MS/s, 50 ms record, 8 bits). The very high frequency resolution allows us to clean the signal of the remaining transmitter lines in the frequency band of study (2-5 MHz). The antennas are classical Butterfly antennas with a modified LONAMOS LNA, adapted for the frequency range, with an East-West horizontal polarization and a vertical one, the SDP being expected to be mainly vertically polarized.

3 Short overview of the main performances

To illustrate the performances of the current experiments, we choose to present the signals of all CODALEMA-3/EXTASIS detectors for the same cosmic ray event in figure 4. From top to bottom and from left to right: time traces of the tripole antenna in 2 frequency bands; time traces of 1 low-frequency antenna, in both polarizations; polarization figures of the Compact Array; map of the entire array, including the scintillator (thin circles) and the standalone antennas (bold circles). For the two latter, size of the circles is proportional to the amplitude of the signal, color scale represents the arrival times on individual detectors. The center plots are the power spectrum densities recorded on the standalone antennas, with a clear contribution on some of them up to 200 MHz. On the right of the figure, the time traces of the Compact Array antennas, and in the lower right corner the polarization figures for the 12 involved autonomous stations (polarization figures are shown without any antenna angular correction).

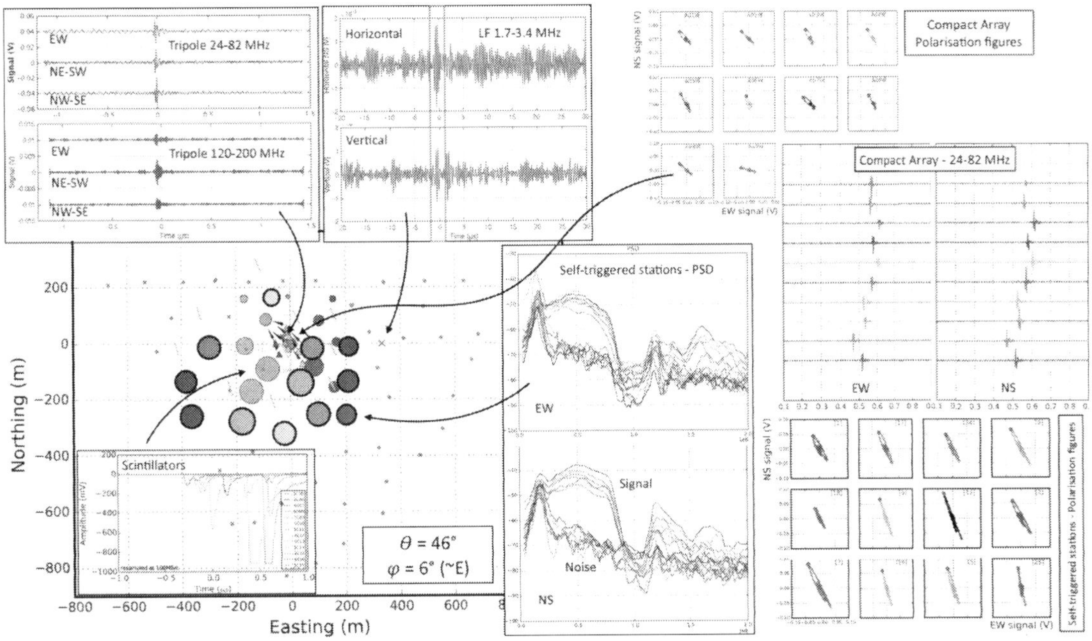

Figure 4. A hybrid event, seen by all the CODALEMA-3/EXTASIS arrays. See text for details.

4 Conclusion

The Nançay radio observatory now offers a variety of instruments dedicated to the radio detection of high energy cosmic rays. The main goal of this instrumental setup is to explore new or neglected ways of exploitation of the radio signal emitted by EAS, to fill the remaining gaps of the current detection method. Taking advantage of the very wide bandwidth of the CODALEMA-3 standalone station, we propose an alternative concept able to compensate for the large detector density currently necessary to reach a good sensitivity to the X_{max}. The EXTASIS project aims at detecting the low frequency counterpart of the regular EAS radio signal, and the sudden death pulse expected to occur at the sudden deceleration of the air shower particles at ground. Last, thanks to a new designed tripole antenna, a study of the total electric field in [20-200] MHz will be performed, in order to verify the standard far-field hypothesis.

References

[1] J. Schulz, for the Pierre Auger collaboration, 34th ICRC, The Hague, The Netherlands (2015).

[2] P. Schellart *et al.* (LOFAR Collaboration), *A&A* **560** A98 (2013).

[3] C. Rivière, PhD thesis, Université Joseph-Fourier - Grenoble I (2009). In french <tel-00526045>.

[4] The Pierre Auger Collaboration, *JINST* **7**, P10011 (2012).

[5] D. Torres-Machado *et al.* (CODALEMA Coll.), 33rd ICRC, Rio de Janeiro, Brazil (2013).

[6] K. de Vries, O. Scholten and K. Werner, *Astropart. Phys.* **45** 23 (2013).

[7] F. Gaté *et al.*, 34th ICRC, The Hague, The Netherlands (2015).

[8] A. Bellétoile *et al.*, *Astropart. Phys.* **69** 50-60 (2015).

[9] B. Revenu *et al.*, 33rd ICRC, Rio de Janeiro, Brazil (2013).

Tunka-Rex: Status, Plans, and Recent Results

F. G. Schröder[1,*], P. A. Bezyazeekov[2], N. M. Budnev[2], O. Fedorov[2], O. A. Gress[2], A. Haungs[1], R. Hiller[1], T. Huege[1], Y. Kazarina[2], M. Kleifges[3], E. E. Korosteleva[4], D. Kostunin[1], O. Krömer[3], V. Kungel[1], L. A. Kuzmichev[4], N. Lubsandorzhiev[4], R. R. Mirgazov[2], R. Monkhoev[2], E. A. Osipova[4], A. Pakhorukov[2], L. Pankov[2], V. V. Prosin[4], G. I. Rubtsov[5], R. Wischnewski[6], and A. Zagorodnikov[2] (Tunka-Rex Collaboration)

[1] *Institut für Kernphysik, Karlsruhe Institute of Technology (KIT), Karlsruhe, Germany*
[2] *Institute of Applied Physics, Irkutsk State University (ISU), Irkutsk, Russia*
[3] *Institut für Prozessdatenverarbeitung und Elektronik, Karlsruhe Institute of Technology (KIT), Germany*
[4] *Skobeltsyn Institute of Nuclear Physics, Lomonossov University (MSU), Moscow, Russia*
[5] *Institute for Nuclear Research of the Russian Academy of Sciences, Moscow, Russia*
[6] *Deutsches Elektronen-Synchrotron (DESY), Zeuthen, Germany*

Abstract. Tunka-Rex, the Tunka Radio extension at the TAIGA facility (Tunka Advanced Instrument for cosmic ray physics and Gamma Astronomy) in Siberia, has recently been expanded to a total number of 63 SALLA antennas, most of them distributed on an area of one square kilometer. In the first years of operation, Tunka-Rex was solely triggered by the co-located air-Cherenkov array Tunka-133. The correlation of the measurements by both detectors has provided direct experimental proof that radio arrays can measure the position of the shower maximum. The precision achieved so far is 40 g/cm^2, and several methodical improvements are under study. Moreover, the cross-comparison of Tunka-Rex and Tunka-133 shows that the energy reconstruction of Tunka-Rex is precise to 15 %, with a total accuracy of 20 % including the absolute energy scale. By using exactly the same calibration source for Tunka-Rex and LOPES, the energy scale of their host experiments, Tunka-133 and KASCADE-Grande, respectively, can be compared even more accurately with a remaining uncertainty of about 10 %. The main goal of Tunka-Rex for the next years is a study of the cosmic-ray mass composition in the energy range above 100 PeV: For this purpose, Tunka-Rex now is triggered also during daytime by the particle detector array Tunka-Grande featuring surface and underground scintillators for electron and muon detection.

1 Introduction

Tunka-Rex started in 2012 as radio extension of the air-Cherenkov array Tunka-133 in Siberia close to lake Baikal [1]. Its original goals have been a cross-calibration of the radio and the air-Cherenkov signal [2], and the demonstration that radio detection can indeed by economic compared to other techniques - at least when not operated as stand-alone detector, but as add on to an existing array [3]. After both original goals have been fulfilled within the first three years of operation, the new primary goal

*e-mail: frank.schroeder@kit.edu

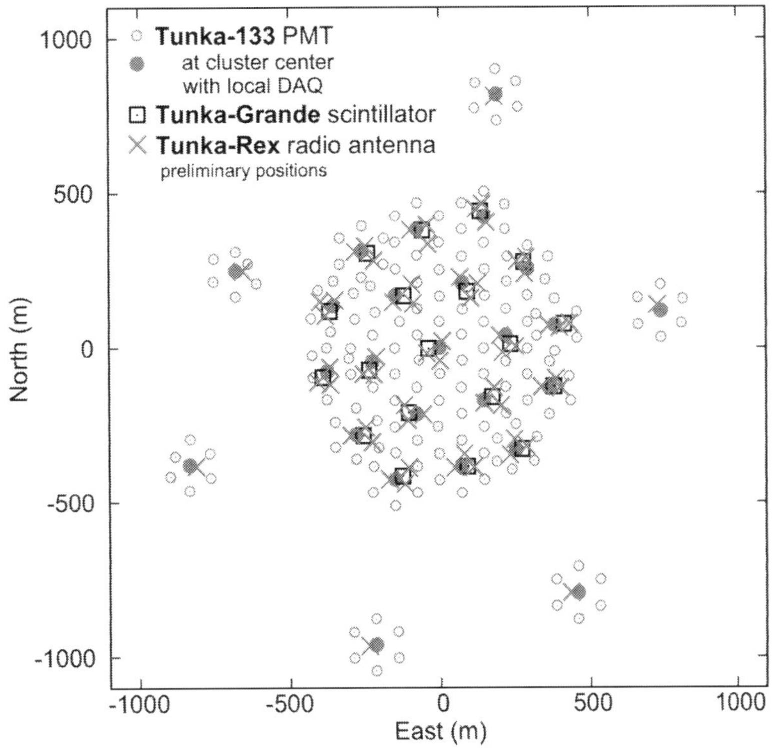

Figure 1. Map of the antenna array Tunka-Rex, the photomultiplier (PMT) array Tunka-133, and the scintillator array Tunka-Grande.

of Tunka-Rex is the determination of the mass-composition in the energy range of $10^{17} - 10^{18}$ where a transition of cosmic rays from galactic to extra-galactic origin is assumed [4]. For this goal, the reconstruction methods for the atmospheric depth of the shower maximum, X_{max}, shall be improved, and the radio measurements of Tunka-Rex shall be analyzed in combination with the co-located Tunka-Grande array of on- and under-ground scintillator detectors [5]. A secondary goal of Tunka-Rex remains the further investigation of the radio signal, e.g., by comparing the predictions of Monte Carlo simulations codes to measurements, and by testing results and methods found by other experiments.

2 Experimental setup

Since this summer Tunka-Rex consists of 63 antenna stations in a hexagonal structure, each consisting of two orthogonal SALLA antennas measuring the radio emission between 30 and 80 MHz with an effective band of 35 – 76 MHz [6]. Apart from six outer satellite stations, the antennas are clustered in triplets with antenna spacings of several 10 m inside of a triplet and about 200 m between the triplets (see figure 1). This structure is unprecedented and future analyses will show whether it lowers the detection threshold and increases the reconstruction accuracy for X_{max} compared to the first years with only one antenna per cluster (see table 1). The reason for the new non-uniform layout is mostly technical. Each Tunka-133 cluster consists of 7 photomultiplier stations for the detection of air-Cherenkov light and features its own local data-acquisition (DAQ) in the center. The first 25 Tunka-Rex antenna stations are attached to this DAQ which provides the trigger during clear nights. Moreover, close

Table 1. Stages of the Tunka-Rex experiment. Measurements take place in 'seasons' from autumn to spring and are interrupted each summer for several reasons.

Season	Number of antennas	trigger by Tunka -133	-Grande
Oct. 2012 - Apr. 2013	18	x	
Sep. 2013 - Mar. 2014	25	x	
Oct. 2014 - Apr. 2015	25	*	
Sep. 2015 - Jun. 2016	44	x	x
from autumn 2016 on	63	x	x

* Data not used due to timing problem.

Table 2. Average contributions to the time uncertainty of Tunka-Rex studied by comparing real events with CoREAS simulations before and after adding background to the simulations.

Source of uncertainty	size (ns)
Synchronization	6
Noise	5
Plane wavefront model	5
total (squared sum)	**9**

to each of the 19 inner cluster centers of Tunka-133 there is one station of the particle-detector array Tunka-Grande consisting of scintillators from the dismantled KASCADE-Grande experiment [7]. Meanwhile two antennas per cluster are attached to the DAQ of Tunka-Grande, triggering Tunka-Rex during daytime and bad weather. Despite the different local DAQs, the configuration is such that for all events all three antennas of a cluster are read out, i.e., no information is lost.

The antennas have been calibrated with the same external reference source used by LOPES [8] and LOFAR [9] with a total accuracy of 22 % dominated by a 17 % uncertainty of the absolute scale [3]. Because a large part of the scale uncertainty is correlated between the experiments, their results can be compared to each other with a relative uncertainty of only 10 %, which enables more precise cross-checks of results by the different experiments [10]. Recently, we also studied the timing accuracy in more detail, since a relative accuracy as good as a nanosecond would be required to exploit timing information for $X_{\max}$ reconstruction [11]. A previous analysis with a constant-wave reference beacon similar to the ones used at LOPES [12] or AERA [13] revealed that the relative timing is indeed stable to better than a nanosecond during single nights [14]. However, due to daily recalibration of the clock system, there are jumps between nights of several nanoseconds, and moreover the offsets between different clusters are only known to 5 ns accuracy. Furthermore, background (= internal + external noise) contributes significantly to the uncertainty of measured pulse times, such that both, the synchronization and the noise uncertainty, are comparable in size to the residuals between the real wavefront and the shower plane (see table 2). This situation is sufficient for a $1-2^\circ$ resolution on the arrival direction, but significant efforts are still necessary to make the timing of Tunka-Rex accurate enough for a measurement of the radio wavefront. Thus, the current standard methods for reconstruction of the shower energy and of $X_{\max}$ rely fully on amplitude measurements [15, 16], but do not yet exploit other observables like the pulse time.

3 Analysis results

In the first two seasons 178 events have been measured with sufficient signal-to-noise ratio in at least 3 antennas, and with the arrival directions reconstructed by Tunka-133 and Tunka-Rex in agreement [2]. For about half of the events, the absolute amplitude has been compared to CoREAS simulations [3], and LOPES measurements [10]. In both cases, Tunka-Rex measurements are in agreement within respective uncertainties of about 20 % and 10 %. Furthermore, the primary energy and $X_{\max}$ have been reconstructed from the radio measurements and compared to the air-Cherenkov measurements, again, finding agreement. The derived precision for the radio measurement is about 15 % for the shower energy, and about 40 g/cm^2 for $X_{\max}$, the latter though only for a subset of high-quality events [2]. As shown at this conference, the shower energy can even be reconstructed from a single amplitude measurements at one antenna station, when using the shower axis provided by the host experiment Tunka-133 [17]. While the accuracy is slightly worse in this case, this lowers the energy threshold and might be an interesting option for future hybrid analyses combining Tunka-Rex with muon measurements by Tunka-Grande [18].

4 Conclusion

Tunka-Rex has shown that a sparse and cost-effective radio array can compete with more expensive arrays in several aspects, in particular with respect to the precision of the shower energy. For $X_{\max}$, Tunka-Rex has brought experimental evidence that the radio technique can indeed be used to measure the distance to the shower maximum, complementing earlier evidence by LOPES [19]. However, the $X_{\max}$ of Tunka-Rex is still two times worse than that of optical detection methods or of the dense radio array LOFAR [20], but we expect that the recently increased density of the array as well as the optimization of our reconstruction methods will improve the precision of Tunka-Rex. In addition, the combination of radio measurements with the electron and muon measurements by Tunka-Grade ought to increase the total accuracy on the mass composition for the energy range about 100 PeV.

Acknowledgements

The construction of Tunka-Rex was funded by the German Helmholtz association and the Russian Foundation for Basic Research (grant HRJRG-303). Moreover, this work has been supported by the Helmholtz Alliance for Astroparticle Physics (HAP), by Deutsche Forschungsgemeinschaft (DFG) grant SCHR 1480/1-1, and by the Russian grant RSF 15-12-20022.

References

[1] V.V. Prosin, et al. - Tunka-133 Coll., Nucl. Inst. Meth. A **756**, 94 (2014)
[2] P.A. Bezyazeekov, et al. - Tunka-Rex Coll., JCAP **01**, 052 (2016)
[3] P.A. Bezyazeekov, et al. - Tunka-Rex Coll., Nucl. Inst. Meth. A **802**, 89 (2015)
[4] W.D. Apel, et al. - KASCADE-Grande Coll., Physical Review D **87**, 081101(R) (2013)
[5] N. M. Budnev, et al. - TAIGA Coll., JINST **9**, C09021 (2014)
[6] Pierre Auger Coll. et al., JINST **7**, P10011 (2012)
[7] W.D. Apel, et al. - KASCADE-Grande Coll., Nucl. Inst. Meth. A **620**, 202 (2010)
[8] W.D. Apel, et al. - LOPES Coll., Astropart. Phys. **75**, 72 (2016)
[9] A. Nelles et al., JINST **10**, P11005 (2015)
[10] W.D. Apel, et al. - LOPES and Tunka-Rex Colls., Physics Letters B **763**, 179 (2016)
[11] W.D. Apel, et al. - LOPES Coll., JCAP **09**, 025 (2014)
[12] F.G. Schröder et al., Nucl. Inst. Meth. A **615**, 277 (2010)
[13] Pierre Auger Coll. et al., JINST **11**, P01018 (2016)
[14] F.G. Schröder, et al. Tunka-Rex Coll., Nucl. Inst. Meth. A **824**, 652 (2016)
[15] D. Kostunin et al., Astropart. Phys. **74**, 79 (2016)
[16] D. Kostunin, et al. - Tunka-Rex Coll., EPJ Web of Conf. **this issue**, arXiv:1611.09127 (2016)
[17] R. Hiller, et al. - Tunka-Rex Coll., EPJ Web of Conf. **this issue**, arXiv:1611.09614 (2016)
[18] F.G. Schröder, Prog. Part. Nucl. Phys. **93**, 1 (2017), arXiv:1607.08781
[19] W.D. Apel, et al. - LOPES Coll., Physical Review D **85**, 071101(R) (2012)
[20] S. Buitink et al. -LOFAR Coll., Phys. Rev. D **90**, 082003 (2014)

DOI: 10.1051/epjconf/201713501004

Tunka-Rex: energy reconstruction with a single antenna station

R. Hiller[1,★], *P. A. Bezyazeekov*[2], *N. M. Budnev*[2], *O. Fedorov*[2], *O. A. Gress*[2], *A. Haungs*[1], *T. Huege*[1], *Y. Kazarina*[2], *M. Kleifges*[3], *E. E. Korosteleva*[4], *D. Kostunin*[1], *O. Krömer*[3], *V. Kungel*[1], *L. A. Kuzmichev*[4], *N. Lubsandorzhiev*[4], *R. R. Mirgazov*[2], *R. Monkhoev*[2], *E. A. Osipova*[4], *A. Pakhorukov*[2], *L. Pankov*[2], *V. V. Prosin*[4], *G. I. Rubtsov*[5], *F. G. Schröder*[1], *R. Wischnewski*[6], and *A. Zagorodnikov*[2] (Tunka-Rex Collaboration)

[1] *Institut für Kernphysik, Karlsruhe Institute of Technology (KIT), Karlsruhe, Germany*
[2] *Institute of Applied Physics, Irkutsk State University (ISU), Irkutsk, Russia*
[3] *Institut für Prozessdatenverarbeitung und Elektronik, Karlsruhe Institute of Technology (KIT), Germany*
[4] *Skobeltsyn Institute of Nuclear Physics, Lomonossov University (MSU), Moscow, Russia*
[5] *Institute for Nuclear Research of the Russian Academy of Sciences, Moscow, Russia*
[6] *Deutsches Elektronen-Synchrotron (DESY), Zeuthen, Germany*

Abstract. The Tunka-Radio extension (Tunka-Rex) is a radio detector for air showers in Siberia. From 2012 to 2014, Tunka-Rex operated exclusively together with its host experiment, the air-Cherenkov array Tunka-133, which provided trigger, data acquisition, and an independent air-shower reconstruction. It was shown that the air-shower energy can be reconstructed by Tunka-Rex with a precision of 15% for events with signal in at least 3 antennas, using the radio amplitude at a distance of 120 m from the shower axis as an energy estimator. Using the reconstruction from the host experiment Tunka-133 for the air-shower geometry (shower core and direction), the energy estimator can in principle already be obtained with measurements from a single antenna, close to the reference distance. We present a method for event selection and energy reconstruction, requiring only one antenna, and achieving a precision of about 20%. This method increases the effective detector area and lowers thresholds for zenith angle and energy, resulting in three times more events than in the standard reconstruction.

1 Introduction

After almost being forgotten for about 30 years, the radio detection technique resurfaced during the last decade due to the advances in digital technology [1, 2]. It has beneficial properties for the energy measurement, like low systematic uncertainties from hadronic processes in the air-shower development, and the possibility to measure the signal on an absolute scale [3]. In recent years, sophisticated methods were developed by the latest generation of radio experiments to optimize the energy reconstruction [4–7]. One of the current radio experiments is Tunka-Rex, the radio extension of Tunka-133 [8–10]. Tunka-Rex is devoted to demonstrating the competitiveness of an economic radio

★e-mail: roman.hiller@kit.edu

detector. In this work we present a novel method for reconstructing the air-shower energy with a single antenna station. This optimizes the efficiency for energy reconstruction and allows for a minimal density of radio antennas. Details on the overview given here can be found inf Ref. [11].

2 Single antenna energy reconstruction

The Tunka-Rex energy estimator S_{120} represents the reconstructed electric field, corrected for azimuthal asymmetry and geomagnetic angle, at a distance of 120 m from the shower axis [5]. If there is an antenna close to this distance, S_{120} can be estimated by this single measurement already. The systematic uncertainty on this measurement certainly increases with higher distance from the shower axis, and a question answered here is at which distance this uncertainty becomes the limiting factor. Using a single antenna to determine the energy of an air shower minimizes the threshold on energy and zenith angle and increases the effective detector area, maximizing the capabilities of the radio part of a hybrid detector for air showers.

As a single antenna has no sensitivity to fine features of the lateral distribution function (LDF) of radio amplitudes, we use the simplest model for the LDF, the exponential function, to extrapolate the signal in a single antenna to a distance of 120 m from the shower axis

$$S_{120}^{1\mathrm{A}} = S_{\mathrm{meas}}^{1\mathrm{A}} \cdot \exp(\eta \cdot (r - 120\,\mathrm{m})). \tag{1}$$

η is the slope and $S_{\mathrm{meas}}^{1\mathrm{A}}$ is the amplitude of the reconstructed electric field, corrected for azimuthal asymmetry, of the antenna at distance r from the shower axis. Although being the simplest among the common models for the LDF, it was demonstrated that a resolution of 15%-20% can be reached with this model, when taking the average azimuthal asymmetry of the signal into account [11]. No other, more complicated model so far shows significantly better resolution [4, 5, 7].

The shower core and direction, which are also required to apply Eq. 1, are reconstructed from the other component of the hybrid detector, which for Tunka-Rex is Tunka-133. For the slope η, we use the following parametrization with the zenith angle θ and $\eta_0 = 8.2 \cdot 10^{-3}\,\mathrm{m}^{-1}$:

$$\eta = \eta_0 \cdot \cos\theta, \tag{2}$$

where η varies from event to event about $\sigma_\eta = 3 \cdot 10^{-3}\,\mathrm{m}^{-1}$ in Tunka-Rex, which does not limit the accuracy of the energy determination with a single antenna for most events.

In case multiple antennas with signal are available in an event, for consistency, only the one with the smallest estimated uncertainty is used, where with $\Delta r = r - 120\,\mathrm{m}$ the uncertainty is estimated to

$$\sigma_{\mathrm{tot}} = \exp(\eta \cdot \Delta r) \cdot \sqrt{\sigma_{\mathrm{m}}^2 + \left(S_{\mathrm{meas}}^{1\mathrm{A}} \cdot \Delta r \cdot \sigma_\eta\right)^2}, \tag{3}$$

The two contribution in the uncertainty arise due to noise (σ_{m}) and shower-to-shower fluctuations of the slope and both increase with higher extrapolated distances.

While reducing the minimum number of antennas from 3 to 1 lowers the threshold for air-shower detection, it also increases the risk of false positive events due to background. There are several possibilities to circumvent this, starting by simply increasing the signal-to-noise threshold. However, as the analysis is tailored for radio detectors in hybrid mode anyway, we decided to use the full information available from Tunka-133. Therefore, a model for the footprint of radio amplitudes was developed, and for each Tunka-133 event the distance r_{th} from the shower axis is calculated, at which the radio amplitude is expected to pass the detection threshold. For a detailed description of this model we refer to Ref.[11] and give only the formula here:

$$r_{\text{th}}(E, \theta, \alpha) = r_0 + \frac{1}{\eta_0 \cos\theta} \ln \frac{E \cdot \sqrt{\sin^2\alpha + \varepsilon^2}}{\kappa S_{\text{th}}} \tag{4}$$

α is the geomagnetic angle and the remaining parameters and their meaning are given in Tab. 1. Only antennas within a distance from the shower axis smaller than r_{th} are considered for energy reconstruction. The energy, geomagnetic angle and zenith angle needed to calculate r_{th} are again determined by Tunka-133. However, unlike the shower geometry, the energy reconstruction of Tunka-133 is exclusively used for the purpose of selecting events (and antennas) entering the radio analysis.

Table 1. Parameters for the footprint of the radio signal above threshold from Eq. 4 for Tunka-Rex. They parameters were partly measured and partly determined from CoREAS simulations [11].

Parameter	Value	Origin
r_0 reference distance	120 m	Simulation
η_0 LDF slope	$8.2 \cdot 10^{-3}\,\text{m}^{-1}$	Measurement
ε charge-excess contribution	0.085	Simulation
κ radio signal strength	$884\,\frac{\text{EeV}}{\text{V/m}}$	Simulation
S_{th} detection threshold	$90\,\mu\text{V/m}$	Measurement

3 Performance on data of first two seasons

To test the performance of the single antenna energy reconstruction we use data from the first and second season of Tunka-Rex (Oct. 2012 - April 2014) and look at events with energies reconstructed by Tunka-133 above $10^{16.5}$ eV and zenith angles below 50°. After removing background from the trace, the electric field is reconstructed using the incoming direction reconstructed by Tunka-133. Then, the signal amplitude S is determined as the peak of a Hilbert envelope in the signal window and the noise N as the RMS of a noise window shortly before the signal window [8]. Only antennas with a signal-to-noise ratio (SNR = S^2/N^2) above 10 and within r_{th} are further considered, i.e., both the measured SNR and the SNR expected from the footprint model are required to be above 10.

649 events pass these requirements, more than three times that in the standard analysis, requiring three antennas with signal [8]. In Fig. 1 (left), the distribution of shower cores of this selection and of the standard reconstruction are shown. The highest gains in event number are from the outer clusters, where the antenna density of Tunka-Rex decreases dramatically. Furthermore, the energy threshold decreases from $10^{16.9}$ eV to $10^{16.7}$ eV and many events with zenith angles below 30° pass the cuts, which in the standard analysis only happens for the highest energies. Thus, dropping the requirement for 3 antennas increases the event number at low and high energies, by increasing the efficiency and effective detector area.

In Fig. 1 (right), the single-antenna estimator for the energy $S_{120}^{1\text{A}}$ is shown versus the Tunka-133 energy. While $S_{120}^{1\text{A}}$ shows a clear correlation with the energy, there are a few outliers, almost exclusively from antennas further than 200 m from the shower axis, where systematic uncertainties from the LDF parametrization start to dominate. Removing these events, 606 events are left. The resolution obtained from the spread of the relative difference between the two energy estimators is 25%. Assuming no correlation between them and 15% uncertainty for the energy from Tunka-133 [12], a resolution of 20% is obtained for $S_{120}^{1\text{A}}$.

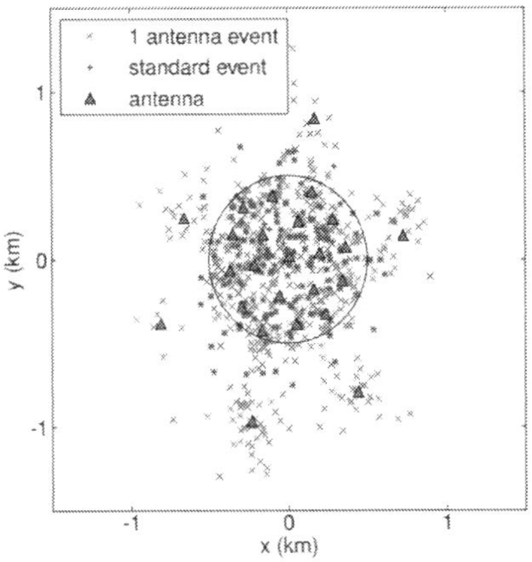

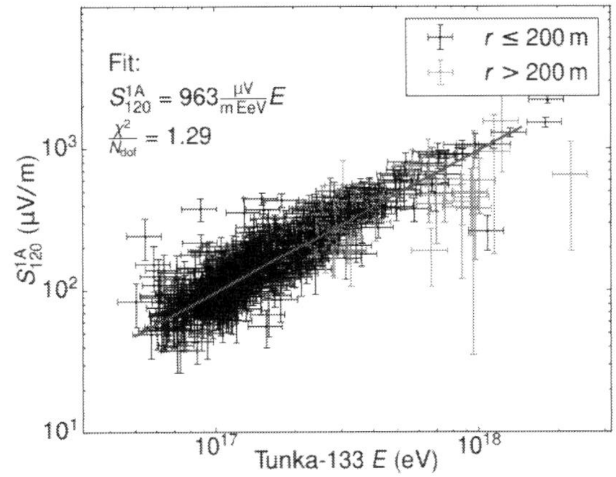

Figure 1. Left: Core distribution of events from standard and single-antenna analyses. Events of the latter extend beyond the borders of the inner, dense array. **Right**: The energy estimator of the singe-antenna analysis S^{1A}_{120} versus Tunka-133 energy. Events far from the shower axis suffer from high systematic uncertainties due to the extrapolation of the LDF.

4 Conclusion

We presented a method for the reconstruction of the air-shower energy with the radio part of a hybrid detector using data from a single antenna station only. With Tunka-Rex an energy resolution of 20% can be reached with this method, obtaining more than 3 times more events than with the standard selection requiring three antennas. Minimizing the required antenna density, this approach achieves the maximum efficiency possible with an economic radio extension, reaching down to the limit of amplitude based methods: the sensitivity of the individual antennas.

Acknowledgments

The construction of Tunka-Rex was funded by the German Helmholtz association and the Russian Foundation for Basic Research (grant HRJRG-303). Moreover, this work has been supported by the Helmholtz Alliance for Astroparticle Physics (HAP), by Deutsche Forschungsgemeinschaft (DFG) grant SCHR 1480/1-1, and by the Russian grant RSF 15-12-20022.

References

[1] F.G. Schröder, Prog. Part. Nucl. Phys. **93**, 1 (2017)
[2] T. Huege, Physics Reports **620**, 1 (2016)
[3] W.D. Apel et al. (Tunka-Rex, LOPES), Phys. Lett. **B763**, 179 (2016), 1610.08343
[4] W.D. Apel, et al. - LOPES Coll., Physical Review D **90**, 062001 (2014)
[5] P.A. Bezyazeekov, et al. - Tunka-Rex Coll., JCAP **01**, 052 (2016)
[6] D. Kostunin et al., Astropart. Phys. **74**, 79 (2016)
[7] Pierre Auger Coll., Physical Review Letters **116**, 241101 (2016)
[8] P.A. Bezyazeekov, et al. - Tunka-Rex Coll., Nucl. Inst. Meth. A **802**, 89 (2015)
[9] F.G. Schröder, et al. - Tunka-Rex Coll., EPJ Web of Conf. **this issue** (2016), arXiv:1611.09604
[10] D. Kostunin, et al. - Tunka-Rex Coll., EPJ Web of Conf. **this issue** (2016), arXiv:1611.09127
[11] R. Hiller, PhD Thesis, Karlsruhe Inst. Tech., Germany (2016), doi:10.5445/IR/1000053456
[12] V.V. Prosin, et al. - Tunka-133 Coll., Nucl. Inst. Meth. A **756**, 94 (2014)

Towards a cosmic-ray mass-composition study at Tunka Radio Extension

D. Kostunin[1,⋆], *P. A.* Bezyazeekov[2], *N. M.* Budnev[2], *O.* Fedorov[2], *O. A.* Gress[2], *A.* Haungs[1], *R.* Hiller[1], *T.* Huege[1], *Y.* Kazarina[2], *M.* Kleifges[3], *E. E.* Korosteleva[4], *O.* Krömer[3], *V.* Kungel[1], *L. A.* Kuzmichev[4], *N.* Lubsandorzhiev[4], *R. R.* Mirgazov[2], *R.* Monkhoev[2], *E. A.* Osipova[4], *A.* Pakhorukov[2], *L.* Pankov[2], *V. V.* Prosin[4], *G. I.* Rubtsov[5], *F. G.* Schröder[1], *R.* Wischnewski[6], and *A.* Zagorodnikov[2] (Tunka-Rex Collaboration)

[1] *Institut für Kernphysik, Karlsruhe Institute of Technology (KIT), Karlsruhe, Germany*
[2] *Institute of Applied Physics, Irkutsk State University (ISU), Irkutsk, Russia*
[3] *Institut für Prozessdatenverarbeitung und Elektronik, Karlsruhe Institute of Technology (KIT), Germany*
[4] *Skobeltsyn Institute of Nuclear Physics, Lomonossov University (MSU), Moscow, Russia*
[5] *Institute for Nuclear Research of the Russian Academy of Sciences, Moscow, Russia*
[6] *Deutsches Elektronen-Synchrotron (DESY), Zeuthen, Germany*

Abstract. The Tunka Radio Extension (Tunka-Rex) is a radio detector at the TAIGA facility located in Siberia nearby the southern tip of Lake Baikal. Tunka-Rex measures air-showers induced by high-energy cosmic rays, in particular, the lateral distribution of the radio pulses. The depth of the air-shower maximum, statistically depends on the mass of the primary particle, is determined from the slope of the lateral distribution function (LDF). Using a model-independent approach, we have studied possible features of the one-dimensional slope method and tried to find improvements for the reconstruction of primary mass. To study the systematic uncertainties given by different primary particles, we have performed simulations using the CONEX and CoREAS software packages of the recently released CORSIKA v7.5 including the modern high-energy hadronic models QGSJet-II.04 and EPOS-LHC. The simulations have shown that the largest systematic uncertainty in the energy deposit is due to the unknown primary particle. Finally, we studied the relation between the polarization and the asymmetry of the LDF.

1 Introduction

In the last years the community made a large step forwards to understand the radio emission from extensive air-showers produced by high-energy (greater than 100 PeV) cosmic rays [1]. Using simulations made with the CoREAS software [2], several parameterizations describing lateral distributions of the radio amplitudes have been developed. The contributions from the two main mechanisms of the generation of the signal (geomagnetic [3], and Askaryan or charge-excess [4]) are either explicitly or implicitly included into all of these modern parameterizations. Basically, they can be grouped by experiments applying them (in order of increasing number of free parameters): Tunka-Rex (4 parameters, asymmetry ε is explicitly included into parameterization) [5]; LOFAR, and later AERA (7

⋆e-mail: dmitriy.kostunin@kit.edu

parameters, describing two gaussian functions shifted respective to each other) [6]; and the straightforward LOFAR approach (full simulation of radio footprints for each antenna position) [7].

The asymmetry caused by interference between the Askaryan effect and the geomagnetic effect has already been treated within all modern approaches, and has been measured by means of the polarization study [8, 9]. In the present paper we will discuss its impact on the lateral distribution and the connection of the asymmetry with the shower maximum.

First, we study the model-independent behavior of the radio emission, then we check the influence and uncertainties given by different hadronic models. Finally, we discuss possible improvements for the reconstruction of the shower maximum.

2 Model-independent considerations and model-dependent uncertainties

Let us study the LDF using the following assumptions: the distribution of the electrons behaves as Gaisser-Hillas function and the density of the Earth's atmosphere falls exponentially with increasing altitude, namely we use the CORSIKA parameterization of the standard atmosphere [10]. The simple form of the amplitude of the radio signal $\mathcal{E}_\nu(r > r_c)$ beyond the Cherenkov bump r_c with frequency ν at distance r from the shower axis is [11]:

$$\mathcal{E}_\nu(r > r_c) = \kappa \int\limits_{h_1^\nu(r,n_r)}^{h_2^\nu(r,n_r)} \frac{N(h)}{h} dh \propto \exp\left(-\frac{(r/r_1)^{\alpha_1}}{h_{\max}} f_{\rm int}(h_{\max}, \ldots)\right), \quad h_{1,2}^\nu(r, n_r) = \left(\frac{r}{r_{1,2}(\nu, n_r)}\right)^{\alpha_{1,2}(\nu,n_r)}, \quad (1)$$

where κ is the normalization coefficient (the dependence on geomagnetic angle α_g has already been taken into account), $N(h)$ is the number of electrons at the altitude h, and $h_{1,2}^\nu(r, n_r)$ are the integration limits depending on the distance to shower axis r and refractive index n_r. In other words, this integral describes what parts of the air-shower contribute to the signal $\mathcal{E}_\nu(r)$. The curves denoting the behavior of $h_{1,2}^\nu(r, n_r)$ are presented in Ref. [12]. One can see from Eq. (1), that the height of the shower maximum $h_{\max}$ is encoded in the slope of the LDF. Simplifying $f_{\rm int} = 1$ one still conserves the correlation between $\mathcal{E}$ and $h_{\max}$ (see Ref. [12]), what means, that the radio signal has a high sensitivity to the position of the shower maximum. On the other side, this one-dimensional slope method does not give additional information on the type of primary particle.

To study uncertainties given by hadronic interactions and shower-to-shower fluctuations, we performed simulations with the recently released CORSIKA v7.5. Both, QGSJET-II.04 and EPOS-LHC yield almost the same radio amplitude with the difference less than one given by shower-to-shower fluctuations. They both show a shift of 12% between proton and iron in the radio energy deposit, which is much larger than the effect of shower-to-shower fluctuations for these particles: 5% and 1.3% for proton and iron, respectively. This uncertainty can be reduced by taking into account information from muon detectors, in our case we plan to combine measurements of Tunka-Rex and Tunka-Grande [13, 14].

3 Hints from the charge-excess asymmetry

In the work [5] it is shown that the charge-excess asymmetry has a non-trivial dependence on the distance to the shower axis, particularly, the asymmetry features a local maximum depending on the distance to the shower maximum (see Fig. 1). This structure was obtained by analyzing the polarization of CoREAS simulations at individual antenna positions. In the present section we discuss the structure of the asymmetry by means of a completely different approach: we try to extract it by

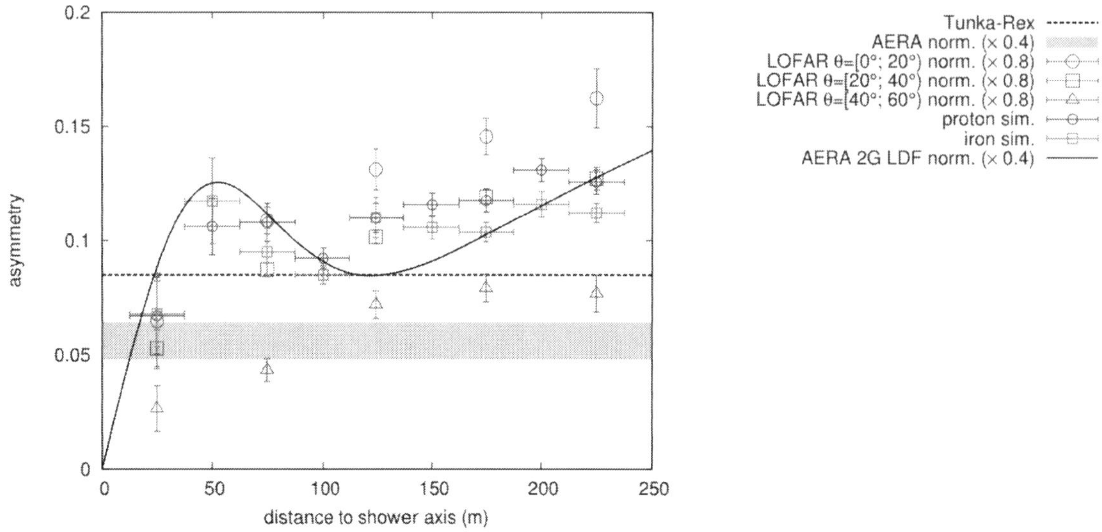

Figure 1. Askaryan asymmetry $\varepsilon(r)$ (Eq. 4) normalized to a geomagnetic field in Tunka Valley as a function of distance to the shower axis. Points indicate polarization measurements be LOFAR [9] (green) and CoREAS simulations [5] (blue and red). Black solid line indicates LDF asymmetry $\varepsilon(r)$ from Eq. (4). The blue band is the polarization measurements by AERA [8] (with uncertainties), the dashed line is the theoretical prediction for Tunka-Rex [5].

combining the Tunka-Rex [5] and AERA [15] LDF parameterizations of the amplitude $\mathcal{E}$ depending on the coordinate $\boldsymbol{r}$ in the shower plane with $\hat{e}_x$ the direction of the geomagnetic Lorentz force.

Let us write down the AERA parameterization:

$$\mathcal{E}_{2G}(\boldsymbol{r}) = A\left[\exp\left(\frac{-(\boldsymbol{r} + C_1\,\hat{\mathbf{e}}_x)^2}{\sigma^2}\right) - C_0\,\exp\left(\frac{-(\boldsymbol{r} + C_2\,\hat{\mathbf{e}}_x)^2}{(C_3 e^{C_4\sigma})^2}\right)\right], \tag{2}$$

parameters of which are described in Ref. [15], and the Tunka-Rex parameterization:

$$\mathcal{E}_\phi(r, \phi_g) = \mathcal{E}_0(r)\sqrt{\sin^2\alpha_g + \varepsilon^2(r) + 2\varepsilon(r)\cos\phi_g}\,, \tag{3}$$

with parameters described in Ref. [5]. One can see, that the asymmetry term $\varepsilon(r)$ is in Eq. (2) only implicit. The asymmetry appears along the Lorentz force, thus we define $\boldsymbol{r}_x = (r, 0)$:

$$\varepsilon(r) = \sin\alpha_g \frac{\mathcal{E}_{2G}(r) - \tilde{\mathcal{E}}_{2G}(r)}{\mathcal{E}_{2G}(r) + \tilde{\mathcal{E}}_{2G}(r)}, \quad \begin{cases} \mathcal{E}_{2G}(r) = \mathcal{E}_{2G}(\boldsymbol{r}_x) \\ \tilde{\mathcal{E}}_{2G}(r) = \mathcal{E}_{2G}(-\boldsymbol{r}_x) \end{cases}. \tag{4}$$

We take $C_0 - C_4$ from Ref. [15] for a zenith angle of $\theta = 45°$, a typical geomagnetic angle for this zenith of $\alpha_g = 45°$ and a width of the main gaussian $\sigma = (-2a_2(E,\theta))^{-1/2} = 152.32$ m, where $E = 1$ EeV, and $a_2(E,\theta)$ is defined in Ref. [16]. By this calculation, the Askaryan asymmetry of the polarization can be directly compared with the Askaryan asymmetry of the lateral distribution. The comparison is presented in Fig. 1. One can see, that both definitions are in good agreement, which leads to an interesting conclusion: the asymmetry (or charge-excess) information can be extracted from the more simple measurement of the total radio amplitude, instead of precise measurements of the components of the electrical field. Measuring the amplitude asymmetry requires higher number of stations per events, but lower signal-to-noise ratios. Moreover, as it was shown in Ref. [5], the behavior of the asymmetry is connected to the distance to shower maximum, i.e. an accurate measurement of the asymmetry by either means should be sensitive to the mass composition. Finally, the asymmetry contains information not only on the total number of the charge particles, but also on the dynamics of their creation.

4 Summary

In the present work we have studied ideas how to better reconstruct the primary mass using radio measurements. Since radio detectors feature calorimetric measurements, the main property related to mass composition is the shower maximum. We have shown with simple calculations, that the Gaisser-Hillas distribution of the particle cascade produces an exponential-like lateral distribution of the amplitudes far from the shower axis, and the position of the shower maximum is directly encoded in the slope of this exponent. However, radio is sensitive only to the electromagnetic component of air-showers, which leads to an additional uncertainty of 6% in energy estimation due to unknown primary particle.

As an alternative to the slope method we have studied other properties of the LDF. We found that the asymmetry, which is slightly sensitive to shower maximum, can be equally well determined from amplitude and polarization measurements. We now plan to introduce this study to our standard analysis of the recently upgraded Tunka-Rex array.

Acknowledgements

The construction of Tunka-Rex was funded by the German Helmholtz association and the Russian Foundation for Basic Research (grant HRJRG-303). Moreover, this work has been supported by the Helmholtz Alliance for Astroparticle Physics (HAP), by Deutsche Forschungsgemeinschaft (DFG) grant SCHR 1480/1-1, and by the Russian grant RSF 15-12-20022.

References

[1] F.G. Schröder, Prog. Part. Nucl. Phys. **93**, 1 (2017), 1607.08781
[2] T. Huege, M. Ludwig, C. James, AIP Conf.Proc. **1535**, 128 (2013), 1301.2132
[3] F.D. Kahn, I. Lerche, Proceedings of the Royal Society of London. Series A **289**, 206 (1966)
[4] G.A. Askaryan, Soviet Physics JETP **14**, 441 (1962)
[5] D. Kostunin et al., Astropart. Phys. **74**, 79 (2016)
[6] A. Nelles, S. Buitink, H. Falcke et al., Astropart. Phys. **60**, 13 (2014)
[7] S. Buitink et al., Phys. Rev. **D90**, 082003 (2014)
[8] Pierre Auger Coll., Phys. Rev. D **89**, 052002 (2014)
[9] P. Schellart et al., JCAP **1410**, 014 (2014)
[10] D. Heck et al., FZKA Report 6019, Forschungszentrum Karlsruhe (1998)
[11] H.R. Allan, Progress in Elementary Particle and Cosmic Ray Physics **Vol. 10**, 171 (1971)
[12] D. Kostunin, et al. - Tunka-Rex Coll., EPJ WoC **ISVHECRI 2016 proc.** (2016), 1701.07165
[13] P.A. Bezyazeekov, et al. - Tunka-Rex Coll., Nucl. Inst. Meth. A **802**, 89 (2015)
[14] F.G. Schröder, et al. - Tunka-Rex Coll., EPJ Web of Conferences **this issue** (2016)
[15] Pierre Auger Coll., Phys. Rev. D **93**, 122005 (2016)
[16] P.A. Bezyazeekov, et al. - Tunka-Rex Coll., JCAP **01**, 052 (2016)

Results and Perspectives of the Auger Engineering Radio Array

Christian Glaser[1,⋆] for the Pierre Auger Collaboration[2]

[1] *RWTH Aachen University, III. Physikalisches Institut A, Aachen, Germany*
[2] *Full author list: http://auger.org/archive/authors_2016_06.html*

Abstract. The Auger Engineering Radio Array (AERA) is an extension of the Pierre Auger Cosmic-Ray Observatory. It is used to detect radio emission from extensive air showers with energies beyond 10^{17} eV in the 30 - 80 MHz frequency band. After three phases of deployment, AERA now consists of more than 150 autonomous radio stations with different spacings, covering an area of about 17 km^2. It is located at the same site as other Auger low-energy detector extensions enabling combinations with various other measurement techniques. The radio array allows different technical schemes to be explored as well as cross-calibration of our measurements with the established baseline detectors of the Auger Observatory. We report on the most recent technological developments and give an overview of the experimental results obtained with AERA. In particular, we will present the measurement of the radiation energy, i.e., the amount of energy that is emitted by the air shower in the form of radio emission, and its dependence on the cosmic-ray energy by comparing with the measurement of the the well-calibrated Auger surface detector. Furthermore, we outline the relevance of this result for the absolute calibration of the energy scale of cosmic-ray observatories.

1 The Auger Engineering Radio Array

The Auger Engineering Radio Array (AERA) is part of the Pierre Auger Observatory [1] and measures extensive air showers with energies above 10^{17} eV. The Pierre Auger Observatory is the world's largest detector for ultra-high energy cosmic rays. It is located in western Argentina and covers an area of 3000 km^2. Its two baseline detectors are a surface detector (SD) array of more than 1600 water-Cherenkov detectors and 27 fluorescence telescopes (FD) that overlook that array from four different sites. In addition, underground scintillators are installed at the AERA site that allow for a separate measurement of the muonic air-shower component.

In its current stage of expansion, AERA consists of 153 autonomously operating radio detector stations covering an area of 17 km^2. Pictures of the two station designs deployed at AERA are shown in Fig. 1. Each station is solar powered and equipped with a battery to run 24 h each day. A fence is put around each station to protect it from animals living in the Argentinian Pampa and the stations are equipped with GPS receivers for accurate timing. In addition, a method was develop to calibrate the relative timing of the radio stations to better than 2 ns using a reference beacon and radio signals from commercial airplanes [2, 3].

⋆e-mail: glaser@physik.rwth-aachen.de

DOI: 10.1051/epjconf/201713501006

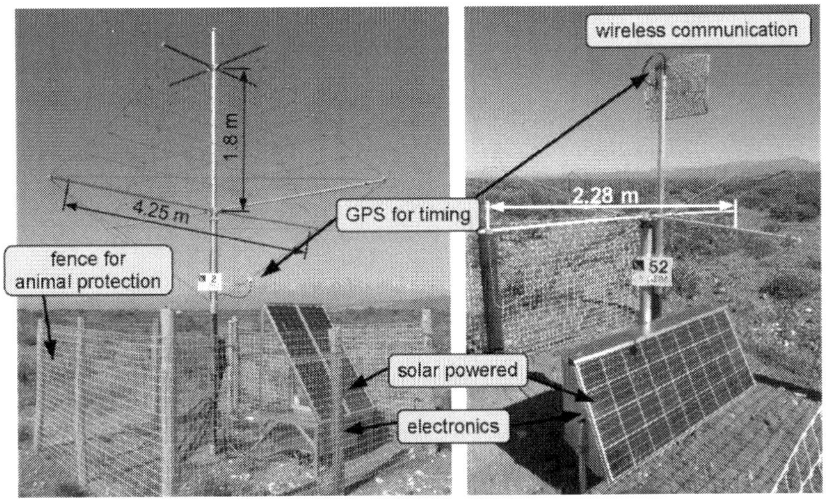

Figure 1. Pictures of the two station designs deployed at AERA, left: log-periodic dipole antenna (LPDA), right: Butterfly antenna.

The first 24 stations that were installed in 2011 are equipped with log-periodic dipole antennas (LPDAs). The antenna response to cosmic-ray signals was previously simulated and measured extensively to an accuracy of 12.5% [4, 5]. The stations are placed on a triangular grid with a distance of 144 m covering an area of 0.6 km^2 and are connected via optical fibers to the data acquisition. All other AERA stations are equipped with a Butterfly antenna and were installed in 2013 (100 additional stations on a 250 m and 375 m grid) and 2015 (29 additional stations on a 750 m grid). Instead of an optical fiber network the stations are equipped with a wireless communication system operating in the 5 GHz band. All stations operate in the band from 30 to 80 MHz and the radio signal is digitized with 180 MHz or 200 MHz depending on the type of digitizing hardware.

A map of the array is presented in Fig. 2. The AERA stations are running with slightly different types of electronics. Most of the stations (filled triangles) are equipped with a ring buffer large enough to store the measured radio signals for 7 s to be able to receive an external trigger from the SD and FD. The rest of the stations which are equipped with a Butterfly antenna contain small scintillators for trigger purposes.

2 Scientific Potential

The different antenna and electronics designs implemented in AERA and the location of AERA next to additional detectors of the Pierre Auger Observatory where air showers are measured with three independent detection techniques create a large engineering and scientific potential.

AERA can be used to explore the optimal setup to measure cosmic rays. Two different antenna designs are tested in the harsh environment of the Argentinian Pampa. The different detector spacings allow the determination of the optimal layout for different energy regimes, e.g., the 750 m spaced stations are located directly next to SD stations to test the potential of a large-scale extension of a surface particle detector with radio antennas. The different types of station hardware allow the implementation of different trigger schemes such as self-triggering directly on the radio signal using advanced signal processing in an FPGA, external triggering or an internal particle trigger using the small scintillators integrated into the station design. The latter setup is especially interesting for a possible future stand-alone radio array.

The Auger surface detector, which is calibrated with the fluorescence detector, provides an accurate measurement of the cosmic-ray energy and is used to develop and cross-calibrate the energy

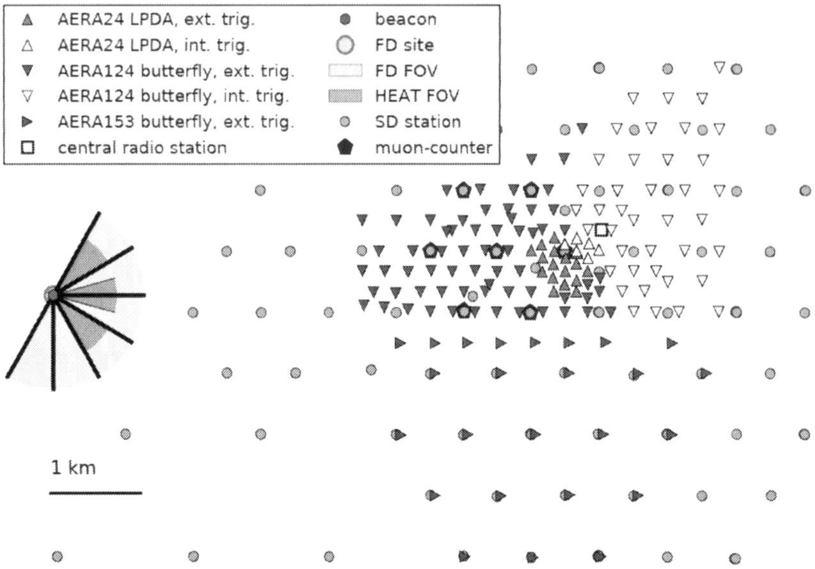

Figure 2. Map of AERA within the Pierre Auger Observatory. The radio detector stations (triangles) equipped with different antennas and digitizing hardware are surrounded by surface detector (SD) stations (gray filled circles) and underground muon counters (black pentagons). AERA is in the field of view of the Coihueco and HEAT fluorescence telescopes. Fig. adapted from [11].

measurement with AERA (see Sec. 4 or [6–8]). Furthermore, the position of the shower maximum X_{max}, which is an estimator of the primary-particle type of the cosmic ray, is directly measured with the fluorescence telescopes. Coincident measurements of AERA with FD are thus used to evaluate the performance of the different techniques used at AERA to reconstruct the cosmic-ray particle type. Different methods are developed at present to derive the cosmic-ray coposition from the radio measurements [9].

In addition to a reconstruction of X_{max} using solely radio data, the AERA measurement can be combined with the other detectors. The radio signal measures the electromagnetic component of the air shower. Hence, a coincident measurement of the muonic shower component using underground particle detectors is sensitive to the cosmic-ray particle type. In case of horizontal air showers, the electromagnetic part of the air shower is completely absorbed in the atmosphere so that the SD measures only the muonic shower component whereas AERA still measures the electromagnetic part of the shower as the atmosphere is transparent to radio waves. Hence, a combined radio and particle measurement of horizontal air showers is sensitive to the cosmic-ray particle type [10].

3 In-situ Calibration of Antenna Response

The most challenging part of the detector calibration is the absolute calibration of the antenna response to cosmic-ray radio signals. To be unaffected by near-field effects, a calibrated signal source needs to be placed at ~20 m distance at any place around the antenna. Several calibration campaigns were performed and the measurement setup was continuously improved. The first campaign used a weather balloon to lift the calibrated signal source and measured the antenna response with an accuracy of 12.5% [4, 8]. The disadvantages of this method are the high costs of each balloon and the need for a windless environment. Therefore a new method that uses a remotely piloted drone to lift the signal source was developed. This setup was completed with an optical camera system to accurately determine the position of the signal source during the flight. Using this new method an accuracy of 9.3% was achieved for the LPDA antenna [5, 12].

DOI: 10.1051/epjconf/201713501006

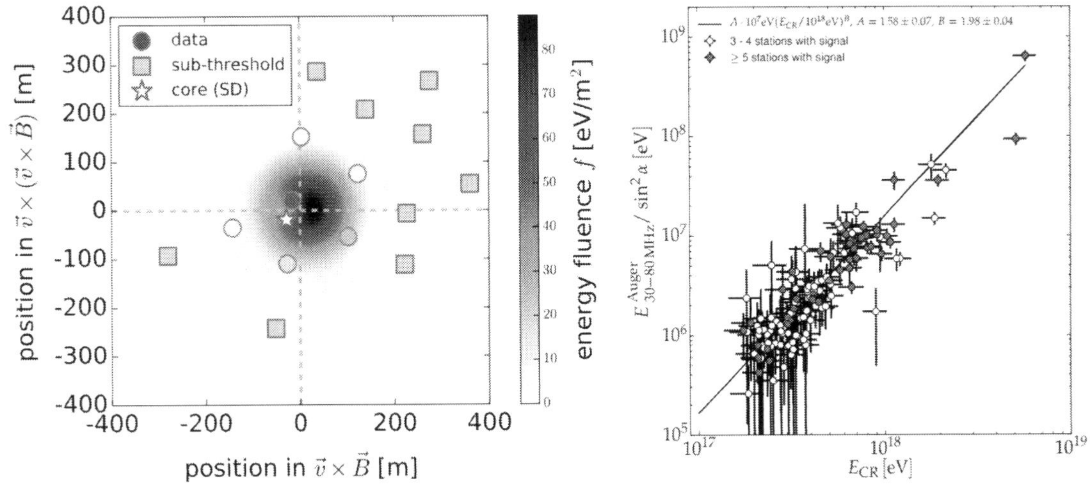

Figure 3. (left) Energy fluence for an extensive air shower with an energy of 4.4×10^{17} eV and a zenith angle of 25° as measured in individual AERA radio detectors (circles filled with color corresponding to the measured value) and fitted with the azimuthally asymmetric, two-dimensional signal distribution function (background color). The fit is performed in the plane perpendicular to the shower axis, with the x axis oriented along the direction of the Lorentz force for charged particles propagating along the shower axis **v** in the geomagnetic field **B**. The best-fitting impact point of the air shower is at the origin of the plot, slightly offset from the one reconstructed with the Auger surface detector [core (SD)]. (right) Correlation between the normalized radiation energy and the cosmic-ray energy E_{CR} as determined by the Auger surface detector. Open circles represent air showers with radio signals detected in three or four radio detectors. Solid circles denote showers with five or more detected radio signals. Figures and Caption from [7].

4 Measurement of the Radiation Energy of Extensive Air Showers

To measure the radiation energy, i.e., the amount of energy that is transferred from the cosmic ray during the air shower development into radio emission in the frequency band of AERA (30 - 80 MHz), we first unfold the detector response from the measured voltage traces. Thereby we obtain the electric field of the cosmic-ray radio signal from which we calculate the energy fluence, i.e., the energy in the radio signal per unit area, at each radio station. A two-dimensional function [8, 13] is fitted to the signal distribution and integrated to obtain the radiation energy of the air shower (cf. Fig. 3 left).

We correct the radiation energy for the different emission strengths at different geomagnetic angles α by dividing the radiation energy by $\sin^2 \alpha$. This corrected radiation energy correlates with the energy of the cosmic ray. Comparing the radiation energy with a measurement of the cosmic-ray energy by the surface detector we found that the radiation energy scales quadratically with the cosmic-ray energy, and we measured a radiation energy of $15.8 \pm 0.7(\text{stat}) \pm 6.7(\text{sys})$ MeV for a 1 EeV air shower arriving perpendicular to the geomagnetic field (cf. Fig. 3 right).

We can generalize our measurement to be applicable at any place on Earth by normalizing the radiation energy to the geomagnetic field strength. Then this corrected radiation energy is the same at any experiment and our result can be used by others to calibrate against the accurate energy scale of the Pierre Auger Observatory. More details can be found in [6–8].

5 Independent Determination of Cosmic-Ray Energy Scale from First Principles

The measurement of the radiation energy can also be used for an independent determination of the cosmic-ray energy scale. The radiation energy released by air showers can be calculated from first principles. The radio emission originates only from the well-understood electromagnetic part of the air shower and is due to (de)acceleration of charges which is described by classical electrodynamics. Such a calculation is implemented in modern Monte-Carlo simulations [14, 15] which can be used for an accurate prediction of the radiation energy [7, 16, 17].

Compared to a measurement of fluorescence light, the measurement of radiation energy has the advantage that it is hardly influenced by environmental conditions. The atmosphere is transparent to radio waves in the frequency range of AERA and also the amount of radiation energy released by an air showers is barely influenced by the current state of the atmosphere [16].

6 Conclusions

The Auger Engineering Radio Array is the world's largest radio detector for high-energy cosmic rays consisting of 153 autonomously-operating radio stations covering an area of 17 km^2. Its location within the Pierre Auger Observatory creates a large scientific potential as air showers are measured with four independent detection techniques simultaneously. The radiation energy of air showers was measured and compared with the accurate cosmic-ray energy information of the Auger surface detector. Furthermore, the radiation energy can be calculated from first principles for an independent determination of the cosmic-ray energy scale.

References

[1] A. Aab et al. (Pierre Auger Collaboration), Nucl. Instrum. Meth. A **798**, 172 (2015)
[2] A. Aab et al. (Pierre Auger Collaboration), J. Inst. **11**, P01018 (2016)
[3] T. Huege for the Pierre Auger Collaboration, this conference (2016)
[4] P. Abreu et al. (Pierre Auger Collaboration), J. Instrum. **7**, P10011 (2012)
[5] R. Krause for the Pierre Auger Collaboration, Nucl. Instrum. Meth. A, Proc. 14th, Vienna Conference on Instrumentation (VCI), Vienna, Austria (2016)
[6] C. Glaser for the Pierre Auger Collaboration, Proceedings of Science (2015), 34th ICRC, The Hague, The Netherlands
[7] A. Aab et al. (Pierre Auger Collaboration), Phys. Rev. Lett. **116**, 241101 (2016), 1605.02564
[8] A. Aab et al. (Pierre Auger Collaboration), Phys. Rev. D **93**, 122005 (2016), 1508.04267
[9] F. Gaté for the Pierre Auger Collaboration, this conference (2016)
[10] O. Kambeitz for the Pierre Auger Collaboration, this conference (2016)
[11] J. Schulz for the Pierre Auger Collaboration, Proceedings of Science (2015), 34th ICRC, The Hague, The Netherlands
[12] F. Briechle for the Pierre Auger Collaboration, this conference (2016)
[13] A. Nelles et al., Astropart. Phys. **60**, 13 (2015)
[14] J. Alvarez-Muñiz, W.R. Carvalho, E. Zas, Astropart. Phys. **35**, 325 (2012), 1107.1189
[15] T. Huege, M. Ludwig, C.W. James, AIP Conf. Proc. **1535**, 128 (2013)
[16] C. Glaser, M. Erdmann, J.R. Hörandel, T. Huege, J. Schulz, submitted to JCAP (2016), 1606.01641
[17] C. Glaser, M. Erdmann, J.R. Hörandel, T. Huege, J. Schulz, this conference (2016)

EPJ Web of Conferences **135**, 01007 (2017) DOI: 10.1051/epjconf/201713501007
ARENA 2016

X_{max} reconstruction from amplitude information with AERA

Florian Gaté[1,★] for the Pierre Auger Collaboration[2]

[1] *SUBATECH - Ecole des Mines de Nantes - CNRS/IN2P3 - Université de Nantes*
[2] *Full author list: http://auger.org/archive/authors_2016_06.html*

Abstract. The standard method to estimate the mass of a cosmic ray is the measurement of the atmospheric depth of the shower maximum (X_{max}). This depth is strongly correlated with the mass of the primary because it depends on the interaction cross section of the primary with the constituents of the atmosphere. Measuring the electric field, emitted by the secondary particles of an extensive air shower (EAS), with the Auger Engineering Radio Array (AERA) in the 30-80 MHz band allows the determination of the depth of shower maximum on the basis of the good understanding of the radio emission mechanisms. The duty cycle of radio detectors is close to 100%, making possible the statistical determination of the cosmic-ray mass composition through the study of a large number of cosmic rays above 10^{17} eV. In this contribution, X_{max} reconstruction methods based on the study of the radio signal with AERA are detailed.

1 Sensitivity of the radio emission to the nature of cosmic rays

The AERA antennas [1] record the electric field for two horizontal polarizations (North-South & East-West directions). The arrival direction of the shower is calculated from the relative arrival time of the pulses at several stations. In the far-field approximation, the vertical component is calculated form the measured horizontal projection of the electric field and the arrival directions. The maximum signal received by an antenna is calculated as the maximum of the Hilbert envelope of the three polarizations. The radio emission is characterized by the lateral distribution function (LDF) of the received electric field. It can be calculated as the maximum electric field recorded by the triggered antennas, or the energy fluence, as a function of their position with respect to the shower axis. The topology of the electric field depends on the mass of the primary, as shown in Fig. 1. Light nuclei are

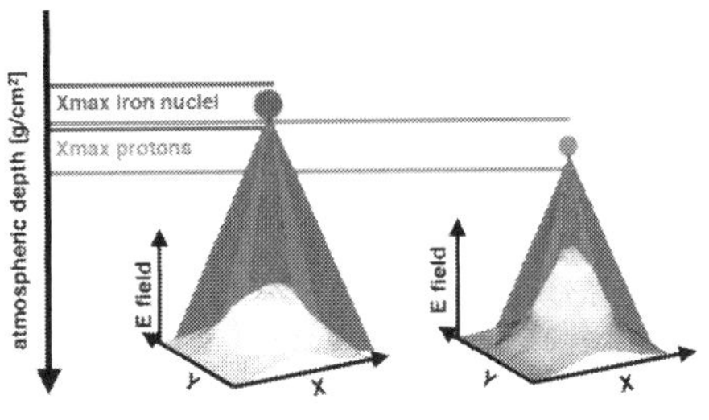

Figure 1. Dependence of the foot print of the radio emission on the mass of the primary. Two LDF are simulated with SELFAS, on the left induced by an iron nucleus, on the right induced by a proton. The vertical axis represent the atmospheric depth and the horizontal lines account for the relative X_{max} depth distributions for proton induced showers and iron nucleus induced showers.

★e-mail: florian.gate@subatech.in2p3.fr

more likely to interact at lower altitude than heavier nuclei of the same energy. As the electric field emission is strongly beamed towards the direction of propagation of the shower, the LDF is narrower in the case of light nuclei compared to heavier nuclei. Thus, the radio signal is correlated to the mass information, that is estimated through the reconstruction of X_{max}, the atmospheric depth at which the number of secondary particles reaches its maximum. The features of the electric field that are known to be correlated to X_{max} are the amplitude [2–4], the shape of the radio wave front [5] and the spectral index of the frequency spectrum [6].

2 Reconstruction methods

Four methods have been developed to perform a reconstruction of the depth of the shower maximum from AERA data. They are based on a comparison of the amplitude (or energy fluence) of the detected electric field to a prediction model. To reconstruct one detected shower, two sets of events are simulated with the experimentally reconstructed arrival direction, initiated by protons and iron nuclei. The total set is composed of a higher fraction of protons to account for the larger first interaction depth and X_{max} fluctuations of the light nuclei. The LDFs corresponding to different X_{max} values are computed by a simulation code and the agreement with the experimental data is tested as a function of the simulated X_{max}, with a χ^2 test. The reconstructed X_{max} is the minimum value of the function $\chi^2 = f(X_{max})$, as shown in Fig. 2. The methods (noted A, B, C, D) are detailed in the next section.

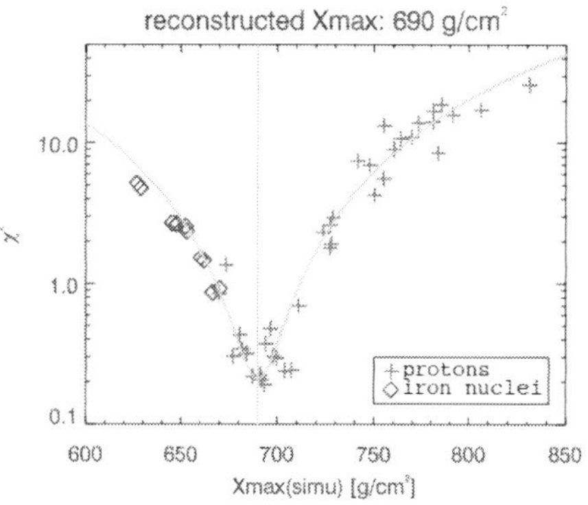

Figure 2. Agreement between a detected LDF at AERA and the simulated LDFs (method A) as a function of their respective X_{max} depths. The values are fitted by a square function represented by the curve. The X_{max} depth reconstructed by the radio method for this event is given by the minimum of the square function (i.e. the value that gives the best agreement), highlighted on the plot by the vertical line. Some χ^2 values are below unity because only errors on the detected electric field have been considered.

A - SELFAS: The electric field induced by 40 protons and 10 iron nuclei with an arbitral energy of 1 EeV is computed by SELFAS [7] on a dense array. No assumption is made on the core position or X_{max}. The core position is estimated by shifting the simulated core position over the frame of AERA until the best agreement with the detected LDF is found. This procedure is followed for all 50 simulations. A scaling parameter is added in the χ^2 test in order to compare only the shapes of the LDFs and the electric field amplitude is assumed to be linearly proportional the energy of the primary.

B - Parametrized model: Instead of computing the electric field, this method uses a parametrized model of electric field emission in the shower frame [8]. The C_i parameters are fixed at values giving the best results for the AERA experiment from CoREAS simulations. The parameter σ accounts for the width of the radio footprint on the ground. The parametrization allows the estimation of the geometrical distance to the shower maximum based on σ. Finally, knowing the arrival direction and an atmospheric model, the distance is converted to the corresponding crossed atmospheric depth X_{max}.

$$u(\mathbf{r}) = A\left[\exp\left(\frac{-(\mathbf{r} + C_1\mathbf{e}_{\mathbf{v}\times\mathbf{B}} - \mathbf{r}_{\mathrm{core}})}{\sigma^2}\right) - C_0\exp\left(\frac{-(\mathbf{r} + C_2\mathbf{e}_{\mathbf{v}\times\mathbf{B}} - \mathbf{r}_{\mathrm{core}})}{(C_3e^{C_4\sigma})^2}\right)\right] \quad (1)$$

C - CoREAS: This method is similar to method A but, the electric field is computed by CoREAS [9]. The primaries are simulated with the energy and arrival direction calculated from the data of the surface detectors (SD). The reconstruction is also based on the comparison of the simulated and measured radio-emitted energy, deposited on the ground by the secondary particles.

D - ZHAireS: The electric field is computed by ZHAireS [10] only along a line of antennas. It uses a superposition model [11] to estimate the electric field around the shower axis from the line of antennas. Indeed, in the shower frame, the amplitudes of the geomagnetic and Askaryan contributions are circularly symmetric around the shower axis. The computing time is considerably reduced. This method also uses the SD energy as a prior.

3 Comparison to fluorescence detection measurements

The methods described in Section 2 are applied on a high quality set of hybrid (radio + FD) showers. The official quality cuts of the fluorescence detector (FD) are applied, at least 5 radio stations must have trigged and the zenith angle must not exceed 55°. The total number of events is 32. However, the number of events reconstructed by each method varies slightly. The combination of the event characteristics (energy of the primary, mass, X_{max}) may not be reproducible with the hadronic interaction model used by one of these methods. Multiple X_{max} solutions can also be produced for a single event. The correlation plots between the reconstructed X_{max} with the radio methods and the FD measurements are presented in Fig. 3 and Fig. 4. The reconstructed values are in good agreement with the FD measurements. The maximum systematic deviation is of 40 g/cm^2 (method A).

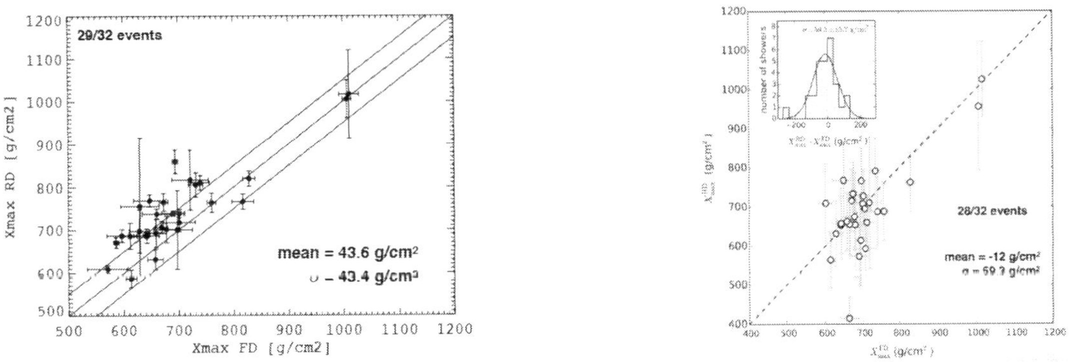

Figure 3. Scatter plots obtained with method A [12] (left) and method B [13] (right). The reconstructed X_{max} are displayed as a function of the FD measurements. The lines represent a one-to-one correlation, surrounded by two lines accounting for ±50 g/cm^2 on the left plot.

4 Conclusion and outlook

The distribution of the deviation of the reconstructed X_{max} from the FD measurements for the set of hybrid showers (ΔX_{max}) are fitted by a gaussian function and the mean difference and standard deviation at 1 σ confidence level are summarized in Table 1 for each method. The offsets could be related to the underlying simulation codes. The reconstruction of X_{max} from the radio signal will be a decisive asset for the estimation of the mass composition of cosmic rays due to the high duty cycle of the radio stations. The next steps are the determination of the composition from radio without any bias and the combination of the strength of the radio methods. To be efficient, the reconstruction needs, to

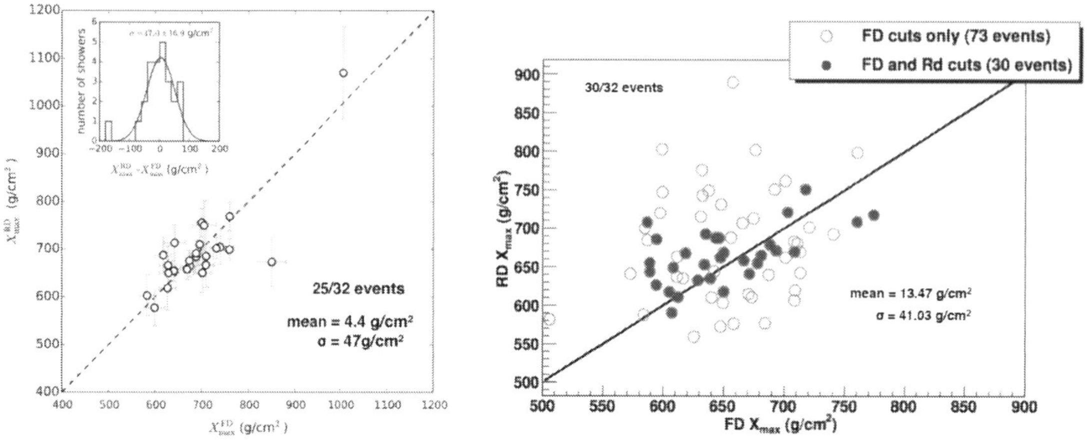

Figure 4. Scatter plots obtained with method C [13] (left) and method D [14] (right).

be precise and with a fast calculation time. The validity of the methods can only be proven through the comparison with the FD measurements. The influence of the air density and refractivity at the time of the measurements on the deviations are under study. The X_{max} resolution from radio methods is currently close to 40 g/cm^2.

Table 1. Summary of the requirements and results of the radio methods

method	A	B	C	D
requirements	RD direction	RD direction	SD direction SD energy	SD direction SD energy
calculation time (one event)	8 hours	- - - - - - - - - -	1 week	1 hour
simulated antennas	168	0	160	60
number of showers	40 p + 10 Fe	0	20 p + 10 Fe	30 p + 30 Fe
mean($X_{max,RD} - X_{max,FD}$) (g/cm^2)	43.6	−12	4.4	13.47
σ ($X_{max,RD} - X_{max,FD}$) (g/cm^2)	43.4	59.3	47	41.03

References

[1] J. Schulz for the Pierre Auger Collaboration, Proc. of the 34th ICRC, PoS(ICRC2015) 615, (2015)
[2] S. Buitink et al. LOFAR Coll., Phys. Rev. D, (2014) 90
[3] W.D. Apel, et al. LOPES Coll., Physical Review D (2012) 85, 10.1103/PhysRevD.85.071101
[4] P.A. Bezyazeekov, et al. Tunka-Rex Coll., JCAP (2016) 01
[5] W.D. Apel, et al. LOPES Coll., JCAP (2014) 09
[6] S. Grebe et al. Pierre Auger Coll., AIP Conference Proceedings, (2013) 1535
[7] V. Marin, B. Revenu, Astropart. Phys. 35 (2012) 733-741
[8] A Nelles et al, Astropart. Phys. 60 (2015) 13
[9] Huege, T., Ludwig, M, James, C.W., AIP Conf. Proc. (2013) 128-132
[10] J. Alvarez-Muñiz, W. R. Carvalho Jr., E. Zas, Astropart. Phys. 35 (2012) 325-34
[11] J. Alvarez-Muñiz, W. R. Carvalho Jr., H. Schoorlemmer, E. ZasAstropart, Phys. 59 (2014) 29
[12] F .Gaté, PhD thesis, https://tel.archives-ouvertes.fr/tel-01427261.
[13] J. Schulz, PhD thesis, http://repository.ubn.ru.nl/handle/2066/151781.
[14] W. R. de Carvalho Jr., private communication.

EPJ Web of Conferences **135**, 01008 (2017) DOI: 10.1051/epjconf/201713501008
ARENA 2016

Updates from the ANITA Experiment

Cosmin Deaconu for the ANITA collaboration[1,*]

[1] *Kavli Institute for Cosmological Physics, University of Chicago*
5640 South Ellis Ave, Chicago, Illinois, United States of America

Abstract.
The ANtarctic Impulsive Transient Antenna (ANITA) collaboration deploys balloon-borne interferometric antenna payloads that fly at 37 km above Antarctica. The primary goal is detection of Askaryan emission from cosmogenic neutrinos interacting in the Antarctic ice. In addition, ANITA has proven sensitive to radio signals from extended air showers. Here, we provide a review of the results of previous missions, with a special focus on recent results, and a preview of our upcoming mission.

1 Introduction

Cosmogenic neutrinos[1] from the interaction of the Ultra-High-Energy Cosmic Ray (UHECR) protons with the cosmic microwave background[2] are expected to exist at EeV energies but have yet to be detected. The low flux and small interaction cross-sections pose formidable challenges, but successful detection would provide new information both about astrophysical sources of UHECR's as well as the highest energy measurement of neutrino-nucleon interactions.

Economic instrumentation of large target material volumes may be accomplished by exploiting the Askaryan effect[3]. The interaction of an extremely-high-energy neutrino in a dielectric, such as ice, produces a moving charge excess resulting in electromagnetic emission that becomes coherent at wavelengths larger than the transverse size of the avalanche (<~ GHz). This emission has been confirmed in laboratory beamline tests[4] and is not greatly attenuated in media such as glacial ice, allowing large acceptance volumes with few antennas.

The ANtarctic Impulse Transient Antenna (ANITA) long-duration interferometric balloon payload scans the Antarctic continent for impulsive radio-frequency emission produced by neutrino interactions in the ice. Antarctica is the ideal location for the payload due to its preponderance of ice, comparatively-few anthropogenic sources of background, and high-altitude air currents suitable for long-duration ballooning. At the 37 km float altitude, roughly one million cubic kilometers of ice are visible at a time, but the large distance from the interactions implies a high threshold (~EeV).

As a broadband impulse radio detector, ANITA is also sensitive to radio-emission from extended air showers (EAS), which is primarily synchrotron emission from charge-separation by the Earth's magnetic field[5]. This allows ANITA to directly measure UHECR's of sufficient energy and also provides sensitivity to another channel for detection of cosmogenic neutrinos – a ν_τ may interact in the Earth producing a τ that can escape and hadronically decay in the atmosphere, creating an upward-going EAS.

*e-mail: cozzyd@kicp.uchicago.edu

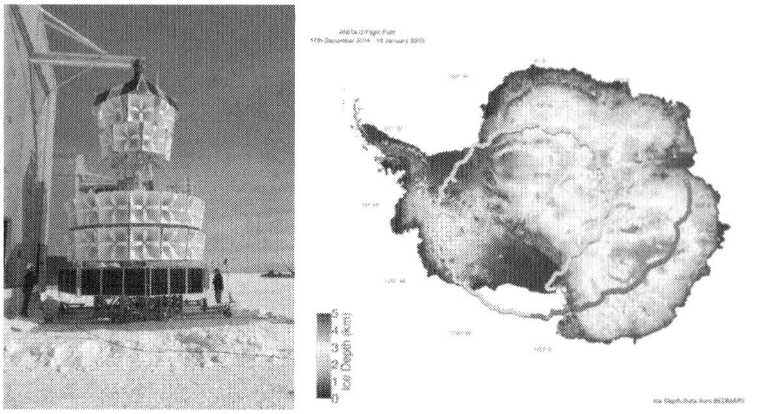

Figure 1. The ANITA-3 payload (left) and flight path (right). The vertical axis denotes the ice depth and the change in color of the flight path occurs at the second revolution around McMurdo

2 Instrument

As a balloon payload in a remote environment, ANITA must be capable of running with minimal interaction and comply with severe weight and power consumption limitations. ANITA's sensitive band is roughly 200-1300 MHz. ANITA payloads consist of 32-48 dual-polarization high-gain antennas, tilted slightly downward towards the horizon to maximize acceptance. The signal from antennas is amplified and then split into digitization and trigger circuits.

LAB4 switched-capacitor arrays[6] are used for digitization due to their their high sampling rate (2.6 GSa/s) and low power requirements. However, this comes at the cost of slow readout and complicated calibration for the analog buffers. Multiple chips are used for each channel to minimize the effect of the long-readout time on deadtime.

Tunnel diode square-law detectors form the basis of the trigger. Onboard field-programmable gate arrays combine individual channel level crossings into global triggers that result in digitization. The exact trigger strategy has varied from flight to flight. Thresholds are dynamically set so that the instrument is taking as much data as possible without significant deadtime. A maximum sustained global trigger rate of $\sim$ 50 Hz is achievable.

So far there have been three flights of the ANITA payload. The most recent flight, ANITA-3 (Fig. 1), flew in the 2014-2015 season for 22 days with 48 antennas. The flight had to be terminated relatively quickly due to wind patterns that threatened to send the payload off the continent. Unlike previous flights which had split the signal into multiple frequency bands and required signal above threshold in each band, a full-band trigger in both vertical and horizontal polarizations was used. This was expected to improve sensitivity at low signal-to-noise ratio, but also proved much more susceptible to triggering on man-made continuous-wave (CW) emissions. These unwanted triggers resulted in high dead-time on the side of the payload pointing North. The analysis from ANITA-3 is still ongoing as of writing.

The fourth flight of ANITA is expected this Austral winter. Upgrades include dynamic notch filtering to combat CW from satellites, a higher-altitude flight, improved amplifiers, and new trigger logic. Phase shifters will be used to convert linear to circular polarization and coincidence between LCP and RCP will be required. As a result, ANITA will have equal acceptance to linearly polarized radiation in any polarization plane and be less likely to trigger on the mainly circularly polarized signals from satellites. As in ANITA-3, a GPU-based online event prioritizer will be used to select

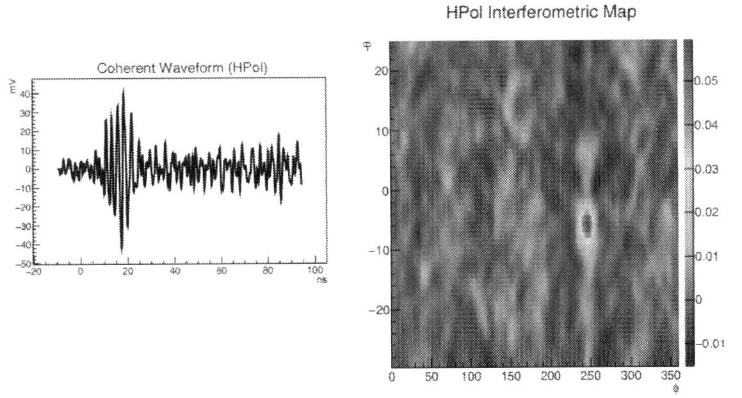

Figure 2. An example coherently-summed waveform (left) and interferometric map (right) from a horizontally-polarized calibration pulser during the ANITA-3 flight.

which events will be telemetered to ground in case the full dataset cannot be recovered from the payload.

3 Calibration and Analysis

In order to help characterize the instrument, high-voltage pulsers transmit impulsive signals to the instrument from various sites on the Antarctic continent. This allows calibration of the antenna positions as they are in flight and allows estimation of trigger efficiency and pointing resolution as a function of signal-to-noise ratio. For ANITA-3, the Hi-Cal trailing balloon provided a balloon-borne high-voltage impulse calibration signal for a portion of the flight. Both the direct and ice-reflected signals from Hi-Cal can be measured. A similar balloon-based calibration is planned to be repeated for ANITA-4.

ANITA analyses have generally followed a similar procedure. First, the waveforms from each event are filtered to remove anthropogenic CW contamination from satellites and bases. Then the likely primary direction of the signal by using an interferometric method [7] considering cross-correlations between antennas. Fig. 2 shows an example interferometric map from a calibration pulser event. Cuts are then applied to remove remaining CW, events likely to be thermal noise (where the signals between antennas will be uncorrelated) and other pathological events, leaving only candidate physics events and anthropogenic activity. Spatial clustering is used to group remaining events into known-bases or apparent hot-spots of activity. Any spatially isolated events are considered physics candidates. The isolated anthropogenic background is estimated from the number of small-event clusters.

4 Science Results

A cosmogenic neutrino search seeks primarily vertically-polarized isolated impulsive emission that may be a result of Askaryan emission. This preferred plane of polarization is expected from geometric acceptance considerations and transmission at the ice-air interface. ANITA-2 has the best current limits on cosmogenic neutrinos in the 10^{21} eV range[8].

Due to nearly vertical magnetic field in the Antarctic region, synchrotron emission from EAS is expected to be primarily horizontally polarized. ANITA-1 measured 16 such events[5]. Because ANITA-2 only triggered on vertical polarization, it was much less sensitive to EAS and only detected

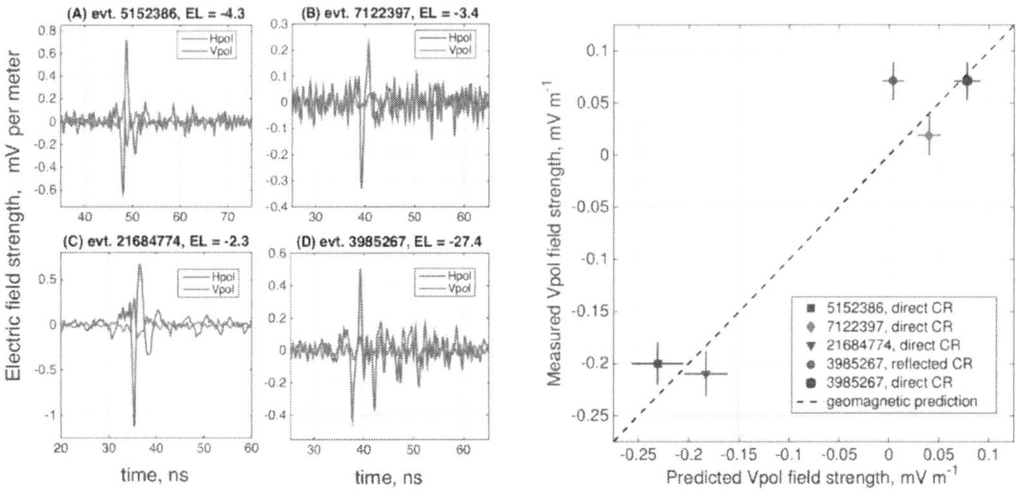

Figure 3. (Left) Coherently-summed waveforms from the three apparent Earth-skimming extended air showers from ANITA-1 and 2, as well as the non-inverted EAS candidate coming from the Earth. The instrument response has been deconvolved. (Right) The predicted versus measured vertical-polarization component of the events, including two assumptions from the apparent upward event.

three. Due to ANITA's high altitude, most EAS development occurs below the payload and the emission measured is reflected off the ice, which results in an apparent incoming direction from the ground and signal polarity inversion. However, ANITA has also detected nearly horizontal showers where the shower misses the Earth entirely; in these cases the signal comes from above the horizon there is no signal inversion.

The geomagnetic origin of these events can be further tested by comparing the measured polarization plane with the expected plane defined by the Earth's local magnetic field, the cosmic ray direction, and, in the case of reflected events, the appropriate Fresnel coefficients. Agreement between prediction and measurement is confirmed for the events from ANITA-1 and 2.

An additional isolated horizontally-polarized event from ANITA-1 [9] has a non-inverted signal shape characteristic of the Earth-skimming EAS's, but appears to come from well below the horizon (Fig. 3). In addition, the geomagnetic polarization plane analysis is also more consistent with no reflection off the ice. This event is challenging to interpret. While it is impossible to rule out anthropogenic origin, only 1.6 events were expected to pass clustering cuts. Only a few percent of all anthropogenics have a shape with as high a cross-correlation coefficient with the average EAS waveform shapes as this event. Moreover, only half would have the correct polarity and most anthropogenics are not expected to have polarization angle so close to the geomagnetic prediction. None of these factors, other than clustering, were used in the event selection.

A potential physics explanation is an upward-going EAS from the hadronic decay of a τ produced by the interaction of a EeV-scale ν_τ in the crust or ice. However, the elevation angle is such that the chord through the Earth would be roughly 5500 km. The projected Standard-Model cross-section for neutrinos at these energies through the Earth predicts an attenuation factor on the order of 10^{-4} for this much material[10]. If this event truly is a τ-induced air shower, it would be in significant tension with the Standard Model. The extremely-high-energy neutrino cross-section reduction on the order of 3-5 would be required to plausibly accommodate such an event. If this really were the case, the acceptance calculations that previous neutrino limits are based on would need to be updated. ANITA-4 is moving to a polarization-plane-blind trigger partially for this reason.

5 Conclusion

The ANITA program has had three successful flights. The analysis of the third flight is still ongoing. Due to triggering on horizontally-polarized events, it is expected to measure additional EAS events, perhaps shedding additional light on the unexpected event from ANITA-1. ANITA-4 is to fly soon, which will address many of the shortcomings of the previous flight. ANITA acknowledges generous support from NASA, the Columbia Scientific Balloon Facility, the National Science Foundation, and the Department of Energy, High Energy Physics Division.

References

[1] V.S. Berezinsky, G.T. Zatsepin, Phys. Lett. **B28**, 423 (1969)
[2] K. Greisen, Phys. Rev. Lett. **16**, 748 (1966)
[3] G.A. Askar'yan, Sov. Phys. JETP **14**, 441 (1962), [Zh. Eksp. Teor. Fiz.41,616(1961)]
[4] P.W. Gorham et al. (ANITA), Phys. Rev. Lett. **99**, 171101 (2007), hep-ex/0611008
[5] S. Hoover et al. (ANITA), Phys. Rev. Lett. **105**, 151101 (2010), 1005.0035
[6] G.S. Varner, L.L. Ruckman, P.W. Gorham, J.W. Nam, R.J. Nichol, J. Cao, M. Wilcox, Nucl. Instrum. Meth. **A583**, 447 (2007), physics/0509023
[7] A. Romero-Wolf et al., Astropart. Phys. **60**, 72 (2015)
[8] P.W. Gorham et al. (ANITA), Phys. Rev. **D82**, 022004 (2010), [Erratum: Phys. Rev.D85,049901(2012)], 1011.5004
[9] P.W. Gorham et al., Phys. Rev. Lett. **117**, 071101 (2016), 1603.05218
[10] E. Bugaev, T. Montaruli, Y. Shlepin, I.A. Sokalski, Astropart. Phys. **21**, 491 (2004), hep-ph/0312295

Cosmic Ray Mass Measurements with LOFAR

Stijn Buitink[1,⋆], *Antonio* Bonardi[2], *Arthur* Corstanje[2], *J. Emilio* Enriquez[2], *Heino* Falcke[2,3,4], *Jörg R.* Hörandel[2,4], *Pragati* Mitra[1], *Katie* Mulrey[1], *Anna* Nelles[5], *Jörg Paul* Rachen[2], *Laura* Rossetto[2], *Pim* Schellart[2], *Olaf* Scholten[6,7], *Satyendra* Thoudam[2], *Gia* Trinh[6,7], *Sander* ter Veen[3], and *Tobias* Winchen[1]

[1]*Vrije Universiteit Brussel, Pleinlaan 2, 1050 Brussels (Belgium)*
[2]*Department of Astrophysics/IMAPP, Radboud University, P.O. Box 9010, 6500 GL Nijmegen (the Netherlands)*
[3]*Astron, Oude Hoogeveensedijk 4, 7991 PD Dwingeloo (The Netherlands)*
[4]*Nikhef, Science Park 105, 1098 XG Amsterdam (The Netherlands)*
[5]*University of California Irvine, Irvine, CA 92697, (USA)*
[6]*KVI-CART, Zernikelaan 25, 9747 AA Groningen (The Netherlands)*
[7]*University of Groningen, Zernikelaan 25, 9747 AA Groningen (The Netherlands)*

Abstract. In the dense core of LOFAR individual air showers are detected by hundreds of dipole antennas simultaneously. We reconstruct X_{max} by using a hybrid technique that combines a two-dimensional fit of the radio profile to CoREAS simulations and a one-dimensional fit of the particle density distribution. For high-quality detections, the statistical uncertainty on X_{max} is smaller than 20 g/cm^2. We present results of cosmic-ray mass analysis in the energy regime of 10^{17} - $10^{17.5}$ eV. This range is of particular interest as it may harbor the transition from a Galactic to an extragalactic origin of cosmic rays.

1 Introduction

Radio detection of extended air showers is a powerful and rapidly developing technique [1] that is suitable for measuring the atmospheric depth of the shower maximum X_{max} [2, 3]. Arrays with a high density of radio antennas, like LOFAR [4] and the future SKA [5], are capable of accurately measuring the mass composition of cosmic rays below the ankle, where a transition of a Galactic to an extragalactic component is expected to occur [6, 7].

LOFAR is a digital radio telescope constructed in the North of the Netherlands with satellite stations across Europe. It consists of thousands of dipole antennas, sensitive in the frequency range of 10 – 240 MHz. It produces extremely detailed air shower radio data, by using the dense core region, or *superterp*, where 384 antennas are located within a circle of 320 m diameter. A particle array, LORA [8], has been installed in the core and is used for triggering and reconstruction. Each antenna contains a ring buffer which is read-out in case of a trigger. The full waveform is stored for offline analysis [9].

The vast amount of data per shower and the complicated radiation pattern require new reconstruction techniques. By fitting two-dimensional radiation profiles to the data, LOFAR has achieved a

⋆e-mail: Stijn.Buitink@vub.ac.be

resolution on X_{max} below 20 g/cm^2 [3] which is comparable to the resolution of fluorescence detection. An analysis of the first set of high-quality LOFAR events has yielded mean X_{max} values that are in agreement with existing data based on other techniques and is indicative of a strong contribution of light nuclei in the energy range of 10^{17} - $10^{17.5}$ eV [10].

2 Simulations

At LOFAR, showers are reconstructed using dedicated sets of CORSIKA [11] simulations, i.e. for each measured shower we generate enough proton showers and iron showers to span the range of possible values for X_{max} . Such simulations are computationally expensive, so we aim to minimise the number of simulated showers per events, while still ensuring that (a) the complete range of naturally occurring X_{max} values is covered, and (b) the reconstruction is accurate enough.

We achieve this by using CONEX to quickly generate hundreds of showers [12]. For these showers only the first few interactions are simulated in CORSIKA, after which the rest of the shower development is based on parametrizations. Since the value of X_{max} critically depends on these first interactions the set of CONEX showers contains a realistic distribution of X_{max} values. Moreover, it is possible to select a subset of these showers for subsequent full CORSIKA simulation. We choose this subset by first selecting the two proton showers with the highest and lowest value for X_{max} . Then, we select nine more proton showers with intermediate values, (nearly) equally spaced between the extreme values. Since the range of X_{max} values for iron showers is much narrower, we select a total of five iron showers using the same algorithm.

Finally we add additional shower simulations close to the estimated X_{max} of the measured shower. This estimation is given by fitting a parametrization of the radio signal [13] to the data, that has an uncertainty of ~ 40 g/cm^2. We select 11 additional showers around this expected value.

The subset acquired in this way is then simulated in CORSIKA including the CoREAS plugin for calculation of the radio emission [14]. The radio emission is simulated for antennas that form a star-shaped pattern in the shower plane. An antenna model is applied to the simulated electric fields to calculate the measured signals in the two dipoles of the LOFAR low band antennas. The measured pulse power is defined as the total power in a window of 55 ns centered around the pulse maximum, summed over the two polarisations. Finally, a two-dimensional map of the pulse power is created by interpolation of the star-shaped pattern. A detailed description of this procedure can be found in [3].

A GEANT4 simulation of the LORA scintillators is used to calculate the lateral distribution of the deposited energy. So, for each measured shower a dedicated set of simulations is produced, and from each simulation in a set a two-dimensional radio profile and a one-dimensional particle profile are produced.

3 Reconstruction

The radio and particle profiles are fitted to the data simultaneously. The fit has four free parameters: two for the shower core position and scaling parameters for both the radio and the particle profiles. Inclusion of the particle data in the fit does not have much effect on the fitted core position in practice. However, it serves two important purposes: as an additional consistency check and to determine the energy scale. The analysis presented here does not yet include the absolute calibration of the antennas [15] so the scaling parameter for the radio power has an arbitrary value. In effect, the ratio of two scaling parameters also has an arbitrary value. It should however be the same constant for

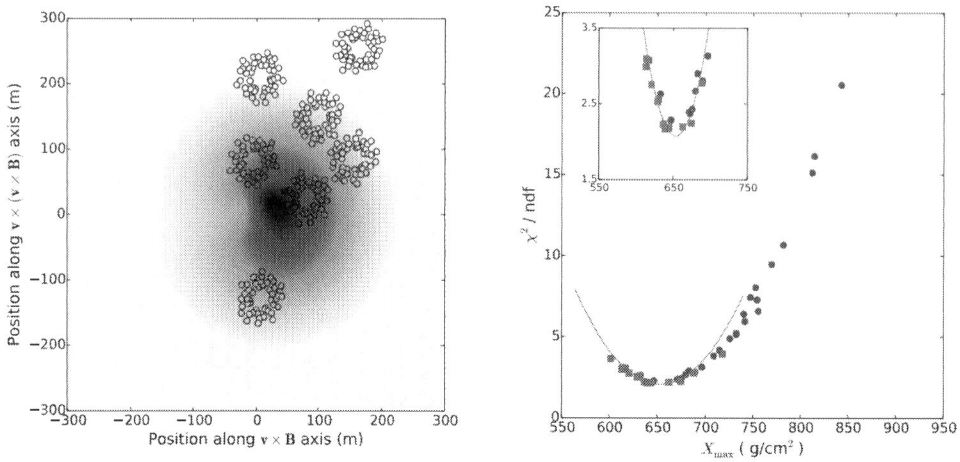

Figure 1. Left: event reconstruction. The small circles indicate LOFAR antenna positions (grouped in larger circles) projected into the shower plane. Their color reflects the measured power of the radio pulse. The background is a two-dimensional radio map from interpolated CoREAS simulations. The fit quality is high when the colors inside the circles blend into the background. Right: reconstruction of X_{max}. Every data point corresponds to a single simulated shower. Together they form the dedicated simulation set produced for this particular shower. The reduced χ^2 of the fit is plotted against X_{max} and a parabola is fitted to find the reconstructed X_{max} . The inlay is a zoomed in version of the minimum of the curve demonstrating the sharpness of the minimum. Purple squares correspond to iron shower and blue circles to proton shower.

all measured showers. Therefore, the width of the distribution of this parameter is a measure for the energy resolution of the reconstruction[1].

The reduced χ^2 of the fit is very sensitive to X_{max} as is seen in Fig. 1. The minimum of the χ^2-curve is found by fitting a parabola to the points near the lowest value. The curve is not smooth because shower-to-shower fluctuations are not just variations in X_{max} . Other parameters that will vary include the width of the longitudinal profile, the shape of the tail of the shower, and the lateral distribution of the shower. These variations cause the jitter in the data points. If an infinite amount of simulations could be produced, the data points are expected to form a band rather than a line.

The reconstructed value for X_{max} found in this method is the atmospheric depth corresponding to the atmosphere used in the simulation. For this analysis we used the standard US atmosphere which is also a good average for northern Europe. For each measurement we construct a more accurate atmosphere using the GDAS database [16]. A first order correction is made by assuming that the reconstructed X_{max} corresponds to the correct altitude h above sea level. The corrected value for X_{max} is then found by evaluating the atmospheric depth of the GDAS profile at altitude h. However, it was realised that the change in refractive index (which also depends on the humidity profile) also affects the reconstructed X_{max}. This effect is included as a systematic effect in the present analysis, although it can be corrected for in the future [17].

[1]In fact this is a conservative choice since it includes the energy resolution of both the radio and particle fit, instead of relying on the most accurate of the two.

4 Event selection

For this analysis we have selected showers that were detected by at least four LOFAR antenna groups (each containing 48 active antennas). At trigger level, we require that 16 out of the 20 scintillator detectors find a signal. Both conditions introduce a possible bias towards light or heavy primaries in the sample. Heavy nuclei interact higher in the atmosphere, so the particle density at the ground becomes smaller, especially for more inclined showers. This leads to reduced trigger probability. On the other hand, protons and light nuclei penetrate deeper into the atmosphere and produce showers closer to the ground. Therefore, the size of the radio footprint is reduced and the probability of having a detectable signal in at least four antenna groups is reduced.

To ensure that our final event sample is bias-free, we require that *all* of simulated showers in the dedicated simulation set of a particular event (a) produce a trigger in the LORA array and (b) have a detectable radio signal in at least four antenna groups. This requirement removes only a few events above 10^{17} eV, but below that energy the amount of removed events rapidly increases, mainly because of the trigger efficiency. Composition measurements below 10^{17} eV become possible by changing the triggering strategy in the future, for example by triggering on radio-only [18] or a combination of radio and particles.

A quality cut is imposed in order to only include showers with relatively small uncertainty in the reconstruction on the core, the energy, and $X_{\max}$. In fact, those three uncertainties are strongly correlated and a single cut on the precision of the core suffices. The uncertainties are calculated using the dedicated set of simulations. They therefore depend on the arrival direction, core position and energy of the measured shower, but not on its value for $X_{\max}$. Thus, no additional bias on composition is introduced by this cut.

The final event sample counts 118 showers recorded before January 2015. The mean energy resolution of the sample is 32% and the resolution on $X_{\max}$ is 17 g/cm^2. The reduced χ^2 values of the fits range from 0.9 to 2.9.

Several systematic effects have been investigated. As mentioned above, humidity in the atmosphere increases the index of refraction and shifts the reconstructed $X_{\max}$. Since the simulations were performed for dry air this introduces an asymmetric systematic uncertainty of 10 g/cm^2. The magnitude of the effect is expected to depend on zenith angle. Indeed, separating the event sample in two groups above or below 32 degrees zenith angle, yields a ± 8 g/cm^2 difference in mean $X_{\max}$. The choice of hadronic interaction model also introduces another uncertainty of 5 g/cm^2 in the reconstruction method[2]. The total systematic uncertainty on $X_{\max}$ is +14/-10 g/cm^2 and on the energy is 27%. More details are given in [3] and [10].

5 Results

The mean value of $X_{\max}$ as a function of energy is shown in the left panel of Fig. 2. The error bars indicate 1σ statistical uncertainties and the shaded region shows the systematic uncertainty. The values are in agreement with other experiments using different techniques (fluorescence detection and non-imaging cherenkov). There is some tension with the data from the low-energy extension of the Pierre Auger observatory [19] towards $10^{17.5}$ eV although the results still agree within systematic uncertainties.

[2]It is important to note that this is the uncertainty when *reconstructing* $X_{\max}$. When *interpreting* $X_{\max}$ in terms of primary masses there is an additional difference between models of order 10 g/cm^2.

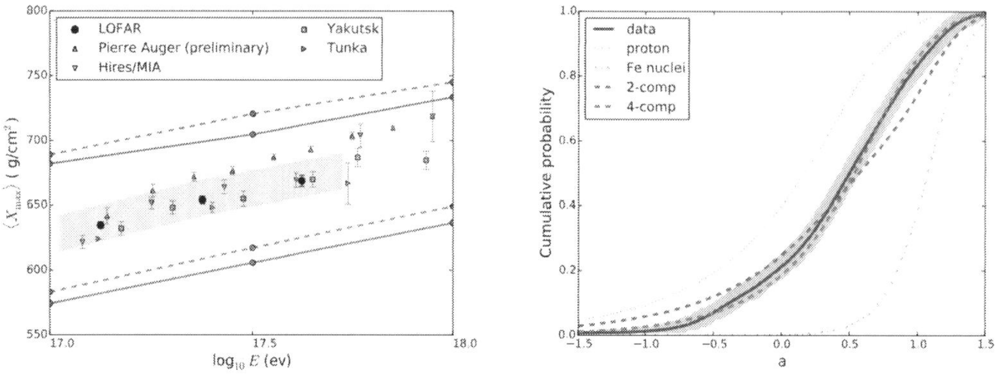

Figure 2. Left: mean $X_{\max}$ as a function of energy. The LOFAR results are in agreement with results based on different techniques within systematic uncertainties. The green lines are for pure proton composition and the red lines for pure iron (dashed: EPOS-LHC, solid: QGSJETII). Right: cumulative distribution of a for LOFAR data (blue; with 1σ uncertainty band) and different compositions. A four-component model is required to reproduce the data.

Further information on the composition can be extracted by studying the shape of the distribution of $X_{\max}$. For each shower, we calculate:

$$a = \frac{\langle X_{\text{proton}} \rangle - X_{\text{shower}}}{\langle X_{\text{proton}} \rangle - \langle X_{\text{iron}} \rangle} \tag{1}$$

where X_{shower} is the reconstructed $X_{\max}$, and $\langle X_{\text{proton}} \rangle$ and $\langle X_{\text{iron}} \rangle$ are mean values predicted by the hadronic interaction code QGSJETII.04.

The cumulative probability density function (CDF) is shown in the right panel of Fig. 2. The yellow curves at the extremes represent pure proton (centered at a=0, wide distribution) and pure iron (centered at a=1, narrow distribution) composition.

We fit theoretical CDFs based on composition with two or four mass components to the data. The test statistic in the fit is the maximum deviation between the data and the model CDFs. The p-value is given by the probability of observing this deviation, or a larger one, assuming the fitted composition model. The two-component model consists of proton and iron nuclei, and has only one free parameter: the mixing ratio. The best fit is found for a proton fraction of 62%, but it describes the data rather poorly with a p-value of 1.1×10^{-6}.

The fit improves when using a four-component model (p+He+N+Fe), yielding a p-value of 0.17. While the best fit is found for a Helium fraction of 80%, the fit quality deteriorates only slowly when replacing helium by protons. Although there is a large uncertainty in the contribution of individual elements, the total fraction of light element (p+He) is well constrained. It has a best fit value of 0.8 and lies within the range [0.38,0.98] at a 99% confidence level based on QGSJETII.

 DOI: 10.1051/epjconf/201713501009

6 Conclusion

It has been shown that the radio technique can be used to measure mass composition of cosmic rays and is competitive with other experiments. The first composition study with LOFAR yields a relatively high fraction of light particles just above 10^{17} eV. The energy range can still be extended to lower and higher energies by expanding the triggering array and improving trigger strategies. Accurate composition measurements in the 10^{16}-10^{18} eV range are of crucial importance for understanding the transition of galactic to extragalactic origin of cosmic rays [7]. In the future, SKA will offer the opportunity of exploiting this technique with increased area, antenna density and frequency bandwidth, leading to extremely accurate measurements [5].

References

[1] T. Huege, Physics Reports **620**, 1-52 (2016).
[2] W. Apel *et al.* [LOPES collaboration] Phys. Rev. D **90**, 062001 (2014).
[3] S. Buitink *et al.* [LOFAR collaboration], Phys. Rev. D **90**, 082003 (2014).
[4] M. P. van Haarlem *et al.* [LOFAR collaboration], Astron. & Astrophys. **556**, A2 (2013).
[5] T. Huege *et al.*, these proceedings (2016); A. Zilles *et al.*, these proceedings (2016).
[6] K.-H. Kampert and M. Unger, M., Astropart. Phys. **35**, 660 (2012).
[7] S. Thoudam *et al.*, arXiv: 1605.03111 (2016).
[8] S. Thoudam *et al.*, [LOFAR collaboration], Nucl. Instr. and Meth. in Phys. Res. A **767**, 339 (2014).
[9] P. Schellart *et al.* [LOFAR collaboration], Astron. & Astrophys. **560**, A98 (2013).
[10] S. Buitink *et al.* [LOFAR collaboration], Nature **531**, 70-73 (2016).
[11] D. Heck, Report FZKA **6019** (1998).
[12] T. Bergmann *et al.*, Astropart. Phys. **26**, 420-432 (2007).
[13] A. Nelles *et al.*, J. Cosm. Astropart. Phys. **5**, 18 (2015).
[14] T. Huege, M. Ludwig, and C. James, AIP Conference Proceedings **1535**, 128-132 (2012).
[15] A. Nelles *et al.* [LOFAR collaboration], JINST **10**, P11005 (2015).
[16] NOAA Air Resources Laboratory (ARL), Global Data Assimilation System (GDAS1) Archive Information, Tech. rep. (2004).
[17] A. Corstanje *et al.* [LOFAR collaboration], these proceedings (2016).
[18] A. Bonardi *et al.* [LOFAR collaboration], these proceedings (2016).
[19] A. Porcelli *et al.* [Pierre Auger Collaboration], PoS (ICRC2015) 420; arXiv:1509.03732 (2015).

A study of radio frequency spectrum emitted by high energy air showers with LOFAR

Laura Rossetto[1,⋆], *Antonio* Bonardi[1], *Stijn* Buitink[2], *Arthur* Corstanje[1], *J. Emilio* Enriquez[1], *Heino* Falcke[1,3,4], *Jörg R.* Hörandel[1,4], *Pragati* Mitra[2], *Katie* Mulrey[2], *Anna* Nelles[5], *Jörg P.* Rachen[1], *Pim* Schellart[1], *Olaf* Scholten[6,7], *Satyendra* Thoudam[1], *Gia* Trinh[6,7], *Sander* ter Veen[2], and *Tobias* Winchen[2]

[1]*Department of Astrophysics/IMAPP, Radboud University, P.O. Box 9010, 6500 GL Nijmegen (The Netherlands)*
[2]*Vrije Universiteit Brussel, Pleinlaan 2, 1050 Brussels (Belgium)*
[3]*Astron, Oude Hoogeveensedijk 4, 7991 PD Dwingeloo (The Netherlands)*
[4]*Nikhef, Science Park 105, 1098 XG Amsterdam (The Netherlands)*
[5]*University of California Irvine, Irvine, CA 92697, (USA)*
[6]*KVI-CART, Zernikelaan 25, 9747 AA Groningen (The Netherlands)*
[7]*University of Groningen, Zernikelaan 25, 9747 AA Groningen (The Netherlands)*

Abstract. The high number density of radio antennas at the LOFAR core in Northern Netherlands allows to detect radio signals emitted by cosmic ray induced air showers, and to characterize the geometry of the observed cascade in a detailed way. We present here a study of the radio frequency spectrum in the 30 – 80 MHz regime, and its correlation with some geometrical parameters of the extensive air shower. An important goal of this study is to find a correlation between the frequency spectrum and the primary particle type. Preliminary results on how the frequency spectrum changes as function of distance to the shower axis, and as function of primary particles mass composition are shown. The final aim of this study is to find a method to infer information of primary cosmic rays in an independent way from the well-established fluorescence and surface detector techniques, in view of affirming the radio detection technique as reliable method for the study of high energy cosmic rays.

1 Introduction

The LOw Frequency ARay (LOFAR) is a radio antenna array consisting of 50 stations spread in Northern Europe with a more dense core in Northern Netherlands [1]. Measurements of cosmic rays are performed mostly by using signals from the Low Band Antennas (LBAs) which operate in the frequency range 10 – 90 MHz. The LOFAR central array is instrumented with 20 scintillator detectors, the so-called LORA array [2], which provides triggers for cosmic ray data acquisition, and allows to reconstruct the arrival direction and the energy of primary particles. The LOFAR array detects cosmic rays in the energy range $10^{16} - 10^{18}$ eV.

⋆e-mail: l.rossetto@astro.ru.nl

Analytical calculations [4, 5] and simulation studies [6] conducted at the beginning of years 2000 already showed a dependence of the radio frequency spectrum on cosmic ray air shower characteristics. Previous measurements of the frequency spectrum performed by LOPES did not show any significant dependence of the frequency spectrum with respect to the distance to the shower axis and to the arrival direction [7]. A recent analysis conducted at the Auger Engineering Radio Array (AERA) of the Pierre Auger Observatory, shows a dependence of the frequency spectrum to the properties of the primary particle [8]. This demonstrates that it is feasible to extrapolate information from the frequency spectrum with the current radio antenna arrays.

The goal of the anaysis described here is to find a reliable method for deriving information of the primary cosmic rays detected at LOFAR by performing a frequency spectrum analysis. Among radio experiments, LOFAR is the one with the most dense number of antennas, thus the most suitable for this kind of analysis. In order to find the best parameters which describe the correlation between primary cosmic ray information and the emitted radio signal, two methods have been applied, and a cross-check between real data and simulations has been done. Regarding real data, cosmic ray radio signals detected by LOFAR since 2011 have been analysed. For the simulation of radio signals, the CoREAS code [9], a plug-in of the CORSIKA particle simulation code, has been used.

2 Analysis and Results

In order to study a correlation between the radio frequency spectrum and the cosmic ray air shower characteristics, two methods have been applied to both simulations and real data: the percentile statistical method and the linear fit method. For each cosmic ray event detected by LOFAR, a set of CORSIKA simulations has been produced by using, as initial input, protons as primary particle, and energy and arrival direction as reconstructed from the real event. In the simulation set the antenna layout has been chosen in order to have a symmetric star-shape around the shower axis, in the plane perpendicular to the shower axis (hereafter, *shower plane*). The total number of antennas is 160, and they are distributed on 8 arms at an angle distance $\alpha = 45^\circ$ between each other in the shower plane. Each antenna is 25 m distant from the previous one, thus covering an area of 500 m radius around the shower axis. Furthermore, the signal intensity in the time-domain is converted to the frequency-domain by applying a Fast Fourier Transform (FFT). The FFT distribution, as function of frequency, has been characterised by using the two following methods:

- **the percentile statistical method** → the FFT distribution has been integrated in the frequency range 30 − 70 MHz, and the statistical percentile indicates the frequency value at which the integrated FFT distribution reaches a certain fraction of the total integral; a preliminary study on simulations has shown that the 50th percentile, also known as *median*, i.e. the frequency value at which the integrated FFT distribution reaches half of the total integral, describes better the slope changing of the FFT distribution as function of distance to the shower axis [10];
- **the linear fit method** → the FFT distribution, in the logarithmic scale, has been fitted with a linear function in the frequency range 30 − 70 MHz; the slope parameter of the fitted function has then been studied as function of distance to the shower axis.

These two analysis methods have been applied to all data detected by LOFAR since 2011 and to the corresponding simulated events. Figure 1 shows the distribution of the 50th percentile frequency as a function of distance to the shower axis in the shower plane for both one simulated event and the corrisponding real event. The detected event has a primary energy of $(1.7 \pm 0.8) \cdot 10^{17}$ eV as reconstructed by the LORA scintillator array, and a value of the atmospheric depth where the cascade reaches its maximum development $X_{max} = (763 \pm 38)$ g/cm^2 as reconstructed by the Lateral Distribution Function method [11]. This real event has been compared with simulations, in particular with

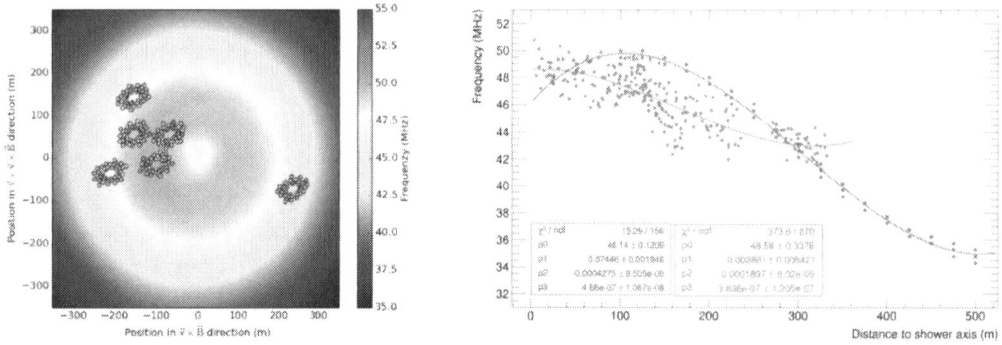

Figure 1. *Left*: comparison between the frequency distribution of the 50th percentile for one simulated event (background) and one real event detected by LOFAR LBAs (circles); the comparison is projected onto the shower plane, and the colour scale refers to the frequency of the 50th percentile. *Right*: distribution of the 50th percentile frequency as function of distance to shower axis in the shower plane for one simulated event (red points) and the corresponding real event (green points); the two distributions have been fitted with a third polynomial function (solid line).

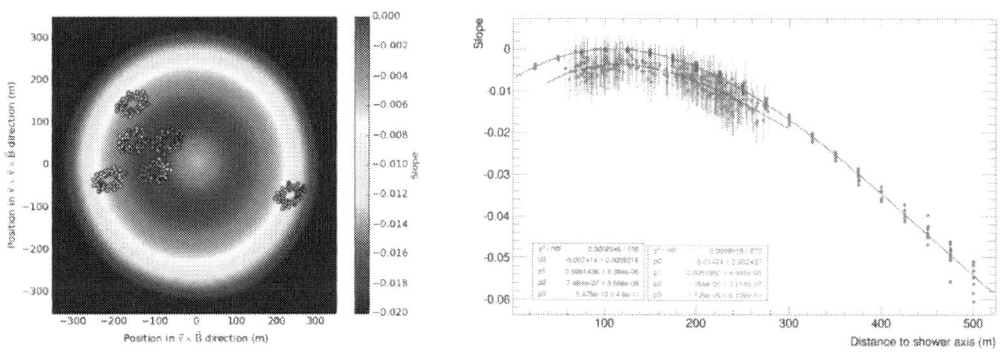

Figure 2. *Left*: comparison between the distribution of the linear fit slope for one simulated event (background) and one real event detected by LOFAR LBAs (circles); the comparison is projected onto the shower plane, and the colour scale refers to the slope of the linear fit. *Right*: distribution of the linear fit slope as function of distance to shower axis in the shower plane for one simulated event (red points) and the corresponding real event (green points); the two distributions have been fitted with a third degree polynomial function (solid line).

the one having the value of X_{max} closest to the real one, i.e. X_{max} = (764 ± 17) g/cm^2, as evaluated from the method described in [12]. Figure 2 shows the distribution of the slope parameter, as obtained from the linear fit method, as function of distance to the shower axis in the shower plane, for both one simulated event and the corrisponding real event. The cosmic ray event and the simulated one are the same event as shown in figure 1 . For both methods, simulations and real data have been fitted with the following third degree polynomial function $p_0 + p_1 \cdot d + p_2 \cdot d^2 + p_3 \cdot d^3$, where d is the distance of each antenna to the shower axis in the shower plane.

A study about the dependence of the frequency spectrum with respect to X_{max} has been performed as well. Figure 3-*left* shows the distribution of the 50th percentile frequency as function of X_{max} for

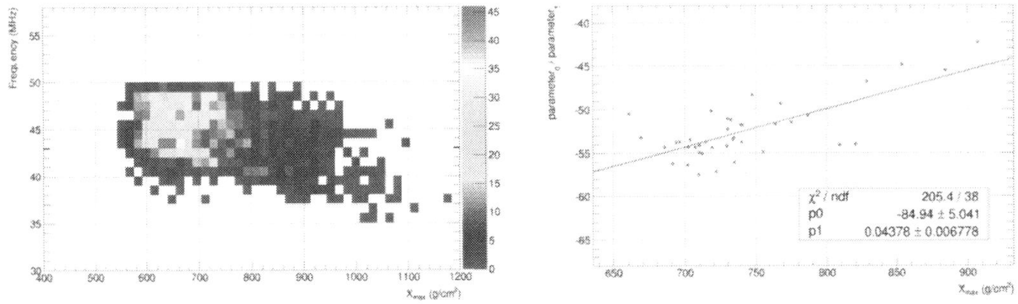

Figure 3. *Left*: distribution of the 50th percentile frequency as a function of X_{max} for simulated events in the energy range $10^{17} - 2 \cdot 10^{17}$ eV; the colour scale refers to the number of events. *Right*: distribution of p_0/p_1 as a function of X_{max}; p_0 and p_1 are the parameters obtained by fitting the slope of the FFT as function of distance to shower axis with a third degree polynomial function.

simulated events in the energy range $10^{17} - 2 \cdot 10^{17}$ eV. Figure 3-*right* shows the distribution of p_0/p_1 as function of X_{max} , where p_0 and p_1 are the parameters obtained by fitting the slope of the FFT as function of distance to shower axis with the third degree polynomial function for different simulated events. Both plots of figure 3 depict a dependence of the frequency spectrum with respect to X_{max} .

3 Conclusions and Outlook

A study of the radio frequency spectrum in the 30 – 70 MHz regime has been performed. Preliminary results depict a very strong correlation between the feature of the frequency spectrum and the distance to the shower axis. This is highlighted by a good agreement between simulations and real event for both analysis methods, as shown in figures 1 and 2 . A correlation between the frequency spectrum and the atmospheric depth of the shower maximum X_{max} has been studied as well. Results performed on simulated data show a clear correlation (figure 3), while this dependence on real data is still under investigation.

References

[1] M.P. van Haarlem et al., *A&A* **556**, A2, 53 (2013)
[2] S. Thoudam et al., *Nucl. Instr. Meth. Phys. Res.* A **767**, 339 (2014)
[3] H.R. Allan, *Prog. in Element. Part. and Cos. Ray Phys.* **10**, 171 – 302 (1971)
[4] H. Falcke and P.W. Gorham, *Astropart. Phys.* **19**, 477 – 494 (2003)
[5] T. Huege and H. Falcke, *A&A* **412**, 19 – 34 (2003)
[6] T. Huege and H. Falcke, *A&A* **430**, 779 – 798 (2005)
[7] A. Nigl et al., *A&A* **488**, 807 – 817 (2008)
[8] S. Jansen, *PhD thesis, Radio for the Masses - Cosmic ray mass composition measurements in the radio frequency domain*, ISBN 978 94 028 0073 9 (2016)
[9] T. Huege, M. Ludwig, and C. James, in proceedings of *AIP Conference* **1535**, 128 – 132 (2012)
[10] L. Rossetto, in proceedings of *ICRC 2015 Conference*, **PoS(ICRC2015)381** (2015)
[11] A. Nelles et al. *JCAP* **05**, 18 (2015)
[12] S. Buitink et al., *Nature* **531**, 70 (2016)

Towards real-time identification of cosmic rays with LOw-Frequency ARray radio antennas

Antonio Bonardi[1,★], *Stijn* Buitink[2], *Arthur* Corstanje[1], *J. Emilio* Enriquez[1], *Heino* Falcke[1,3,4], *Jörg R.* Hörandel[1,4], *Pragati* Mitra[2], *Katie* Mulrey[2], *Anna* Nelles[5], *Jörg Paul* Rachen[1], *Laura* Rossetto[1], *Pim* Schellart[1], *Olaf* Scholten[6,7], *Satyendra* Thoudam[1], *Gia* Trinh[6,7], *Sander* ter Veen[2], and *Tobias* Winchen[2]

[1]*Department of Astrophysics/IMAPP, Radboud University, P.O. Box 9010, 6500 GL Nijmegen (the Netherlands)*
[2]*Vrije Universiteit Brussel, Pleinlaan 2, 1050 Brussels (Belgium)*
[3]*Astron, Oude Hoogeveensedijk 4, 7991 PD Dwingeloo (The Netherlands)*
[4]*Nikhef, Science Park 105, 1098 XG Amsterdam (The Netherlands)*
[5]*University of California Irvine, Irvine, CA 92697, (USA)*
[6]*KVI-CART, Zernikelaan 25, 9747 AA Groningen (The Netherlands)*
[7]*University of Groningen, Zernikelaan 25, 9747 AA Groningen (The Netherlands)*

Abstract. Cosmic rays entering the Earth's atmosphere produce Extensive Air Showers, which emit a radio signal through Geo-magnetic radiation and Askaryan emission. At the present time, one of the biggest challenges for assessing the Radio detection as a valuable technique for Cosmic-ray observation is to identify in real-time the very short (less than 100 ns) radio signals over the background noise. In this work, we present the latest updates on the real-time identification of radio signals from Extensive Air Showers by using the data from LOFAR Low Band Antenna stations, which are sensitive in the 30-80 MHz region.

1 Introduction

When a Cosmic Ray enters the Earth's atmosphere, it starts an Extensive Air Shower (hereafter "EAS") of secondary particles, which emits a radio signal, through Geo-magnetic radiation and Askaryan emission.
It has been recently proven that the primary particle properties, as arrival direction, energy, and mass composition, can be reconstructed by ground-based radio detector arrays [1],[2],[3] with an accuracy similar, if not even better, than current Cosmic-ray experiments. The big advantage of the Radio-detection technique is its very high duty cycle (virtually up to 100%) and its construction and operational costs, which are considerably smaller compared to current experiments based on Fluorescence-detection technique or secondary particles detection. On the other hand, due to the very short duration of the radio signal (less than 100 ns) and the overwhelming background noise of Radio-Frequency Interferences (hereafter "RFI"), the direct detection of EAS with Radio detectors

★e-mail: a.bonardi@astro.ru.nl

 DOI: 10.1051/epjconf/201713501011

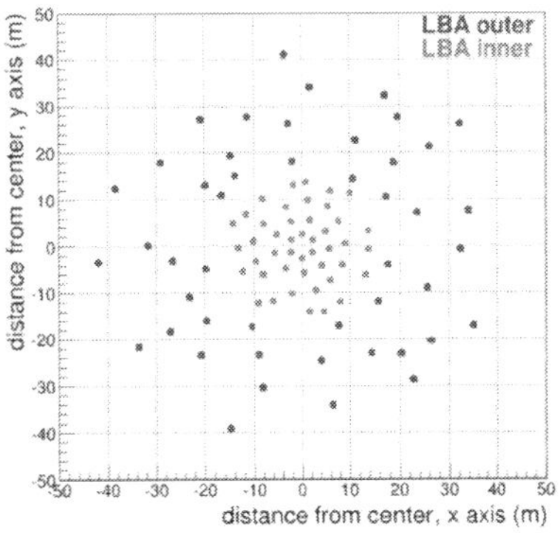

Figure 1. Antenna location in one LOFAR station for LBA inner and LBA outer configuration

only has been very difficult so far. As consequence, particle detectors are usually used in combination with Radio detectors as trigger for the Radio data acquisition, but this procedure sets currently a strong limitation to the exploitation of the Radio detection technique.

The multipurpose LOw-Frequency ARray (LOFAR)[4] is currently the best-suited experiment for developing a realtime Radio-detection algorithm for EASs, thanks to the external trigger provided by the LORA[5] particle detector array and to the very large number of omnidirectional Low-frequency Band Antennas (hereafter "LBA"), each of them equipped with two polarization arms NE-SW and NO-SE oriented and operational in the 30-80 MHz frequency band. LBAs are clustered in units of 96 antennas, each of them called "station". The highest density of stations is the innermost region which is located nearby Exloo (the Netherlands), while remote stations are spread all around Central and Northern Europe. During standard astronomical observation, on each station only half of LBAs are active in the so-called "LBA inner" or "LBA outer" configuration, as shown in Fig. (1). When LORA provides a trigger, the time series ($\sim$ 2.2 ms length, 200 MSample s^{-1}) of each active antenna polarization is downloaded for further offline analysis. At the present time, just the innermost region is covered by the LORA array, significantly affecting the LOFAR Cosmic-ray detection at energies above 10^{18} eV, where a larger covered area is needed.

2 The Radio self-trigger algorithm

The developed algorithm consists of two steps: in the first step Radio pulses are detected ("selection phase"), while in the second one RFI-candidates are rejected ("RFI rejection phase").
During the selection phase, the radio signal of each antenna along both polarizations is compared with its baseline fluctuation: when on a certain antenna the signal of one polarization is above the given threshold, the antenna is considered as "fired" along that polarization. A station is considered "triggered" if at least 24 antennas are fired along the same polarization within a coincidence time of 20 ns (in the LBA inner configuration) or 30 ns (in the LBA outer configuration).
The RFI rejection phase is based on a multi-criteria approach, consisting of evaluating the radio pulse duration, its frequency spectrum, its timing and its arrival zenith angle. Radio pulses satisfying any of the following conditions are threated as RFI and rejected:

- **pulse duration:** the time window where 3 or more antennas are fired (within the coincidence time) is larger than 75 ns (LBA inner) or 300 ns (LBA outer);

Table 1. Trigger performances for LBA inner configuration

Threshold Level	CR triggers	rejected CR triggers	RFI triggers after rejection
3 RMS	395	65%	0
4 RMS	246	36%	0
5 RMS	168	27%	0
10 RMS	51	35%	0

Table 2. Trigger performances for LBA outer configuration

Threshold Level	CR triggers	rejected CR triggers	RFI triggers after rejection
3 RMS	1536	25%	4
4 RMS	899	21%	2
5 RMS	598	22%	2
10 RMS	196	24%	0

- **frequency spectrum:** the ratio between the station mean pulse intensity in the 30-45 MHz frequency band and in the 45-70 MHz band is smaller than 1 or larger than 2;
- **timing:** the pulse occurs less than 5 μs before/after from the next/previous triggered pulse;
- **arrival direction:** its reconstructed arrival zenith angle is larger than 60° (only used with the LBA inner configuration).

3 Analysis and Results

For this study we considered the raw-data from the 4081 recorded events with LBA outer configuration and the 634 events with LBA inner configuration. All the considered events have time, energy, position, and arrival direction successfully reconstructed by LORA, and have at least one LOFAR station with all the antenna channels properly working. Stations with one or more missing read-out channels have been removed.
In the selection criterion four different threshold levels have been considered for the antenna "firing" condition, equal to 3, 4, 5, and 10 times the baseline RMS. The detected radio pulses are considered originated by the same EAS detected by LORA, if they occur within 5 μs before and 12.5 μs after the LORA time stamp. Any pulse falling outside this time window is considered as RFI pulse.
The results obtained for LBA inner and LBA outer are summarized in Table 1 and Table 2.

For both LBA inner and LBA outer, the primary particle energy distributions of all the analyzed events and of the events generating a trigger in at least one station is shown in Fig. (2). The distributions of distance from the EAS core position of all the available stations and of the triggered ones are shown in Fig. (3) . Fig. (2) and Fig. (3) show that for LOFAR stations the energy threshold increases and the effective collecting area slightly decreases by increasing the trigger threshold level.

4 Conclusions and Outlook

For both LBA inner and LBA outer, we obtained a good trigger efficiency together with a very good RFI rejection efficiency, especially for primary energy above $2 \cdot 10^{17}$ eV. If we exclude data acquired in not fair-weather conditions, only one RFI pulse passed our selection and rejection criteria. This was found with 3 RMS threshold level and LBA outer configuration.

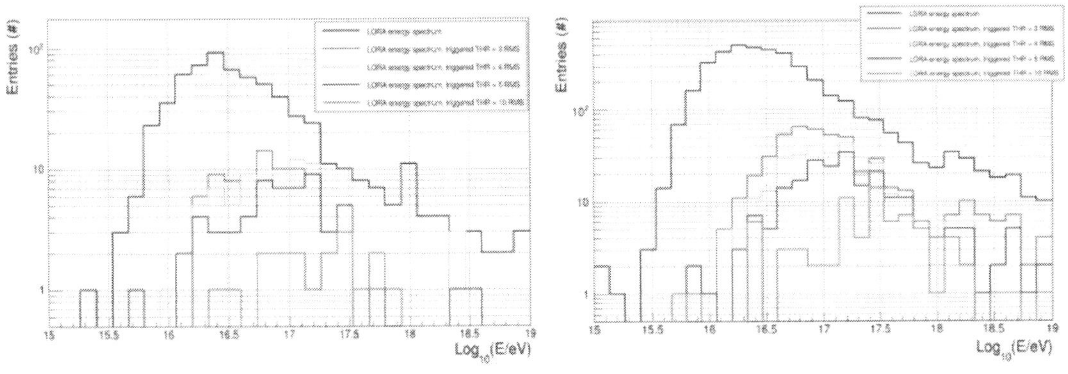

Figure 2. Primary particle energy distributions for LBA inner configuration (left) and LBA outer configuration (right), as reconstructed by LORA. The distribution of the whole set of events is shown together with the distribution of triggered events at the trigger threshold equal to 3, 4, 5, and 10 RMS.

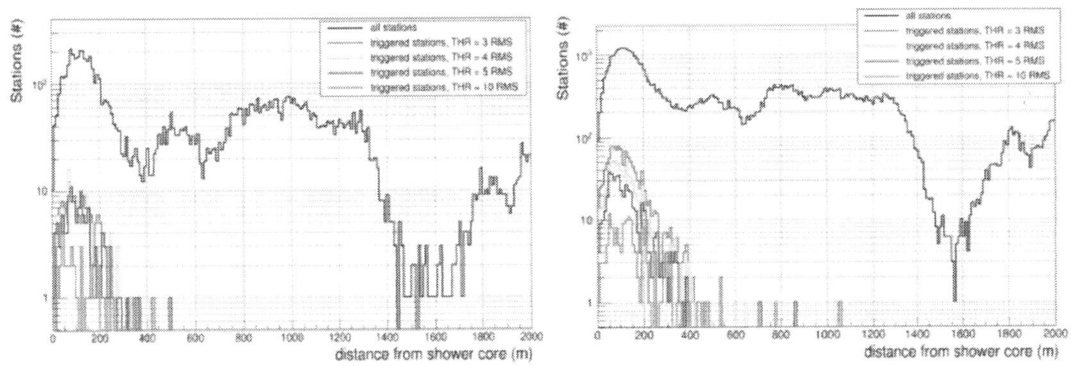

Figure 3. Station distance distribution from the LORA reconstructed EAS core position for LBA inner configuration (left) and LBA outer configuration (right). The distribution of all the available stations is shown together with the distributions of the only triggered stations at the 4 threshold levels.

On the other hand, the available statistics is not enough (~9 s of independent live-time and ~119 s total live-time for LBA outer, ~2 s and ~22 s for LBA inner) for concluding that the desired RFI rejection efficiency has been achieved. Therefore, low threshold acquisitions are planned to be performed on one LOFAR test station for increasing the available statistics. Once the RFI trigger rate will be found smaller than 1 per hour per station, the following step will be the implementation of the self-trigger algorithm to the LOFAR remote stations, so to boost up the cosmic-ray collecting area.

References

[1] Corstanje A. et al. *Astropart. Phys.* **61**, 22 (2014)
[2] Nelles A. et al. *JCAP* **05**, 18 (2015)
[3] Buitink S. et al., *Nature* **531**, 70 (2016)
[4] Van Haarlem M. P. et al., *A&A* **556**, 53 (2013)
[5] Thoudam S. et al., *Nucl. Instr. Meth. Phys. Res.* A **767**, 339 (2014)

DOI: 10.1051/epjconf/201713501012

The influence of the atmospheric refractive index on radio X_{max} measurements of air showers

Arthur Corstanje[1,⋆], *Stijn* Buitink[2], *Antonio* Bonardi[1], *Heino* Falcke[1,3,4], *Jörg R.* Hörandel[1,4], *Pragati* Mitra[2], *Katie* Mulrey[2], *Anna* Nelles[5], *Jörg Paul* Rachen[1], *Laura* Rossetto[1], *Pim* Schellart[1], *Olaf* Scholten[6,7], *Satyendra* Thoudam[1], *Gia* Trinh[6,7], *Sander* ter Veen[3], and *Tobias* Winchen[2]

[1]*Department of Astrophysics/IMAPP, Radboud University, P.O. Box 9010, 6500 GL Nijmegen (the Netherlands)*
[2]*Vrije Universiteit Brussel, Pleinlaan 2, 1050 Brussels (Belgium)*
[3]*Astron, Oude Hoogeveensedijk 4, 7991 PD Dwingeloo (The Netherlands)*
[4]*Nikhef, Science Park 105, 1098 XG Amsterdam (The Netherlands)*
[5]*University of California Irvine, Irvine, CA 92697, (USA)*
[6]*KVI-CART, Zernikelaan 25, 9747 AA Groningen (The Netherlands)*
[7]*University of Groningen, Zernikelaan 25, 9747 AA Groningen (The Netherlands)*

Abstract. The refractive index of the atmosphere, which is $n \approx 1.0003$ at sea level, varies with altitude and with local temperature, pressure and humidity. When performing radio measurements of air showers, natural variations in n will change the radio lateral intensity distribution, by changing the Cherenkov angle. Using CoREAS simulations, we have evaluated the systematic error on measurements of the shower maximum X_{max} due to variations in n. It was found that a 10 % increase in refractivity $(n-1)$ leads to an underestimation of X_{max} between 8 and 22 g/cm^2 for proton-induced showers at zenith angles from 15 to 45 degrees, respectively.

1 Introduction

Measurements of the radio emission from air showers, together with microscopic simulations, form a useful method to infer their defining physical quantities, such as primary energy [1] and the depth of shower maximum X_{max} [2]. The latter is sensitive to the composition of the primary particles.

Using the densely instrumented inner region of the LOFAR radio telescope [3] has allowed to sample the radio lateral intensity distribution ('footprint') in close detail, and to infer X_{max} to a precision of 20 g/cm^2 [2]. This precision is comparable with more common fluorescence measurements (see e.g. [4]), and good for detailed composition studies [5]. In a central ring of 320 m diameter, one can use nearly 300 dual-polarized antennas, operating in the 30 – 80 MHz band. Outside this ring, 18 more stations of 96 antennas are spread over an area of 6 km^2. LOFAR has measured the radio signals from air showers since 2011 [6]. For the X_{max} measurement, the measured radio footprints are compared to simulated footprints, using the CoREAS simulation package [7] for the radio emission, and CORSIKA [8] for the particles in the air showers.

⋆e-mail: a.corstanje@astro.ru.nl

In CORSIKA and CoREAS, the US Standard atmosphere [9] is taken by default as the model atmosphere. It features a constant temperature lapse rate of $L = 6.5\,\mathrm{K/km}$, zero humidity, and a pressure altitude profile given by

$$T(h) = T_0 - Lh, \tag{1}$$

$$p(h) = p_0 \left(\frac{T(h)}{T_0} \right)^{\frac{gM}{LR}}. \tag{2}$$

Standard values are $p_0 = 1013.25\,\mathrm{hPa}$, $T_0 = 288.15\,\mathrm{K}$, $g = 9.80665\,\mathrm{m/s^2}$, $R = 8.31447\,\mathrm{J/(mol\,K)}$ and $M = 0.0289644\,\mathrm{kg/mol}$. For more realistic circumstances, humidity can be added as a non-zero partial pressure of water vapour p_w. The refractivity $N = 10^6\,(n-1)$ is given by [10]

$$N = 77.6890\,\frac{p_d}{T} + 71.2952\,\frac{p_w}{T} + 375463\,\frac{p_w}{T^2}, \tag{3}$$

with p_w and dry-air pressure p_d in hPa, and T in K.

In CoREAS, a default refractivity is taken as $N = 292$ at sea level. Using the above formulas, we obtain refractivity values at altitudes 3 and 5 km as a function of sea-level temperature, Fig. (1) and Fig. (2). Shown is the relative correction in % to the CoREAS default. It is seen that at lower altitudes, N is generally overestimated by about 2 %, with a large spread with humidity at higher temperatures. At higher altitudes, relevant for slanted air showers, N is overestimated by about 4 %. Below we evaluate the effect of a 10 % change in refractivity on measurements of $X_{\max}$; a more accurate treatment of the simulations, by including local atmospheric data from the GDAS database [11] is under investigation.

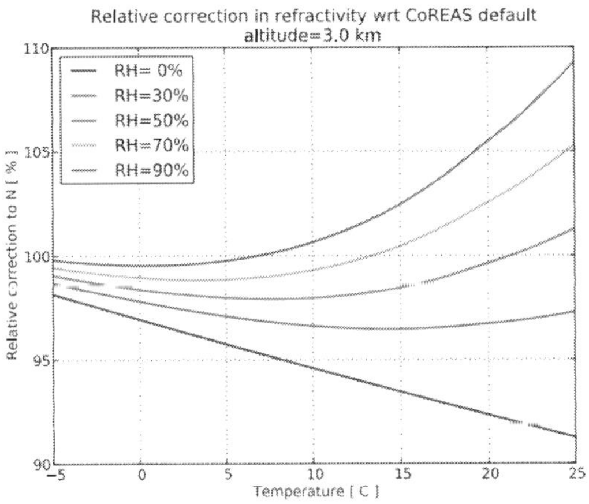

Figure 1. Correction to default refractivity, versus sea-level temperature, at 3 km altitude and relative humidity *RH*.

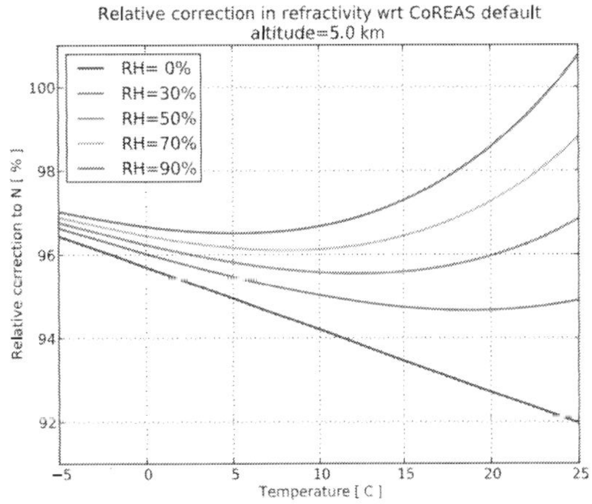

Figure 2. Idem at 5 km altitude.

2 Method

Using CoREAS simulations, we have determined the systematic offset in $X_{\max}$ measurements for a 10 % increase in refractivity. For proton primary particles coming in at four different zenith angles, we have created 50 simulated showers at a default refractivity of $N = 292$ at sea level, and 50 showers at a 10 % higher refractivity. The refractivity at higher altitudes is proportional to air density for dry air, and is taken as such in the simulations, which currently assume dry air.

The two ensembles of showers are compared, by taking one shower from the higher-refractivity ensemble, and fitting the lateral intensity distribution by all 50 from the normal-N ensemble. A mean squared error is used as a fit quality measure. This method is similar to [2], where we compare two simulated ensembles, instead of simulation versus data. As the simulated $X_{\max}$ is known for all showers, the optimum in the fit quality (fitted $X_{\max}$) is compared to the true $X_{\max}$. Due to natural shower-to-shower fluctuations, scatter arises in the fit quality plots; an example is shown in Fig. (4). To lowest order, the fit quality is a quadratic function of $X_{\max}$ near its optimum. Therefore, we fit the points with a parabola. A weighting factor of $1/y^2$ is introduced to put more emphasis on well-fitting showers, increasing precision of the $X_{\max}$-fit by about 20 %.

Averaging fitted minus true $X_{\max}$ over the 50 showers yields the systematic offset in $X_{\max}$. We compare the simulation results with a simplistic model, assuming the size of the radio footprint scales with the Cherenkov angle at the $X_{\max}$ altitude. The altitude-dependent Cherenkov angle α is given by

$$\cos\alpha = \frac{1}{\beta\, n(h)}. \tag{4}$$

The model is depicted schematically in Fig. (3). It is expected to hold qualitatively for higher frequencies, above 100 MHz, as for these frequencies, a clear Cherenkov ring is observed both in simulations and in LOFAR data [12].

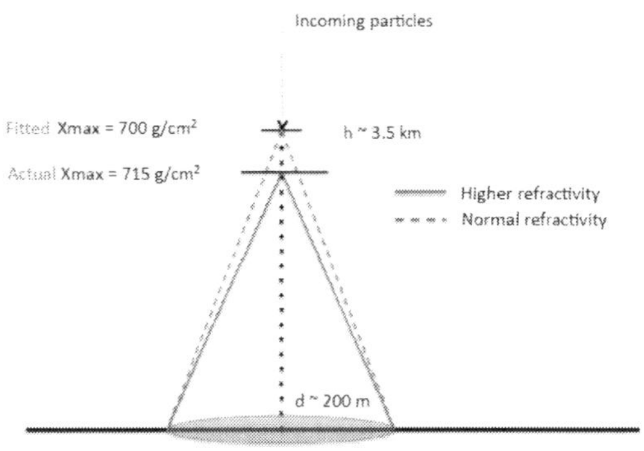

Figure 3. Cherenkov model.

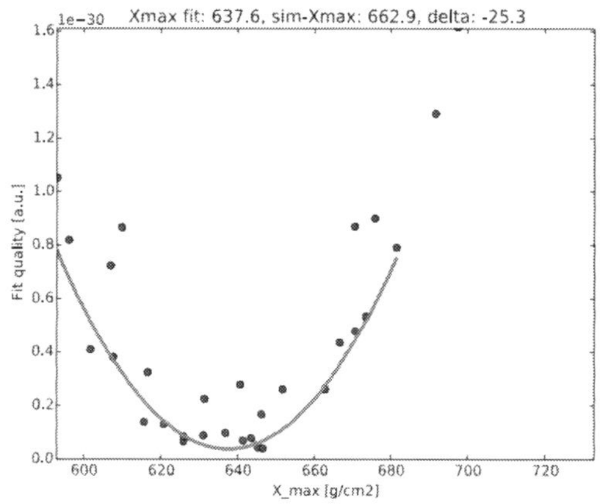

Figure 4. Fit quality as a function of $X_{\max}$.

3 Results

The offset in $X_{\max}$ per 10 % change in refractivity has been evaluated for four zenith angles in the range of 15 to 45 degrees. For each shower in the ensemble, the difference between fitted and simulated $X_{\max}$ is shown in Fig. (5). A quadratic fit to the points is shown.

A modeled offset is calculated from the simplistic assumption that the radio footprint would scale (only) with the size of the Cherenkov cone, starting at the $X_{\max}$ altitude. The offsets thus found are nearly twice as large as those found from full simulations. This indicates that Cherenkov time compression is not the only factor that sets the shape of the radio footprint, as is known for lower frequencies (< 100 MHz).

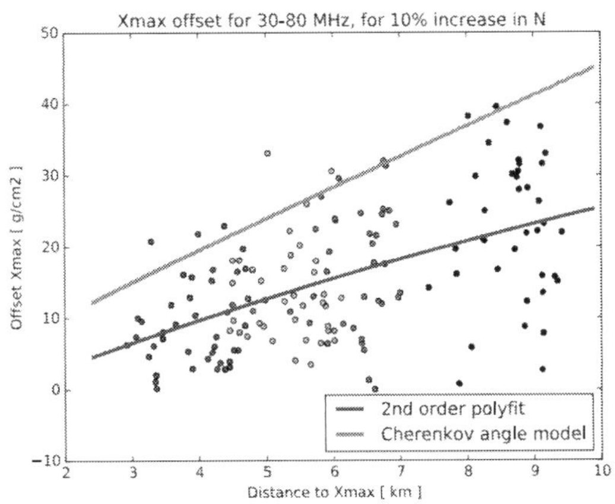

Figure 5. Results for the LOFAR low-band frequency range, 30 – 80 MHz.

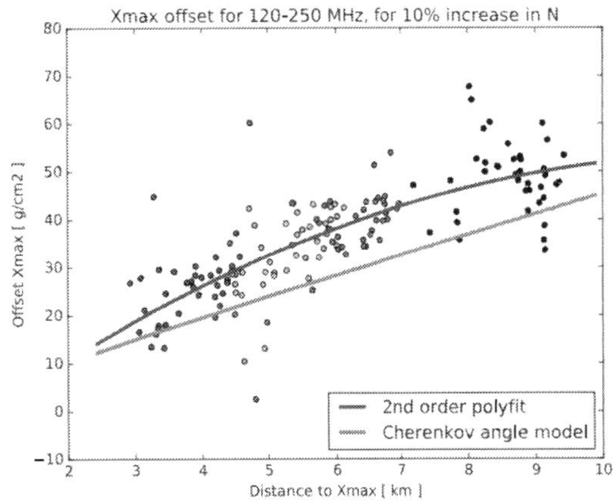

Figure 6. Results for a higher-frequency range, 120 – 250 MHz.

4 Summary

For mass composition analysis, the accuracy of X_{max} measurements is critical. Here, we have simulated the systematic error due to variations in the atmospheric refractive index. For the 30 – 80 MHz band, a 10 % change in refractivity yields changes in inferred X_{max} from 8 to 22 g/cm^2 for zenith angles ranging from 15 to 45 degrees. At higher frequencies, 120–250 MHz, the range is 20 to 45 g/cm^2, and agrees qualitatively with a toy model based on the Cherenkov angle at the X_{max} altitude.

Realistic corrections to the CoREAS default refractivity are on the order of 4 %, leading to a zenith-angle dependent bias of up to about 10 g/cm^2. Including local atmospheric data per air shower into the simulations is a subject of further investigation.

References

[1] A. Aab et al. (Pierre Auger Collaboration), Phys. Rev. D **93**, 122005 (2016)
[2] S. Buitink et al., Phys. Rev. D **90**, 082003 (2014)
[3] M.P. van Haarlem et al., A & A **556**, A2 (2013), 1305.3550
[4] K.H. Kampert, M. Unger, Astroparticle Physics **35**, 660 (2012)
[5] S. Buitink et al., Nature p. 70 (2016)
[6] P. Schellart, A. Nelles, et al. LOFAR collaboration, A & A **560**, A98 (2013), 1311.1399
[7] T. Huege, M. Ludwig, C.W. James, AIP Conference Proceedings **1535**, 128 (2013)
[8] D. Heck et al., *CORSIKA: a Monte Carlo code to simulate extensive air showers.* (1998)
[9] U.S. Government Printing Office (1976)
[10] J. Rüeger, Proceedings of FIG XXII International Congress, 2002 (2002)
[11] National Oceanic and Atmospheric Administration, *Global Data Assimilation System* (2016)
[12] A. Nelles, P. Schellart et al., Astroparticle Physics **65**, 11 (2015)

Nanosecond-level time synchronization of AERA using a beacon reference transmitter and commercial airplanes

Tim Huege[1,*] for the Pierre Auger Collaboration[2]

[1] *Karlsruhe Institute of Technology (KIT), Institut für Kernphysik (IKP), Karlsruhe, Germany*
[2] *Full author list: http://auger.org/archive/authors_2016_06.html*

Abstract. Radio detection of cosmic-ray air showers requires time synchronization of detectors on a nanosecond level, especially for advanced reconstruction algorithms based on the wavefront curvature and for interferometric analysis approaches. At the Auger Engineering Radio Array, the distributed, autonomous detector stations are time-synchronized via the Global Positioning System which, however, does not provide sufficient timing accuracy. We thus employ a dedicated beacon reference transmitter to correct for event-by-event clock drifts in our offline data analysis. In an independent cross-check of this "beacon correction" using radio pulses emitted by commercial airplanes, we have shown that the combined timing accuracy of the two methods is better than 2 nanoseconds.

1 Introduction

With the Auger Engineering Radio Array (AERA) [1], we measure radio emission from cosmic-ray air showers [2, 3] in the 30-80 MHz band. To maximize the event reconstruction potential, a highly accurate time synchronization of the individual radio detectors is required. In particular, interferometric and wavefront analyses rely on a time synchronization to within 1 ns or better. In cabled setups, nanosecond time synchronization can be achieved with established techniques. AERA, however, consists of independent, autonomous detector stations which are only connected wirelessly, making accurate time synchronization very challenging. In this article, we describe how we achieve a time synchronization of the AERA detector stations to within 2 ns using a dedicated reference transmitter, a so-called "beacon", together with an independent cross-check based on the measurement of radio pulses emitted by commercial airplanes. Details of the presented analysis can be found in [4].

2 Time synchronization with a reference beacon

The individual AERA stations each use a GPS clock to synchronize the array timing. Manufacturers state timing accuracy of order 5 ns for their GPS clocks. This, however, is insufficient for interferometric and wavefront analyses. To improve the time synchronization, we thus operate a reference beacon transmitter [5] in the vicinity of the radio detector array. This transmitter emits monochromatic, continuous sine waves at frequencies of 58.887 MHz, 61.523 MHz, 68.555 MHz, and 71.191 MHz, i.e.,

*e-mail: tim.huege@kit.edu

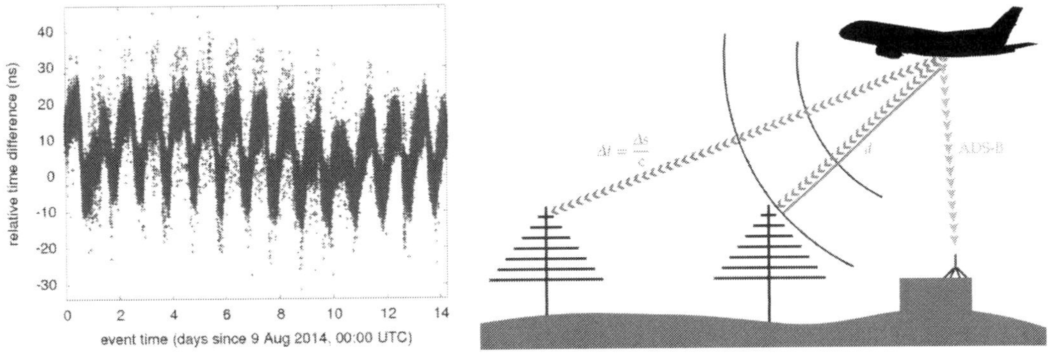

Figure 1. Left: Relative timing between two AERA stations as measured using the beacon phasing. Each dot represents a measured offset with respect to the expectation, interpreted as a time offset between the GPS clocks of the two detector stations. Right: Illustration of the time calibration using commercial airplanes. The airplane broadcasts real-time position information via digital ADS-B packets at 1090 MHz, which are received and interpreted by a dedicated setup in the AERA field in real time. If the airplane also emits pulsed signals in the frequency range of 30–80 MHz recorded by the AERA detector stations, it can be used as a timing calibration source.

within the AERA detection band of 30-80 MHz. Whenever an air-shower event is recorded with AERA, the beacon transmitter signals are thus recorded within the normal data stream. As radio pulses from air showers are broad-band, the very narrow-band signals of the beacon transmitter can be suppressed in an offline analysis [6] with minimal loss of data quality once they have been used for timing calibration.

The phasing of the four frequencies expected at a given detector station with respect to the phasing at an arbitrary reference station can be calculated from the relative positions of the two stations and the position of the beacon transmitter. Any deviations from the expected phasing signify a timing offset between the two compared stations, i.e., a drift in the GPS clock of either or both of the two stations. In fact, an analysis of the measured reference phases reveals that there are very significant drifts on timescales of minutes to days, well beyond the nominal 5 ns timing accuracy of the GPS clocks, as is shown in the left panel of Fig. 1.

Using the beacon phasing, the relative GPS clock drift can be measured and corrected for on a per-event level. However, it was unclear whether the beacon purely measures GPS clock drifts or can be affected by other factors, e.g., environmental conditions that could possibly influence the beacon signal transmission. An independent cross-check of this method was therefore needed.

3 Time synchronization with airplane pulses

To independently cross-check the beacon method, we need a source with a known position which emits pulsed radio signals in the 30-80 MHz band. These signals need to be measurable simultaneously in as many AERA stations as possible. It turns out that some commercial airplanes emit detectable radio pulses in this frequency band. The aircraft position can be determined in real-time from the Automatic Dependent Surveillance Broadcast (ADS-B) messages that modern airplanes transmit digitally at a frequency of 1090 MHz. To intercept this position information, we have installed a "software-defined radio" setup based on a commercial USB receiver for digital terrestrial video broadcasts (DVB-T). The total cost for the setup was below 50 USD, yet it has been running very reliably

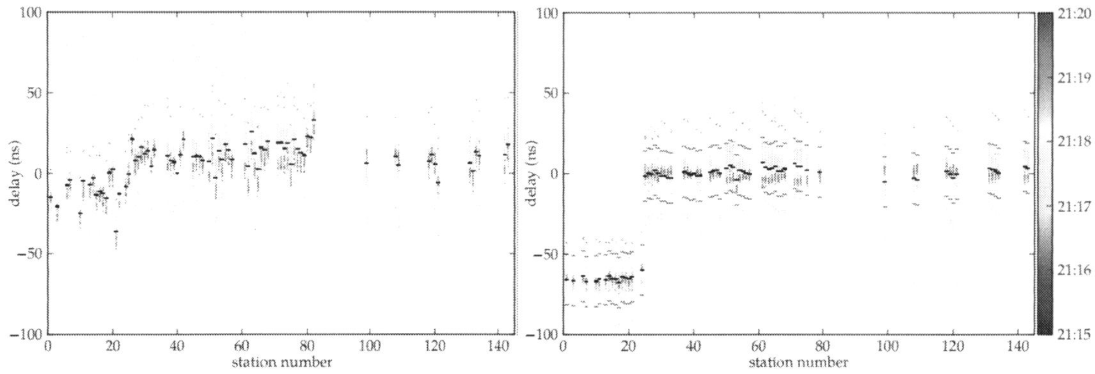

Figure 2. Left: Time offsets determined from AERA and ADS-B data for an individual airplane. Offsets are relative to station 40 chosen as an arbitrary reference. Gaps in the station numbering are related to detector stations not participating in the airplane trigger. The color code denotes the local time at which the airplane pulses were detected. Right: Time offsets when applying the airplane timing calibration after per-event clock-drift correction using the beacon method. The red lines show the range of mean ± 4 · MAD (median absolute deviation), values further out are ignored for the calculation of the mean time offsets.

in the field for years. We use the position information for two purposes: First, we notify our data acquisition in real-time of approaching airplanes and allow repeated triggers from the corresponding direction to pass through — normally, signals like these that are clearly not of cosmic-ray origin are filtered out by our self-trigger algorithms. Second, in an offline analysis, we use the airplane position to calculate the expected arrival times of the pulsed emissions at the individual AERA stations, which we then compare with the measured pulse arrival times. Deviations point to offsets in the timing of the individual AERA stations, in particular arising from their GPS clocks. The concept is sketched in the right panel of Fig. 1.

In the left panel of Fig. 2, we show the time offsets (with respect to arrival time at a chosen reference station) for each individual detector station as determined from the radio pulses emitted by a single airplane that overflew AERA over the course of approximately 5 minutes. Each individual data point marks a measurement for an airplane radio pulse emitted at the indicated local time. There is significant scatter as well as clear outliers in these data for a given station. The outliers can be explained by misidentification of the pulse maximum and thus the arrival time. The visible time-ordering of the data is not fully understood. Nevertheless, the mean of the time offsets can be determined per station. If time synchronization over the whole array were perfect, all means should be at a delay value of zero. However, the mean values scatter by tens of nanoseconds over the various stations, indicating deviations in the time synchronization of the same order as was observed in the beacon reference phasing measurements shown in the left panel of Fig. 1.

In the right panel of Fig. 2, the same data were analyzed. However, the beacon timing correction was applied before determining the timing offsets on the basis of the airplane analysis. If the beacon timing correction of the GPS clocks works as expected, all means should be shifted to the zero line. For stations 25 and higher, this is indeed the case. These are detector stations using Butterfly antennas [7], same as the reference station chosen in this analysis (station 40). Stations 1-24 use logarithmic-periodic dipole antennas (LPDAs) [7] and have mean time offsets of approximately -65 ns. This means that after the beacon timing correction, indeed stations with the same type of antenna have a much better time synchronization than before the beacon correction. However, there is a previously

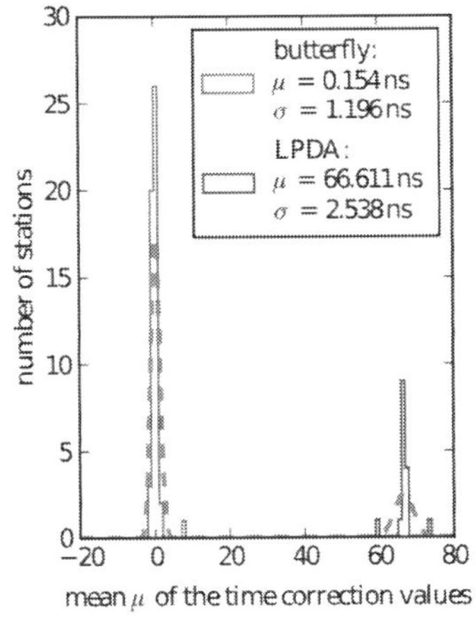

Figure 3. Histogram of the mean time correction values μ determined from the combination of beacon and airplane time synchronization methods. The standard deviations σ of the distributions provide a measure for the average agreement between the two methods. The values stated in the statistics box are from fitted Gaussians (dashed lines).

unknown time difference between the two types of antenna, which can easily be corrected for now that it is known.

A more quantitative view at the distribution of the mean values after beacon correction is given in Fig. 3. The standard deviation of the mean values within the group of detector stations with the same antenna is of order 2 ns. The combined timing accuracy of both the beacon- and airplane methods is thus of order 2 ns. The airplane method is expected to have the larger systematic uncertainties (see, e.g., the time-ordering of data points), so that the beacon method alone might well be more accurate than the combined 2 ns.

4 Conclusion

Nanosecond-level time synchronization of distributed arrays without cable connection is a challenging task. In AERA, we have tackled this challenge with two independent approaches. A reference beacon transmitter continuously emits signals that are recorded in our normal event data stream and can be used to correct GPS clock offsets on a per-event basis. Indeed, offsets of order tens of nanoseconds are visible in our data, i.e., the nominal 5 ns accuracy of the GPS clocks is not reached in practice. To validate the beacon-correction method, we have also determined timing offsets in our detector array on the basis of radio pulses emitted by commercial airplanes, the positions of which can be determined in real-time from ADS-B transmissions. The airplane analysis confirms that the beacon timing correction works and reaches an accuracy of 2 ns or better. In addition, a systematic time offset between the LPDA and Butterfly antennas used in AERA was discovered and is now routinely corrected for.

References

[1] J. Schulz for the Pierre Auger Collaboration, in *Proc. 34th ICRC* (2015), PoS(ICRC2015)615
[2] T. Huege, Physics Reports **620**, 1 (2016)
[3] F.G. Schröder, Prog. Part. Nucl. Phys. (submitted), `arXiv:1607.08781`
[4] A. Aab, P. Abreu, M. Aglietta, E. Ahn, et al., JINST **11**, P01018 (2016)
[5] F.G. Schröder, T. Asch, L. Bähren, et al., Nucl. Instr. Meth. A **615**, 277 (2010)
[6] P. Abreu, M. Aglietta, E.J. Ahn, et al., Nucl. Instr. Meth. A **635**, 92 (2011)
[7] P. Abreu, M. Aglietta, M. Ahlers, et al., JINST **7**, P10011 (2012)

EPJ Web of Conferences **135**, 01014 (2017) DOI: 10.1051/epjconf/201713501014
ARENA 2016

In-situ absolute calibration of electric-field amplitude measurements with the LPDA radio detector stations of the Pierre Auger Observatory

Florian Briechle[1,*] for the Pierre Auger Collaboration[2]

[1] *RWTH Aachen University, III. Physikalisches Insitut A*
[2] *Full author list: http://www.auger.org/archive/authors_2016_06.html*

Abstract. With the Auger Engineering Radio Array (AERA) located at the Pierre Auger Observatory, radio emission of extensive air showers is observed. To exploit the physics potential of AERA, electric-field amplitude measurements with the radio detector stations need to be well-calibrated on an absolute level. A convenient tool for far-field calibration campaigns is a flying drone. Here we make use of an octocopter to place a calibrated source at freely chosen positions above the radio detector array. Special emphasis is put on the reconstruction of the octocopter position and its accuracy during the flights.
The antenna response pattern of the radio detector stations was measured in a recent calibration campaign. Results of these measurements are presented and compared to simulations. It is found that measurements and simulations are in good agreement.

1 Introduction

To be able to do high precision measurements of ultra-high energy cosmic rays with the Auger Engineering Radio Array (AERA) [1, 2] at the Pierre Auger Observatory [3], the radio detector needs to be well-calibrated. The largest challenge in performing such a calibration is to place a signal source at every direction relative to the detector station and at least 20 m away to guarantee far-field conditions. Different calibration methods have been developed and already performed at AERA [4] and other experiments [5–7]. In this contribution, we present a novel and very flexible method using a flying drone to carry the signal source. A drone is a very useful tool, since it can be placed at any point above the array and can carry any piece of equipment as long as it is not too heavy. With the presented method, a higher accuracy can be achieved than in the previous calibrations mostly due to a larger set of statistics and a better handling of environmental conditions.

At AERA, two different antenna types are used. The core consists of a dense array of 24 stations equipped with log-periodic dipole antennas (LPDA). They consist of nine separate dipoles optimized for a frequency range of 30 to 80 MHz. The other stations are equipped with butterfly antennas. However, in this proceeding, only LPDA stations are considered.

Extensive air showers produce short radio pulses. These pulses then induce a signal measurable as voltages at the antenna output. The value which relates the Fourier transformed electric field $\vec{\mathcal{E}}$ to the measured Fourier transformed voltage $\mathcal{U}$ is called the vector effective length (VEL) $\vec{H}$:

*e-mail: briechle@physik.rwth-aachen.de

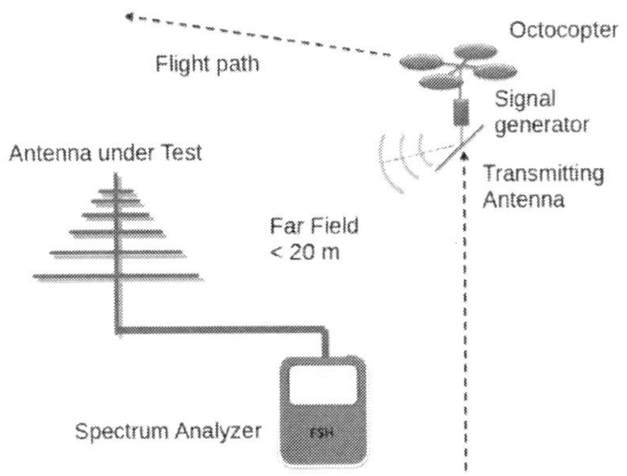

Figure 1. Experimental setup for the calibration measurement.

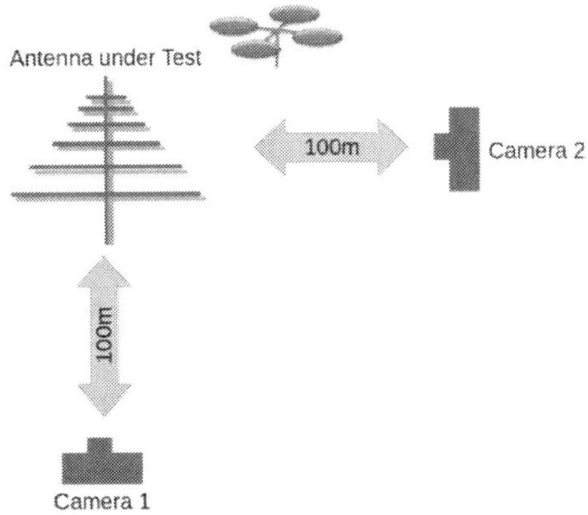

Figure 2. Experimental setup for the optical position determination.

$$\mathcal{U}(f,\theta,\phi) = \vec{H}(f,\theta,\phi)\cdot\vec{\mathcal{E}}(f). \tag{1}$$

$\vec{H}$ depends on the frequency of the signal f and the incoming direction, described by the zenith angle θ and the azimuth angle ϕ. In spherical coordinates, $\vec{H}$ is a superposition of a horizontal H_ϕ and a perpendicular meridional H_θ component. The absolute value of these components is determined in a transmission measurement using a calibrated transmitting antenna. For the absolute value of a VEL component at a specific frequency and direction the following holds

$$|H_i| = \sqrt{\frac{4\cdot\pi\cdot Z_A}{Z_0}}\cdot R\cdot\sqrt{\frac{P_{r,i}}{G_t\cdot P_t}}, \tag{2}$$

where $i = \theta, \phi$ denotes the VEL component, Z_A denotes the impedance of the antenna, $Z_0 = 120\pi\Omega$ is the vacuum impedance, R describes the distance between the antenna under test (AUT) and the transmitting antenna, P_r denotes the power received at the AUT, and G_t and P_t are the gain and power of the transmitting antenna [8].

2 Calibration Setup

A sketch of the setup for the calibration measurements is shown in Fig 1. The AUT is connected to a spectrum analyzer model Rohde & Schwarz FSH 4. This device measures the received power at each frequency. A signal generator and a calibrated transmitting antenna are attached to an octocopter and placed at a distance of at least 20 m to make sure far-field conditions apply. The signal generator RSG-1000 produces continuous signals at every 5 MHz. During the measurement the octocopter then flies straight up to a height of approximately 30 m, flies towards the antenna until it is directly above it. Afterwards, it flies back and finally lands at the starting position. In this way, the complete zenith angle range is covered.

3 Reconstruction of the Octocopter Position

As seen in Eq. (2), the distance R between the transmitting antenna used for the calibration and the AUT is of crucial importance. The positions of the antennas at AERA have been previously measured

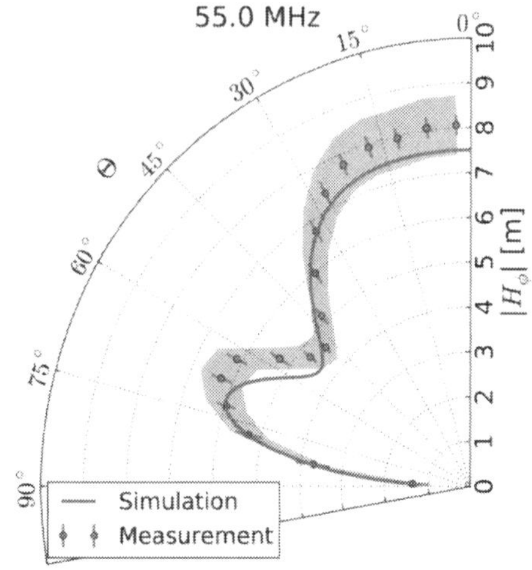

Figure 3. $|H_\phi|$ at 55 MHz as function of θ. The red points denote the measured values, the red error bars the statistical uncertainty and the grey color band the systematical uncertainty. The green line denotes simulated values.

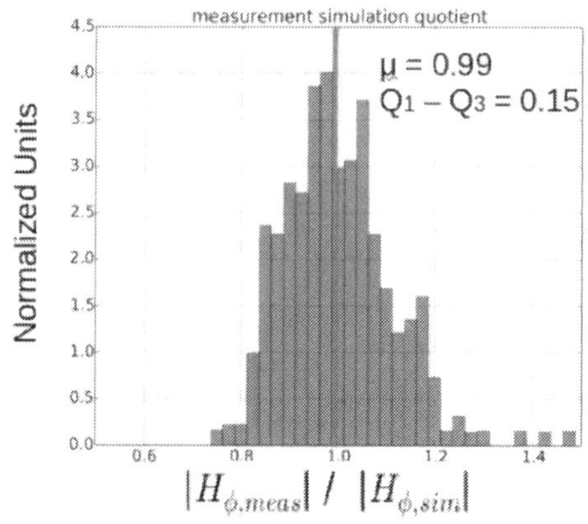

Figure 4. Ratio of measured and simulated $|H_\phi|$. Each entry represents one specific frequency and zenith angle. Q1 and Q3 denote the 25% and 75% quartiles, respectively.

with a differential GPS system and are known to within a few cm. For the position of the octocopter, an internal GPS receiver for the position in the x-y plane and a barometer for the height measurement are built-in. The statistical uncertainty of the position measurement has been found to be better than 1 m. However, the systematical uncertainty is on the order of a few meters. To make a high precision measurement with small systematical uncertainties, another position determination method has been developed. It makes use of two standard digital cameras. The setup is shown in Fig 2. Two cameras are placed on orthogonal axes with a distance of ~100 m to the AUT. Canon Ixus 132 cameras with a resolution of 16 Megapixel are utilized. They are set to an autonomous mode where they take pictures every three seconds.

From these pictures the position of the octocopter during the full flight is reconstructed. In the first step, the pixel tuple of the octocopter in each picture is determined by a "template matching" algorithm. Using this pixel tuple, a vector pointing from each camera to the octocopter is constructed. The point of closest approach of these vectors is calculated and used as the reconstructed octocopter position. For more details on the whole reconstruction process, see [9].

From comparison with a differential GPS and the built-in position sensors, a statistical uncertainty of 1.5 m and a systematical uncertainty of 1 m for this optical method has been found. So, the optical method compared to the built-in sensors has a smaller systematical uncertainty, albeit with a larger statistical uncertainty. Therefore, in the calibration flights the built-in sensors are used, corrected by the optical method. In the x-y plane, the absolute offset between the built-in sensors and the optical method, measured in each separate flight, is added, while in the height, the height measured with the built-in sensors is multiplied by the ratio of the measured height with the built-in sensors and the height measured with the optical method.

4 Results of the Calibration Measurements

Several flights for measuring both $|H_\phi|$ and $|H_\theta|$ over a time of a few weeks have been performed. In Fig. 3, the results of the calibration for $|H_\phi|$ as a function of θ at a frequency of 55 MHz for a fixed azimuth angle $\phi = 0°$ are shown. Each value represents the mean over a zenith angle range of 5°. The error bars denote the statistical spread of the measurement while the error bands denote the systematical uncertainties. These uncertainties are dominated by the spectrum analyzer, which has an uncertainty of 0.5 dB given by the manufacturer. This uncertainty has to be applied to the injected and the received power. For a frequency of 55 MHz, which is the middle of the frequency spectrum of the antennas, and a zenith angle of 45°, from which most of the air shower signals originate, a statistical uncertainty of 4.4% is found. The systematical uncertainty is found to be 8.1%. These values do not vary much over the whole frequency and directional range. Combining these uncertainties gives an overall uncertainty of 9.3%. This is a significant improvement on the previous calibration, which had an uncertainty of 12.5% [4].

The green line in Fig. 3 shows a simulation of the antenna pattern done with the 4NEC2 simulation code. In Fig. 4, the ratios of the simulated and measured VEl for each 5° zenith angle bin and each measured frequency are histogrammed. The mean of the distribution is 0.99, which shows a good agreement between simulation and measurement.

The same analysis has also been done for the meridional component $|H_\theta|$. It is found that the sensitivity is much lower than in the horizontal component.

5 Summary and Outlook

A calibration campaign of the LPDA antennas at AERA using an octocopter has been performed. To determine the position of the octocopter during the flight, a new optical reconstruction method has been developed and evaluated. The absolute vector effective length, which describes the antenna response pattern to an incoming electric field, has been measured with an uncertainty of 9.3%. A good agreement between simulation and measurement was found. A journal article about the whole calibration process is currently in the works.

This calibration is a crucial component in the determination of an absolute energy scale of cosmic rays from first principles. Using the new calibration presented here, the uncertainty of the cosmic-ray energy measurements can be lowered with regard to current values [10, 11]. In the future, these results from AERA will be used to improve the measurements of the whole Pierre Auger Observatory [2].

References

[1] J. Schulz for the Pierre Auger Coll., *Proc. of the 34th ICRC, The Hague, The Netherlands* (2015)
[2] C. Glaser for the Pierre Auger Coll., *Proc. of ARENA 2016, Groningen, The Netherlands* (2016)
[3] A. Aab et al. for the Pierre Auger Coll., Nucl. Instrum. Meth. A **798**, 172 (2015)
[4] P. Abreu for the Pierre Auger Coll., JINST **7**, P10011 (2012)
[5] W.D. Apel, et al. for the LOPES Coll., Astropart. Phys. **75**, 72 (2016)
[6] P.A. Bezyazeekov, et al. for the Tunka-Rex Coll., Nucl. Inst. Meth. A **802**, 89 (2015)
[7] A. Nelles et al., JINST **10**, P11005 (2015)
[8] K. Weidenhaupt, Ph.D. thesis, RWTH Aachen University (2014)
[9] F. Briechle, Master's thesis, RWTH Aachen University (2015)
[10] A. Aab for the Pierre Auger Coll., Phys. Rev. Lett. **116**, 241101 (2016)
[11] A. Aab for the Pierre Auger Coll., Phys. Rev. D **93**, 122005 (2016)

EPJ Web of Conferences **135**, 01015 (2017) DOI: 10.1051/epjconf/201713501015
ARENA 2016

Measurement of horizontal air showers with the Auger Engineering Radio Array

Olga Kambeitz on behalf of the Pierre Auger Collaboration[1,2,★]

[1] *Institut für Kernphysik, Karlsruhe Institute of Technology (KIT), Germany*
[2] *Full author list: http://auger.org/archive/authors_2016_06.html*

Abstract. The Auger Engineering Radio Array (AERA), at the Pierre Auger Observatory in Argentina, measures the radio emission of extensive air showers in the 30-80 MHz frequency range. AERA consists of more than 150 antenna stations distributed over 17 km^2. Together with the Auger surface detector, the fluorescence detector and the underground muon detector (AMIGA), AERA is able to measure cosmic rays with energies above 10^{17} eV in a hybrid detection mode. AERA is optimized for the detection of air showers up to 60° zenith angle, however, using the reconstruction of horizontal air showers with the Auger surface array, very inclined showers can also be measured. In this contribution an analysis of the AERA data in the zenith angle range from 62° to 80° will be presented. CoREAS simulations predict radio emission footprints of several km^2 for horizontal air showers, which are now confirmed by AERA measurements. This can lead to radio-based composition measurements and energy determination of horizontal showers in the future and the radio detection of neutrino induced showers is possible.

1 Horizontal Air Showers

The radio emission of horizontal air showers covers a larger area on ground than the radio emission of vertical showers of the same energy and azimuth angle. CoREAS [1] simulations have predicted the large footprint of horizontal air showers of several km^2, which can be seen in figure 1. This makes it possible to increase the distance between the antenna stations, as is done for the current phase of AERA with up to 750 m distance between the antenna stations. The Pierre Auger Observatory [2] with its 3000 km^2 size provides the optimal conditions to detect horizontal air showers because the SD tanks are not flat like scintillators. The sketch of figure 2 illustrates that with the radio detection of extensive air showers the electromagnetic component of horizontal air showers can be accessed by detecting the radio emission even if the particle shower is absorbed in the atmosphere. The 1.5 km grid of the surface detector water-Cherenkov tanks triggers the externally triggered AERA [3] antenna stations. The self-triggered stations of AERA are only sensitive up to 45° zenith angle and cannot be used for detecting horizontal air showers.

★e-mail: olga.kambeitz@kit.edu

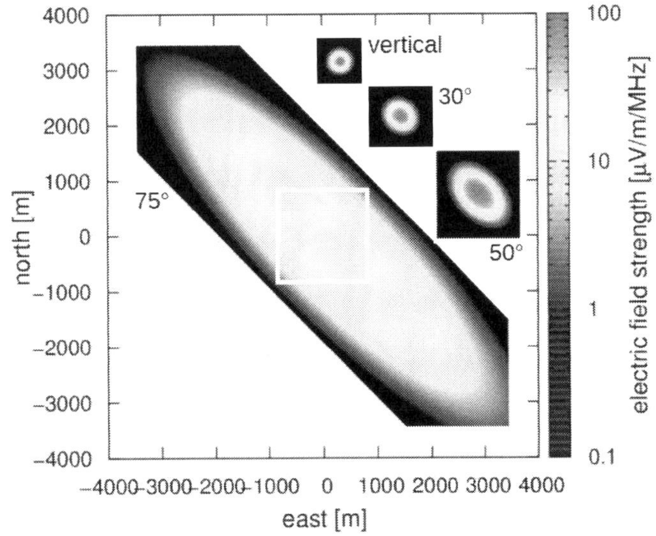

Figure 1. Field strength of the radio emission as a function of ground position for CoREAS simulations of primary proton showers of 10^{18} eV for different zenith angles. The white box marks the change of antenna distance from 40 m to 100 m in the simulation to save computing time. [4, 5]

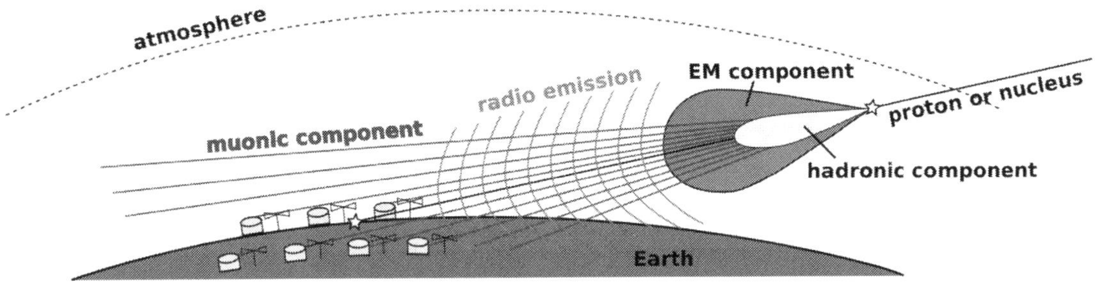

Figure 2. Illustration of a horizontal air shower. The hadronic and electromagnetic component is absorbed and the muonic component and the radio signal can be measured on the ground. [5, 6]

2 Radio Detection of Horizontal Air Sowers with AERA

Data selection of surface-detector-triggered and reconstructed horizontal events of the AERA data sample in the period 1st of January 2012 to 15th of August 2015 results in 427 high quality events. The selection criterion of the zenith angle range from 62° to 80° is illustrated in the left plot of figure 3 and the energy distribution of the detected events can be seen in the right plot of figure 3. Comparing the amplitude of each AERA station of each measured event with CoREAS simulations of proton primaries results in the left plot of figure 4. The mean relative difference for CoREAS simulations with proton primaries is 10% and the RMS is 35%. These systematic shifts and uncertainties need to be further investigated, but the agreement illustrates that signals from horizontal air showers are basically well-described by the simulations.

The AERA horizontal air shower data sample contains 27 events with energy larger than 10 EeV, if no zenith angle cut is applied. These events have higher amplitudes and therefore allow deeper studies of the radio emission also in comparison with simulations. For half of the 427 horizontal showers the core is not contained in AERA and nevertheless a clear radio signal can be measured. One example event of the current AERA set-up can be seen in figure 5. This measured event reveals the potential of the radio detection of horizontal air showers. The event has an energy of 1.13×10^{19} eV, a zenith angle of 74.4° and contains 69 antenna stations above signal-to-noise threshold. The antenna stations with antenna distance of 750 m are contained and the detections of such kind of events is no problem on a 1.5 km grid.

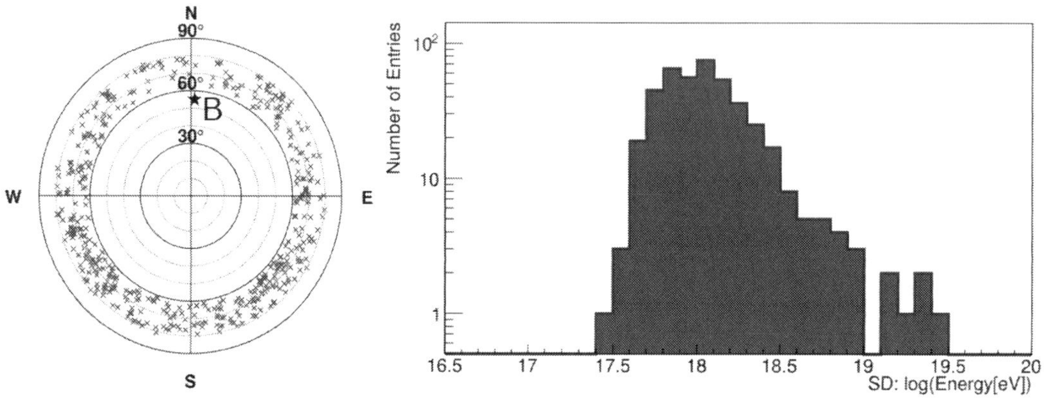

Figure 3. The data sample and data selection of the externally triggered AERA events in the period from 1st of January 2012 to 15th of August 2015 results in 427 high quality events.

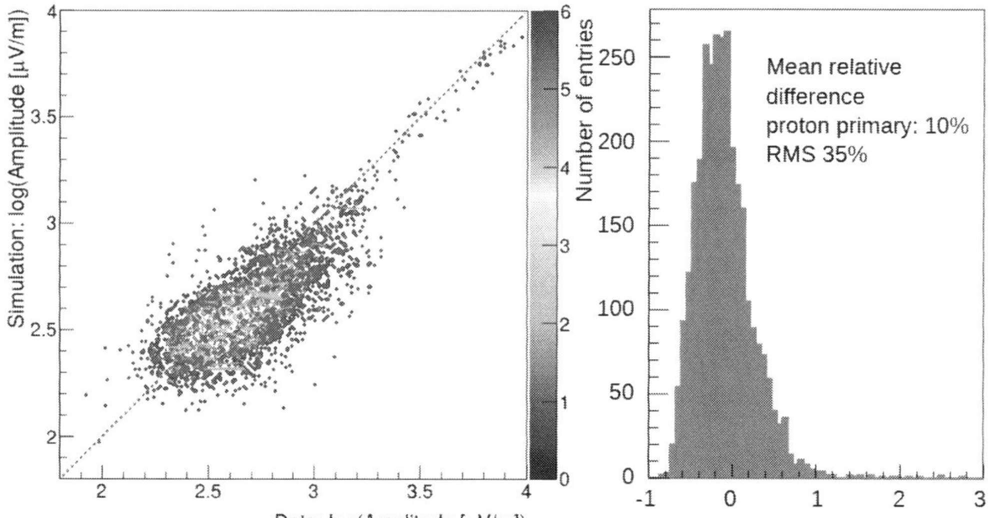

Figure 4. Comparison of measured and simulated events. In 427 horizontal showers 3065 AERA stations are above signal-to-noise threshold. The chosen simulation code is CoREAS. The systematic shifts and uncertainties need to be further investigated.

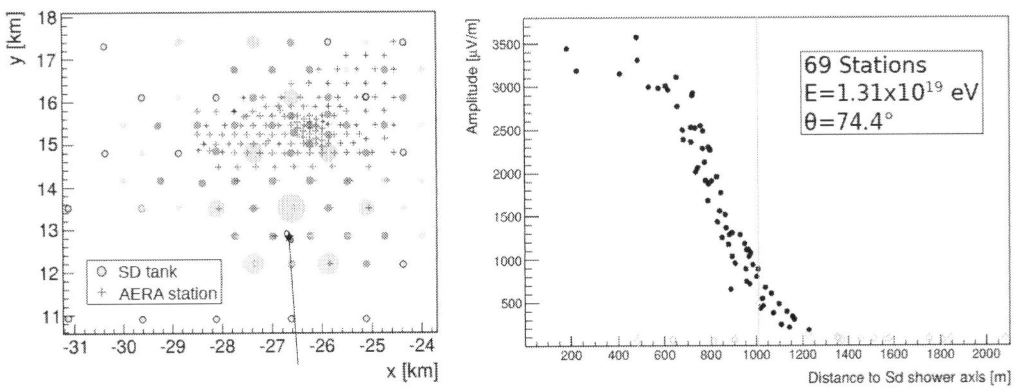

Figure 5. Example event of a horizontal air shower with the current AERA set-up. The grid of the surface detector stations and antenna stations can be seen on the left and on the right the radio lateral distribution can be seen. The color scheme of the SD tanks and the AERA stations on the left indicate the timing information of the triggered tanks and stations and thus the arrival direction of the shower. The size of the circles indicates the signal strength.

3 Conclusion and Outlook

The radio detection of horizontal air showers reveals great potential to be a comprehensive detection method to access the electromagnetic component of the extensive air shower and for horizontal air showers this is only possible with radio detection. The analysis of the AERA data sample results in 427 high quality events in the zenith angle range from 62° to 80°. The large radio footprint covering several km^2 could be detected for the first time with the large-scale radio detector AERA. The radio signal of horizontal air showers is understood within the uncertainties. Sparse antenna arrays are well-suited for horizontal air shower measurements. Now the physics of the radio detection of horizontal air showers for composition measurements and energy determination is in reach and the detection of neutrino induced air showers possible, but more studies are needed in sensitivity and systematics.

References

[1] T. Huege, M. Ludwig, C.W. James, Proc. ARENA 2012, Erlangen, Germany, AIP Conference Proc. 1535, 128, 2013.

[2] A. Aab et al. (Pierre Auger Collaboration), Nucl. Instrum. Meth. A 798, 172 (2015), 1502.01323.

[3] C. Glaser for the Pierre Auger Collaboration, this conference (2016)

[4] T. Huege, Radio detection of cosmic ray air showers in the digital era, Phys.Rept. 620 (2016) 1-52.

[5] O. Kambeitz, Radio Detection of Horizontal Extensive Air Showers, PhD thesis (2016), DOI(KIT):10.5445/IR/1000055758.

[6] O. Kambeitz, Radio Detection of Horizontal Extensive Air Showers with AERA, Proceedings of ARENA 2014, Anapolis, USA, arXiv:1509.08289.

Simulation of the Radiation Energy Release in Air Showers

Christian Glaser[1,⋆], *Martin* Erdmann[1], *Jörg R.* Hörandel[2,3], *Tim* Huege[4], and *Johannes* Schulz[2]

[1] *RWTH Aachen University, III. Physikalisches Institut A, Aachen, Germany*
[2] *IMAPP, Radboud University Nijmegen, Nijmegen, Netherlands*
[3] *Nikhef, Science Park, Amsterdam, Netherlands*
[4] *Karlsruhe Institute of Technology, Institut für Kernphysik, Karlsruhe, Germany*

Abstract. A simulation study of the energy released by extensive air showers in the form of MHz radiation is performed using the CoREAS simulation code. We develop an efficient method to extract this radiation energy from air-shower simulations. We determine the longitudinal profile of the radiation energy release and compare it to the longitudinal profile of the energy deposit by the electromagnetic component of the air shower. We find that the radiation energy corrected for the geometric dependence of the geomagnetic emission scales quadratically with the energy in the electromagnetic component of the air shower with a second order dependency on the atmospheric density at the position of the maximum of the shower development X_{max}. In a measurement where X_{max} is not accessible, this second order dependence can be approximated using the zenith angle of the incoming direction of the air shower with only a minor deterioration in accuracy. This method results in an intrinsic uncertainty of 4% with respect to the electromagnetic shower energy which is well below current experimental uncertainties.

1 Introduction

The measurement of high-energy cosmic rays using short radio pulses emitted by air showers is a quickly evolving field of research [1]. Recently, a new method to measure the cosmic-ray energy using the radiation energy, i.e., the energy that is emitted by the air shower within the frequency band of the detector, was presented [2–5]. In this work, we study the emission of the radiation energy from the theoretical side using Monte Carlo simulations of air showers and the calculation of the radiation from first-principles based on classical electrodynamics. More details of this analysis can be found in [6].

2 Method

We use the CoREAS code [7], which is an extension of the CORISKA [8] software, to simulate the radio emission from air showers. It uses a full MC approach where in principle all shower particles are followed and their contributions to the radiation at a specific observer position are calculated. We simulated 592 air showers with energies between 10^{17} eV and 10^{19} eV, zenith angles between 0°

⋆e-mail: glaser@physik.rwth-aachen.de

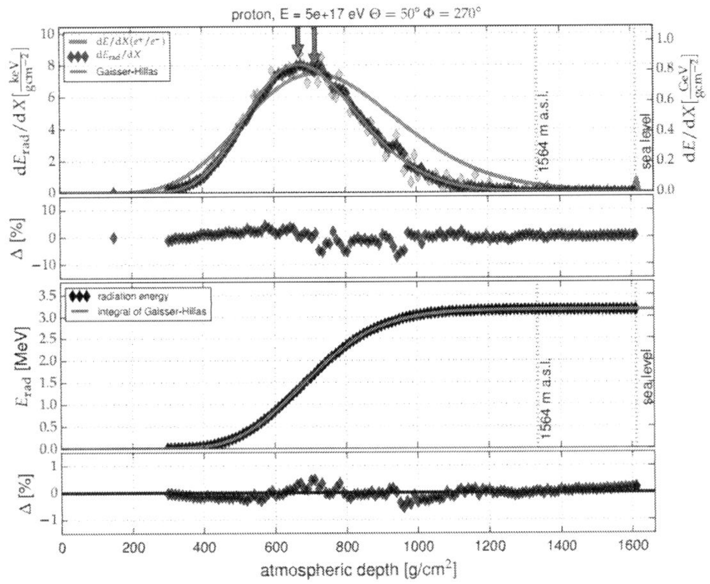

Figure 1. Longitudinal profile of the radiation energy release. The lower diagram plots the total radiation energy measured at a given atmospheric depth. The green line demonstrates a fit of the integral of a Gaisser-Hillas function to the radiation energy. The residuals are also shown. The upper diagram plots the same data and fit, but differentially, i.e., as radiation energy release at a given atmospheric depth. Also shown is the energy deposit of the electromagnetic shower particles as a purple line that needs to be read on the right vertical axis. The arrows indicate the position of the shower maximum. Figure and caption adapted from [6].

and 80° and uniformly distributed azimuth angles. For each combination of energy and geometry we simulate a proton- and an iron-initiated air shower.

As the radiation at each observer position needs to be calculated separately, the radiation energy is no direct outcome of the simulation but needs to be determined via integration of the radio emission footprint. As this would require a large number of sampling points and the computing time scales linearly with the number of observers, we developed a method that exploits the radial symmetry of the geomagnetic and the charge-excess contributions in the radio signal. We found that it is sufficient to simulate the radiation only at positions along the positive $\vec{v} \times (\vec{v} \times \vec{B})$-axis, where $\vec{v}$ is the shower direction and $\vec{B}$ is the geomagnetic field. This is because on this axis the geomagnetic and charge-excess radiation are polarized orthogonally to each other and can therefore be separated. This method does not only allow for an efficient extraction of the radiation energy from air-shower simulations but also allows for a separate determination of the radiation energy resulting from the geomagnetic and the charge-excess radiation.

3 Longitudinal Profile of the Radiation Energy Release

To determine at which stage of the shower development how much radiation energy is emitted, we simulate observers at different atmospheric depths along the shower axis. The resulting longitudinal profile of the radiation energy release of one of the simulated air showers is presented in Fig. 1. The radiation energy rises with increasing atmospheric depth until it reaches its maximum and then remains constant as the atmosphere is transparent to radio waves. This does not imply that the lateral signal distribution remains constant. The signal distribution changes strongly with the distance between observation height and shower maximum. For larger distances the radiation energy is distributed over a larger area resulting in a broader signal distribution with smaller signal strengths.

The longitudinal profile of the radiation energy release can be described with a Gaisser-Hillas function with three free parameters. We compared the profile of the radiation energy release with the longitudinal profile of the energy deposit of the air shower (dE/dX) and found that the radio profile is shifted to smaller atmospheric depths with respect to the dE/dX profile with an average shift of $46\,\mathrm{g/cm^2}$.

4 Correlation with Electromagnetic Shower Energy

To achieve a good correlation of the radiation energy with the electromagnetic shower energy, the dependence of the geomagnetic emission on the angle α between the geomagnetic field and the shower axis needs to be corrected for. As the charge-excess emission is independent of the geomagnetic field we only correct for the geomagnetic contribution to the radiation energy by dividing the radiation energy by $[a^2 + (1 - a^2)\sin^2\alpha]$, where a is defined as the square root of the ratio of the charge-excess and the geomagnetic contribution to the radiation energy normalized to maximum geomagnetic emission ($\sin\alpha = 1$). We use the square root of the ratio to be consistent with previous work on this topic where a was defined as the ratio of electric-field amplitudes [9, 10].

With our method, the geomagnetic and charge-excess contributions to the radiation energy can be extracted separately from the simulated air showers. We found that a is not constant but depends on the atmospheric density in which the shower develops. The larger the atmospheric density in which the shower develops the larger the ratio of charge-excess to geomagnetic radiation energy. We observe values ranging from $a = 0.02$ to $a = 0.5$. This effect can be parametrized with the density at the shower maximum ρ_{X_max} which is primarily a function of the zenith angle and depends to a smaller extend on the variations of the shower maximum X_max. This dependence is described by the following exponential function:

$$a(\rho_{X_\mathrm{max}}) = 0.43\left(e^{1.11\,\mathrm{m^3/kg}\,(\rho_{X_\mathrm{max}} - \langle\rho\rangle)}\right) - 0.24\,, \qquad (1)$$

where we use $\langle\rho\rangle = 0.65\,\mathrm{kg/m^3}$ as the average atmospheric density at the shower maximum.

In addition to the $\sin\alpha$ dependence we found that the radiation energy itself also depends on the atmospheric density in which the shower develops. For shower developments in small density regions we observe an increased radiation energy. This is because the shower development depends on the amount of atmosphere that is traversed (slant depth) whereas the amount of radiation depends on the geometric path length of the shower particles. For showers developing in thin regions of the atmosphere the ratio of geometric path length to slant depth is larger than for showers developing in thick regions resulting in an increased radiation energy.

This dependence can again be parametrized using ρ_{X_max} and we add a second correction term to the radiation energy.

$$S^{\rho}_{\mathrm{RD}} = \frac{E_\mathrm{rad}}{a(\rho_{X_\mathrm{max}})^2 + (1 - a(\rho_{X_\mathrm{max}})^2)\sin^2\alpha}\,\frac{1}{(1 - p_0 + p_0\,\exp[p_1(\rho_{X_\mathrm{max}} - \langle\rho\rangle)])^2}\,. \qquad (2)$$

The correlation between the corrected radiation energy S^{ρ}_{RD} and the electromagnetic shower energy is presented in Fig. 2. We describe the correlation with a function of the form $S_\mathrm{RD} = A \times 10^7\,\mathrm{eV}\,(E_\mathrm{em}/10^{18}\,\mathrm{eV})^B$ and determine the free parameters A and B as well as the parameters p_0 and p_1 of Eq. (2) in a combined χ^2 fit. We find $A = 1.683$, $B = 2.006$, $p_0 = 0.251$ and $p_1 = -2.95\,\mathrm{m^3/kg}$. Hence, the corrected radiation energy scales quadratically with the electromagnetic shower energy as expected for coherent emission. The intrinsic energy resolution of this method can be inferred from the scatter around the calibration curve and amounts to $\sim 3\%$ which is shown in the inset figure.

In a measurement, X_max is often not accessible or has large experimental uncertainties. We therefore approximated the density at the shower maximum using only the zenith angle of the incoming shower direction and assuming an average value of X_max for all air showers. With this simplification we still obtain an intrinsic uncertainty of only 4% which is well below current experimental uncertainties.

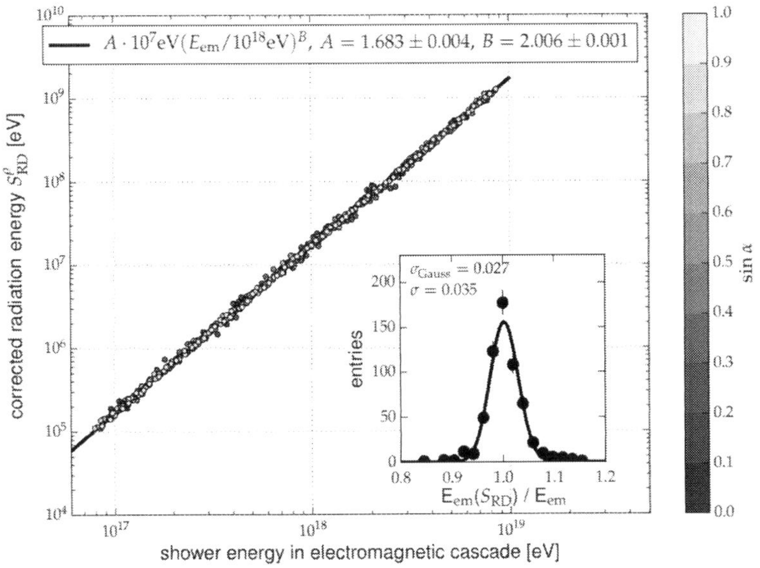

Figure 2. Correlation between the energy in the electromagnetic part of the air shower and the corrected radiation energy (cf. Eq. (2)). The black line shows a power-law fit to the data. The inset figure shows the scatter around the calibration curve. The color scale shows the value of sin α. Figure from [6].

5 Conclusion

In this work, we presented a prediction of the radiation energy emitted by air showers using first-principles calculations based on classical electrodynamics. We determined the longitudinal profile of the radiation energy release. Furthermore, we studied the dependence on the shower energy and found that the radiation energy scales quadratically with the energy in the electromagnetic cascade of the air shower after correcting for the dependence of the geomagnetic emission on the geomagnetic field. In addition, we found that the radiation energy shows a second-order dependence on the atmospheric density in which the shower develops. This dependence can be parametrized using the atmospheric density at the shower maximum resulting in an intrinsic uncertainty of the method of 3%. In a more practical parametrization using only the zenith angle, the method shows an intrinsic uncertainty of 4%. Hence, the results presented here can be used by cosmic-ray radio experiments to improve the precision in the energy reconstruction and to calibrate the energy scale from first-principles calculations.

References

[1] T. Huege, Physics Reports **620**, 1 (2016), 1601.07426
[2] C. Glaser for the Pierre Auger Collaboration, Proceedings of Science (2015), 34th ICRC, The Hague, The Netherlands
[3] A. Aab et al. (Pierre Auger Collaboration), Phys. Rev. D **93**, 122005 (2016), 1508.04267
[4] A. Aab et al. (Pierre Auger Collaboration), Phys. Rev. Lett. **116**, 241101 (2016), 1605.02564
[5] C. Glaser for the Pierre Auger Collaboration, this conference (2016)
[6] C. Glaser, M. Erdmann, J.R. Hörandel, T. Huege, J. Schulz, submitted to JCAP (2016), 1606.01641
[7] T. Huege, M. Ludwig, C.W. James, AIP Conf. Proc. **1535**, 128 (2013)
[8] D. Heck, G. Schatz, T. Thouw, J. Knapp, J. Capdevielle, Report FZKA 6019 (1998)
[9] A. Aab et al. (Pierre Auger Collaboration), Phys. Rev. D **89**, 052002 (2014)
[10] P. Schellart et al., J. Cosmol. Astropart. P. **10**, 014 (2014), 1406.1355

SLAC T-510: Radio emission from particle cascades in the presence of a magnetic field

Katharine Mulrey[1,⋆] for the SLAC T-510 collaboration

[1] *Vrije Universiteit Brussel*

Abstract. Cosmic ray induced particle cascades radiate in radio frequencies in the Earth's atmosphere. Geomagnetic and Askaryan emission provide an effective way to detect ultra-high energy cosmic rays. The SLAC T-510 experiment was the first to measure magnetically induced radiation from particle cascades in a controlled laboratory setting. An electron beam incident upon a dense dielectric target produced a particle cascade in the presence of a variable magnetic field. Antennas covering a band of 30-3000 MHz sampled RF emission in vertical and horizontal polarizations. Results from T-510 are compared to particle-level RF-emission simulations which are critical for reconstructing the energy and composition of detected ultra-high energy cosmic ray air showers. We discuss the experimental set up, the data processing, the systematic errors and the main results of the experiment, which we found in a good agreement with the simulations.

1 Introduction

Radio emission from extensive air showers is a promising way to probe the properties of high energy cosmic rays [1]. Two mechanisms describe radio emission from particle cascades inside a dielectric in a magnetic field. Askaryan radiation was hypothesized in 1962 [2]. It is due to a charge excess in the cascade that builds up from Compton scattering, positron absorption, and other processes. It is linearly polarized in the direction radial to the cascade axis. Askaryan radiation has been measured in previous SLAC experiments with cascades developing in silica sand, rock salt, and ice targets [3] [4] [5]. In the presence of a magnetic field a magnetically induced component of radiation also develops. The charges in the shower form a transverse current due to the Lorentz force. This radiation is polarized in the $\mathbf{v} \times \mathbf{B}$ direction. Magnetically induced radiation is particularly useful for detecting cosmic rays whose cascades develop in the Earth's magnetic field. The detected signal is a sum of both types of radiation. The radio emission from a cosmic ray air shower has to be compared to simulations in order to be interpreted. In particular, the endpoint [6] and ZHS [7] formalisms track individual particles in the cascade and calculate the radiation from each track segment, which is then summed up at the point of an observer. Since there are systematic uncertainties in the hadronic interaction models and the primary mass of a detected cosmic ray is unknown, the only way to directly compare radio emission from a particle cascade to simulations is through a controlled beam experiment. The SLAC T-510 experiment was designed to provide the first measurements of magnetically induced radiation in a controlled laboratory setting and to compare experimental data to simulations [9].

⋆e-mail: kmulrey@gmail.com

2 Experimental Design

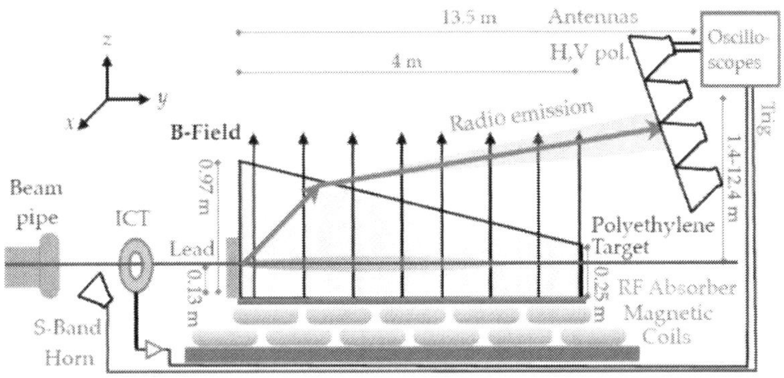

Figure 1. Schematic of the SLAC T-510 experiment [9].

The SLAC T-510 experiment used a target of high density polyethylene (ρ = 0.97 g/cm^3). Electron bunches of 4.35 and 4.55 GeV cascaded in the target, creating a total cascade charge equivalent to a 4×10^{18} eV cosmic ray air shower. Fifteen coils underneath the target provided a variable magnetic field between ±970 Gauss. The top edge of the target was sloped at 9.8° to allow the radiation in the target to escape. In order to make the experiment applicable to cosmic ray air shower detection and to see the effects of both emission mechanisms at a comparable amplitude in one experiment, the ratio of magnetically induced radiation to Askaryan radiation was designed to be roughly 1. Askaryan radiation scales with the shower length (which goes as $1/\rho$) and is weighted by the charge asymmetry. Magnetically induced radiation scales with the shower length as well as the drift velocity of the charges, which goes as the magnetic field strength divided by the dielectric density. In total, the ratio of magnetically induced radiation to Askaryan radiation is B/ρ. In an air shower the magnetic field is on the order of 0.5 Gauss and the density of air is 1.225×10^{-3} g/cm^3 at shower max. These values were scaled to 970 Gauss and 0.97 g/cm^3.

A variable height antenna array consisted of four dually polarized antennas. The experiment was designed so that Askaryan radiation would be detected in only the vertical polarization and magnetic induced radiation would be detected in only the horizontal polarization. Both radiations could then be measured simultaneously and independently. A schematic of the experiment is shown in Figure 1. The antenna array covered distances of 1.4 to 12.4 meters above the beam line and was sensitive to radiation between 300 and 3000 MHz, and results in the 200-1200 MHz range are discussed in the following sections. This frequency band is comparable to 10 - 60 MHz when scaled to air shower frequencies.

3 Results

Data was taken across the Cherenkov cone in both horizontal and vertical polarizations and at magnetic field strengths ranging from -970 to 970 Gauss. The left panel of figure 2 shows the peak voltage from the detected signal plotted against the antenna height above the beam. Vertically polarized emission (in red) remains at the same amplitude regardless of magnetic field strength. Horizontally polarized emission (in blue) is only seen in the presence of a magnetic field. The shape of the Cherenkov cone is clear, and the signal strength peaks at the expected position of the cone, based on the target geometry. The right panel of figure 2 shows the narrowing of the Cherenkov cone with increasing frequency band. The yellow line indicates a band of 300-600 MHz, red 600-900 MHz, and blue 900-1200 MHz.

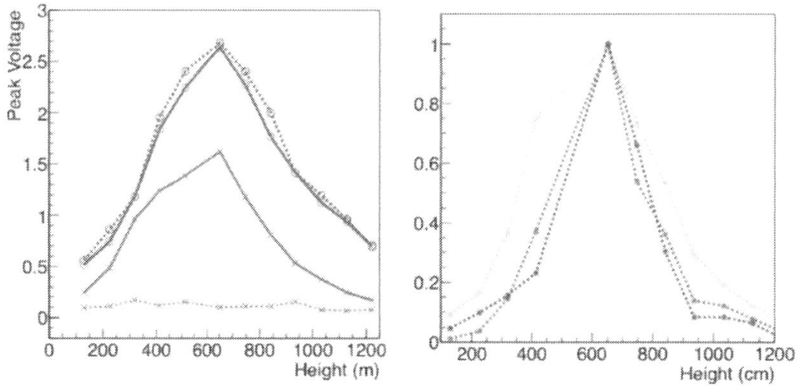

Figure 2. Left: Peak amplitude of the signal at different antenna positions. Vertically polarized emission is in red and horizontally polarized emission is in blue. Solid lines represent data taken with full strength magnetic field and dashed with no magnetic field. Right: Normalized Cherenkov cones for different frequency bands. In yellow is 300-600 MHz, red is 600-900 MHz, and blue is 900-1200 MHz.

The left panel of Figure 3 shows the linearity of the magnetically induced emission with the magnetic field strength. The signal polarity is expected to change with the sign of magnetic field, indicating that the transverse velocity changes direction with magnetic field. The right panel of Figure 3 shows horizontally polarized signals at ±240 G, ±480 G, ±720 G, and ±970 G, with negative magnetic field runs in green and positive in blue. The shape of the signal remains the same and only the polarity changes.

The comparison of T-510 data with simulations is ciritically important and is discussed in [8].

4 Systematic Uncertainties

For the T-510 data to be compared to simulations, the systematic uncertainties in the experiment must be well understood. Quantities used in simulation include cascade energy and charge, detector and target geometry, and magnetic field strength and distribution. The charge bunch that generated the electromagnetic cascade in the target was measured with an integrating current transformer to have a mean charge of 131 pC with 3% variation. The height of the antenna array was measured with laser distance meter to an accuracy of 0.01 cm. The measurement was made relative to a target on the array, for a total uncertainty of 5 cm. The accuracy of the angular geometry of the array was estimated to be known to within 3°. The magnetic field was measured at beam height in a 5 cm × 5 cm grid to 3.64 G precision and was also monitored during the runs at a single point. The RMS of the single point measurement was 72 G, giving a total of 7.4% uncertainty.

The limiting uncertainty was the reflection of the signal off the bottom of the target. A radio frequency absorbing mat was placed under the target during the experiment, but the characteristics of the mat below 1000 MHz are not well known. The direct reflection of the signal is separated from the main signal by less than 1 ns, and cannot be separated in time. Ideas for handling the reflection include doing another run optimized to understand the reflection, or including the reflection in simulations (see [8]). In a follow up experiment, a mat with well known characteristics could be included, or the geometry of the target could be redesigned so that the reflection was separated in time from the main

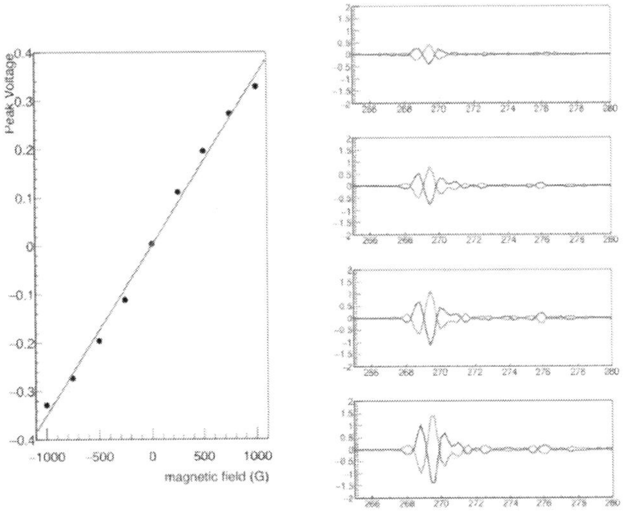

Figure 3. Left: Peak amplitude of the horizontally polarized signal normalized by the vertically polarized signal. Right: Voltage waveform at ±240 G, ±480 G, ±720 G, and ±970 G magnetic fields.

signal. A mirror could also be placed at the bottom of the target so that the reflection has a coefficient of 1. The reflection coefficient could also be measured independently of a second run and subtracted from the T-510 data.

5 Summary

The T-510 experiment provides the first data for radio emission from an electromagnetic cascade developing in the presence of a magnetic field in a controlled laboratory experiment. The radiation shows a distinct Cherenkov cone that narrows with increasing frequency. The magnitude of the magnetically induced radiation increases linearly with magnetic field strength while the Askaryan signal remains the same. The frequency range of detected emission in this experiment is relevant to current cosmic ray experiments. Other than the reflection, the systematic uncertainties are well understood. Comparison of T-510 data to simulations will lower the uncertainty in the interpretation of detected radio emission from cosmic ray air showers.

References

[1] T. Huege, Physics Reports 620, 1-52 (2016)

[2] G. A. Askaryan, JETP 14, 441 (1962).

[3] Saltzberg et al., Phys. Rev. Lett. 86, 2802 (2001).

[4] P. W. Gorham et al., Phys. Rev. D 72, 023002 (2005).

[5] P. W. Gorham et al., Phys. Rev. Lett. 99, 171101 (2007).

[6] C. W. James, H. Falcke, T. Huege and M. Ludwig, arXiv:1007.4146 [physics.class-ph].

[7] J. Alvarez-Muniz, W. R. Carvalho Jr., E. Zas, arXiv:1107.1189 [astro-ph.HE].

[8] A. Zilles, et al., *Modeling of radio emission from a particle cascade in magnetic field and its experimental validation*, these proceedings.

[9] K. Belov et al., Phys. Rev. Lett. 116, 141103 2016.

Modeling of radio emission from a particle cascade in a magnetic field and its experimental validation

Anne Zilles[1,*] for the SLAC T-510 collaboration

[1] *Institut für Experimentelle Kernphysik, Karlsruher Institut für Technologie, 76128 Karlsruhe, Germany*

Abstract. The SLAC T-510 experiment was designed to compare controlled laboratory measurements of radio emission of particle showers to predictions using particle-level simulations, which are relied upon in ultra-high-energy cosmic-ray air shower detection. Established formalisms for the simulation of radio emission physics, the "endpoint" formalism and the "ZHS" formalism, lead to results which can be explained by a superposition of magnetically induced transverse current radiation and charge-excess radiation due to the Askaryan effect. Here, we present the results of Geant4 simulations for the SLAC T-510 experiment, taking into account the details of the experimental setup (beam energy, target geometry and material, magnetic field configuration, and refraction effects) and their comparison to measured data with respect to e.g. signal polarisation, linearity with magnetic field, and angular distribution. We find that the microscopic calculations reproduce the measurements within uncertainties and describe the data well.

1 Introduction

The interpretation of measured data from air-shower events caused by ultra-high energy cosmic rays is based on state-of-the-art Monte-Carlo simulations. Codes like AIRES [1] and CORSIKA [2] are available for the simulation of the development of extensive air-showers. Both tools are based on different hadronic and electromagnetic interaction codes to handle particle interaction for air shower simulations. They have been extended with formalisms to calculate the radio emission by the particle showers. The two microscopic formalisms calculate the electric field of the moving charged particles in the shower while making no assumptions on the emission mechanisms. As the basis for the calculation, they use the positions of the simulated particle tracks and their corresponding times, which are available by design of the particle shower simulation codes. The first approach is called the endpoint formalism [3][4] and is applied in CoREAS [5], a plug-in for CORSIKA. The second, the ZHS formalism [6], is applied in ZHAireS [7], an extension of AIRES.

For the interpretation of the measured radio signals of extensive air showers, the uncertainty in the simulations has to be reduced by validating them in a controlled laboratory environment. The T-510 experiment at the SLAC National Accelerator Laboratory was performed to verify simulations for the radio emission from particle showers by a comparison to data. An overview of the experimental setup of the T-510 experiment as well as the physics motivation are described in [8]. Details on the simulation study can be found in [9].

*e-mail: anne.zilles@kit.edu

2 Comparison of simulations and data and their interpretation

As already published in [10], the time-domain peaks of the magnetic component for the ZHS and endpoints formalisms agree within 3% while the time-domain peaks of the measured data exceed the simulations by 35%. The difference could be shown to be of about the same magnitude as the systematic uncertainty due to the first-order reflection at the bottom surface of the target which we discuss in more detail. We found that uncertainties in the index of refraction of the RF absorbing blanket underneath the target dominate the systematic uncertainty.

2.1 Impact of internal reflections on signal amplitude

Since the frequency-dependent reflection properties of the RF absorbing blanket, on which the target was positioned, are not known, a quantitative modeling of the reflection in the simulation is not possible. Therefore, total reflection at the bottom surface with a reflection coefficient of $R = 1$ and a refractive index of the blanket larger than the one of the target is assumed. Given by the geometry of the target, the reflected signal arrives at the antenna about 1 ns later than the direct signal. Therefore, the direct and the reflected signals overlap and cannot be separated in time. A signal which is first reflected at the top surface and then at the bottom will have a time shift of about 7 ns and with a much lower amplitude.

To predict the polarity of the reflection, the signal can be split up into a horizontal and a vertical component. Due to the geometry, the polarity of the vertical component is inverted for the part of the signal arriving first at the bottom target surface compared to the one directly propagating to the antenna. Here, the horizontal component retains the same polarity as the direct signal, but will experience a sign flip at the boundary to the blanket due to the higher refractive index of the blanket. To predict the impact of the reflection on the radio signal measured at a particular antenna position, a time-delayed signal with equal amplitude but inverted polarity for both components is added to the simulated direct signal. This was done for the antenna position on the Cherenkov cone at a height of 6.52 m above the beam line (see Fig. (1)). A comparison of the measured radio signal in the time domain with the results from the simulations shows that with the inclusion of the reflected signal as explained above a good description of the measured peak amplitude by the simulations is achieved. However, the after-pulse is now predicted to be larger than measured. In addition, a comparison of the frequency spectra leads to the conclusion that the assumption in the modeling of the total reflection for all frequencies is too simple since the simulations including the reflections show a strong suppression of the signal for high frequencies.

2.2 Scaling with magnetic field strength

To check for the expected linear scaling of the electric field strength of the magnetic component on the magnetic field strength, measurements with different magnetic field strengths were performed [10]. In Fig. (2)a the results are compared with those from simulations that include internal reflections. The amplitude of the horizontally polarised component was observed to depend on the magnetic field strength. This supports the conclusion that the transverse current generates the magnetic emission in particle showers. To study the linearity of the scaling, the ratio of the peak amplitudes of the magnetically induced component to that induced by the charge excess during the shower development is shown in Fig. (2)b. As expected the ratio rises with increasing magnetic field strength. However, we observe a flattening of the ratio towards higher magnetic field strengths and a power law fit to the data does not yield an exponent of 1, meaning that the scaling is not linear. The simulations seem not to reproduce the flattening for the studied magnetic field strength. But the uncertainty introduced

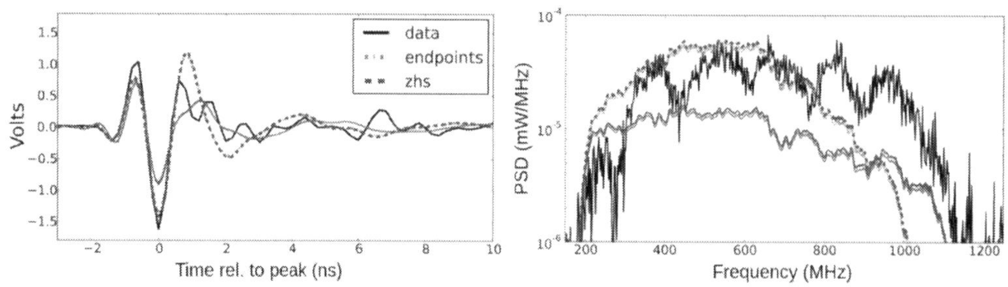

Figure 1. Comparison of the measured radio signal at a height of 6.52 m with simulations including the internal reflection (dashed lines) as described in section 2.1. The thin solid lines represent the traces of the simulated results for the direct signal for comparison. Left: Signals in the time domain. Right: Signals in the frequency domain.

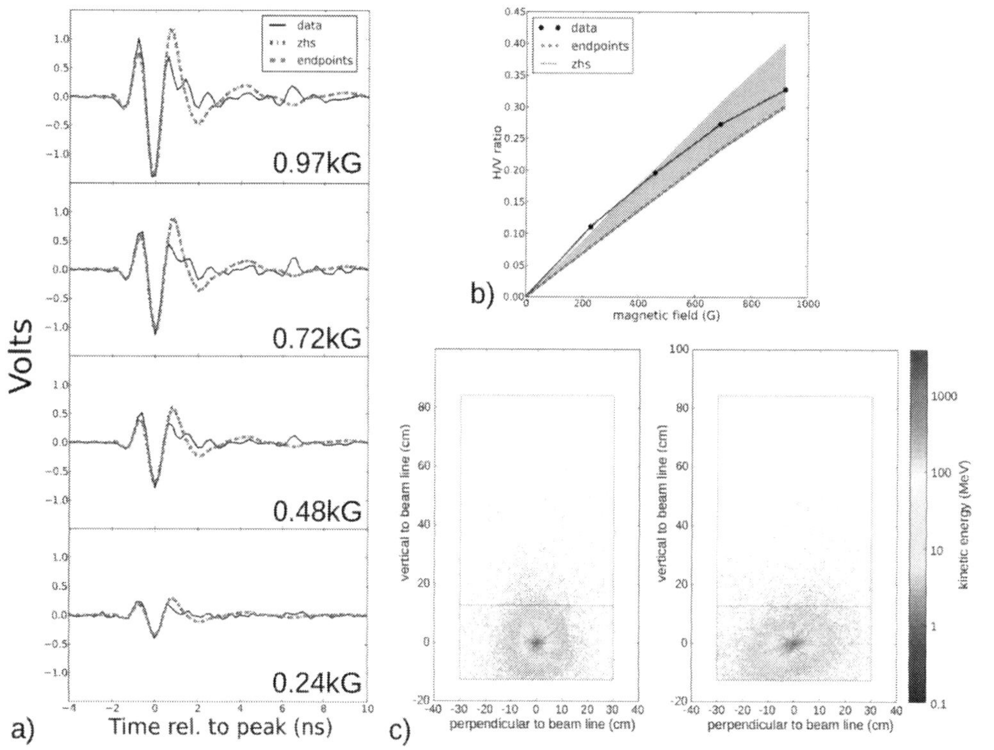

Figure 2. a) Measurements of the scaling of the peak amplitude of the magnetically induced component of the radio signal with increasing magnetic field strength. b) Measured and simulated ratio of the peak amplitudes of the magnetically induced component and the component induced by the charge excess during the shower development. The gray band on top of the simulations indicates the possible impact of the reflections. c) Distribution of particles in the shower for different magnetic field strengths (left: $B = 0\,\mathrm{G}$, right: $B = 1940\,\mathrm{G}$) looking along the beam line. The color code represents the kinetic energy of the particles.

by internal reflection, indicated by a gray band in the plot, contains the measured ratios and even indicates a slight flattening for the higher magnetic field strengths.

There are different possibilities to explain the measured flattening of the ratio of the peak amplitudes of the magnetically induced component with respect to the component induced by the charge excess during the shower development: First, there could be a so far unknown physical effect, like a saturation effect in the drift velocity of the electrons and positrons, which is not reproduced in the simulations. Secondly, to exclude the possibility that high-energy particles escape the target and therefore don not contribute to the radio signal of the shower, the distribution of particles in the shower without any magnetic field is studied (see Fig. (2)). As shown by the color code, no high-energy particles leave the target. But, the particle distribution with an induced magnetic field strength of an exaggerated value of $B = 1940\,\mathrm{G}$ shows that the particle shower is rotated out of the horizontal alignment due to the non-uniform three-dimensional magnetic field. This leads to another possible explanation namely that the geometry of the radio emission, assuming that the magnetically induced radio component is aligned with the horizontal channel of the antenna and the charge excess induced component is aligned with the vertical component, is not valid for high magnetic fields. A rotation of the particle showers would also rotate the electric field vector of the magnetically induced component and thus the Askaryan and magnetic component of the radio emission can not be measured independently in the T-510 experiment. This hypothesis will be tested by a more detailed simulation study.

3 Summary

The SLAC T-510 experiments constitutes the first laboratory benchmark of radio-frequency radiation from electromagnetic cascades under the influence of a magnetic field [10]. The experiment confirmed that simulations based on the first-principle formalisms of electrodynamics can predict the absolute scale of the radio emission accurately to within the systematic uncertainty of the measurements. Internal reflections at the bottom surface of the target are the main limiting uncertainty in this experiment. The magnetically induced radio signal component shows a near-linear scaling with rising magnetic field strength, but a flattening towards higher magnetic field strengths is measured. There are indications that the flattening can be explained by a rotation of the particle shower due the non-uniform three-dimensional magnetic field. A future more-detailed simulation study can answer this question.

References

[1] S. J. Sciutto, arXiv:astro-ph/9911331.
[2] D. Heck, *The Air Shower Simulation Program CORSIKA*, https://web.ikp.kit.edu/corsika/.
[3] C. W. James, H. Falcke, T. Huege and M. Ludwig, arXiv:1007.4146 [physics.class-ph].
[4] M. Ludwig, T. Huege, Astroparticle Physics 34, 438-446, 2011.
[5] T. Huege, M. Ludwig, C. W. James, arXiv:1301.2132 [astro-ph.HE].
[6] J. Alvarez-Muniz, A. Romero-Wolf, E. Zas, arXiv:1002.3873v1 [astro-ph.HE].
[7] J. Alvarez-Muniz, W. R. Carvalho Jr., E. Zas, arXiv:1107.1189 [astro-ph.HE].
[8] K. Mulrey, et al., *SLAC T-510: Accelerator measurements of radio emission from particle cascades in the presence of a magnetic field*, these preceedings.
[9] A. Zilles et al., PoS(ICRC2015)313.
[10] K. Belov et al., Phys. Rev. Lett. 116, 141103 2016.

The Giant Radio Array for Neutrino Detection

The GRAND collaboration: *Olivier* Martineau-Huynh[1,*], *Mauricio* Bustamante[2], *Washington* Carvalho[3], *Didier* Charrier[4], *Sijbrand* De Jong[5], *Krijn D.* de Vries[6], *Ke* Fang[7], *Zhaoyang* Feng[8], *Chad* Finley[9], *Quanbu* Gou[8], *Junhua* Gu[10], *Hongbo* Hu[8], *Kumiko* Kotera[11], *Sandra* Le Coz[10], *Clementina* Medina[1,12], *Kohta* Murase[13], *Valentin* Niess[14], *Foteini* Oikonomou[13], *Charles* Timmermans[5], *Zhen* Wang[8], *Xiangping* Wu[10], and *Yi* Zhang[8]

[1]*LPNHE, CNRS-IN2P3 and Universités Paris 6 & 7, BP200, 4 place Jussieu, 75252 Paris, France*
[2]*Center for Cosmology and AstroParticle Physics, The Ohio State University, Columbus, OH 43210, USA*
[3]*Physics institute, University of São Paulo, Rua do Matão, trav. R, Cid. Universitária, São Paulo, Brazil*
[4]*SUBATECH, CNRS-IN2P3, Université de Nantes, Ecole des Mines de Nantes, Nantes, France*
[5]*Nikhef/Radboud University, Nijmegen, the Netherlands*
[6]*Vrije Universiteit Brussel, Dienst ELEM, B-1050 Brussels, Belgium*
[7]*Department of Astronomy, University of Maryland, College Park, MD, 20742*
[8]*Key Lab of Particle Astrophysics, IHEP, Chinese Academy of Sciences, Beijing 100049, China*
[9]*Oskar Klein Centre and Dept. of Physics, Stockholm University, SE-10691 Stockholm, Sweden*
[10]*National Astronomical Observatory, Chinese Academy of Sciences, Beijing 100012, China*
[11]*IAP, Sorbonne Universités, Paris 6 and CNRS, 98 bis bd Arago, 75014 Paris, France*
[12]*Instituto Argentino de Radioastronomia, CCT La Plata-CONICET, 1894, Villa Elisa C.C. No. 5, Argentina*
[13]*Dept. of Physics, Dept. of Astronomy & Astrophysics, Penn State University, University Park, PA, USA*
[14]*LPC, CNRS-IN2P3, Université Blaise Pascal, BP 10448, 63000 Clermond-Ferrand, France*

Abstract. The Giant Radio Array for Neutrino Detection (GRAND) is a planned array of $\sim 2 \cdot 10^5$ radio antennas deployed over $\sim 200\,000\,\mathrm{km}^2$ in a mountainous site. It aims primarly at detecting high-energy neutrinos via the observation of extensive air showers induced by the decay in the atmosphere of taus produced by the interaction of cosmic neutrinos under the Earth surface. GRAND aims at reaching a neutrino sensitivity of $5 \cdot 10^{-11} E^{-2}\,\mathrm{GeV}^{-1}\,\mathrm{cm}^{-2}\,\mathrm{s}^{-1}\,\mathrm{sr}^{-1}$ above $3 \cdot 10^{16}$ eV. This ensures the detection of cosmogenic neutrinos in the most pessimistic source models, and ~50 events per year are expected for the standard models. The instrument will also detect UHECRs and possibly FRBs. Here we show how our preliminary design should enable us to reach our sensitivity goals, and discuss the steps to be taken to achieve GRAND, while the compelling science case for GRAND is discussed in more details in [1].

1 Detection Method

Cosmic ν_τ can produce taus[1] under the Earth surface through charged-current interactions. Taus may then exit and decay in the atmosphere, generating Earth-skimming extensive air showers (EAS) [2, 3].

*e-mail: omartino@in2p3.fr

[1]Other neutrino flavors can be neglected, as the electron range in matter at these energies is too short and the muon decay length too large.

EAS emit coherent electromagnetic radiations at frequencies of a few to hundreds of MHz, detectable by radio antennas for shower energies $E \gtrsim 3 \cdot 10^{16}$ eV [4, 5]. The strong beaming of the electromagnetic emission, combined with the transparency of the atmosphere to radio waves, will allow the radiodetection of EAS initiated by tau decays at distances up to several tens of kilometers (see Fig. (1)), making radio antennas ideal instruments for the search of cosmic neutrinos. Furthermore, antennas offer practical advantages (e.g. limited unit cost, easiness of deployment) that allow the deployment of an array over very large areas, as required by the expected low neutrino event rate.

Remote sites, with low electromagnetic background, should obviously be considered for the array location. In addition, mountain ranges are preferred, first because they offer an additional target for the neutrinos, and also because mountain slopes are better suited to the detection of Earth-skimming showers compared to flat areas which are parallel to the neutrino-induced EAS trajectories.

GRAND antennas are foreseen to operate in the 30 – 100 MHz band. Below this range, short-wave background prevents detection, while coherence of radio emission fades above it. However, an extension of the antenna response up to 200 or 300 MHz would enable us to better observe the Cherenkov ring associated with the air shower [6], which represents a sizable fraction of the total electromagnetic signal at these frequencies. This could provide an unambiguous signature for background rejection.

2 GRAND layout and neutrino sensitivity

We present here a preliminary evaluation of the potential of GRAND for the detection of cosmic neutrinos, based on the simulated response of a 90 000 antennas setup deployed on a square layout of 60 000 km^2 in a remote mountainous area, the Tianshan mountains in the XinJiang province, China.

Simulation method. We perform a 1D tracking of a primary ν_τ, simulated down to the converted tau decay. We assume standard rock with a density of 2.65 g/cm^3 at sea level and above, while the Earth core is modeled following the Preliminary Reference Earth Model [7]. The simulation of the deep inelastic scattering of the neutrinos is performed with Pythia6.4, using the CTEQ5d probability distribution functions (PDF) combined with [8] for cross section calculations. The propagation of the outgoing tau is simulated using randomized values from parameterisations of GEANT4.9 PDFs for tau path length and proper time. Photonuclear interactions in GEANT4.9 have been extended above PeV energies following [9]. The tau decay is simulated using the TAUOLA package. The radiodetection of neutrino-initiated EAS is simulated in the following way:

- for a limited set of ν_τ showers simulated with ZHaireS [10] at various energies (see Fig. (1)), we determine a conical volume inside which the electric field is above the expected detection threshold of the GRAND antennas (30 μV/m in an agressive scenario, 100 μV/m in a conservative one).

- from this set of simulations, we parametrize the shape (angle at top and height) of this detection cone as a function of energy.

- for each neutrino-initiated EAS in our simulation, we compute the expected cone shape and position, and select the antennas located inside the corresponding volume, taking into account signal shadowing by mountains.

- if a cluster of 8 neighbouring units can be found among these selected antennas, we consider that the primary ν_τ is detected.

Results and implications. Assuming a 3-year observation with no neutrino candidate on this 60 000 km^2 simulated array, a 90% C.L. integral limit of 6.6×10^{-10} GeV^{-1} cm^{-2} s^{-1} can be derived for an E^{-2} neutrino flux in our agressive scenario (1.3×10^{-9} in our conservative scenario). This is a factor ≥ 5 better than than other projected giant neutrino telescopes for EeV energies [11].

This preliminary analysis also shows that mountains constitute a sizable target for neutrinos, with ~50% of down-going events coming from neutrinos interacting inside the mountains. It also appears

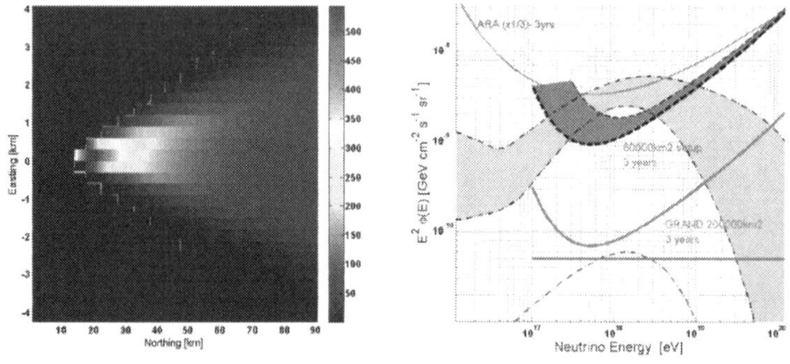

Figure 1. *Left:* Expected radio footprint for a $5 \cdot 10^{17}$ eV horizontal shower induced by a tau decay at the origin. The color coding corresponds to the Efield maximum amplitude integrated over the 30-80MHz range (in μV/m). The sky background level is $\sim 15\mu$V/m in this frequency range. Note the different x and y scales. *Right:* Differential sensitivity of the 60 000 km^2 simulated setup (brown region, top limit: conservative, bottom: aggressive) and of the projected GRAND array (brown thick curve). The integral sensitivity limit for GRAND is shown as a thick line. We also show the expected limit for the projected final configuration of ARA [11] and theoretical estimates for cosmogenic neutrino fluxes [13]: the blue line stands for the most pessimistic fluxes, the gray-shaded region to the "reasonable" parameter range. All curves are for single-flavor neutrino fluxes.

that specific parts of the array (large mountains slopes facing another mountain range at distances of 30 – 80 km) are associated with a detection rate well above the average. By splitting the detector into smaller sub-arrays of a few 10 000 km^2 each, deployed solely on favorable sites, an order-of-magnitude improvement in sensitivity could be reached with only a factor-of-3 increase in detector size, compared to the 60 000 km^2 simulation area. This is the envisioned GRAND setup.

This neutrino sensitivity corresponds to a detection rate of 1 to 60 cosmogenic events per year. Besides, the angular resolution on the arrival directions, computed following [12], could be as low as 0.05° for a 3 ns precision on the antenna trigger timing, opening the door for neutrino astronomy.

3 Background rejection

A few tens of cosmogenic neutrinos per year are expected in GRAND. The rejection of events initiated by high-energy particles other than cosmic neutrinos should be manageable [1]. The event rates associated to terrestrial sources (human activities, thunderstorms, etc.) are difficult to evaluate, but an estimate can be derived from the results of the Tianshan Radio Experiment for Neutrino Detection (TREND). TREND [12] is an array of 50 self-triggered antennas deployed over a surface $\gtrsim 1$ km^2 in a populated valley of the Tianshan mountains, with antenna design and sensitivity similar to what is foreseen for GRAND. The observed rate of events triggering six selected TREND antennas separated by $\sim$800 m over a sample period of 120 live days was found to be around 1 day^{-1}, with two-thirds of them coming in bursts of events, mostly due to planes. Direct extrapolation from TREND results thus leads to an expected event rate of $\sim$ 1 Hz for GRAND for a trigger algorithm based on coincident triggers on neighbouring antennas and a rejection of events bursts.

Amplitude patterns on the ground (emission beamed along the shower axis and signal enhancement on the Cherenkov ring [6]), as well as wave polarization [14] are strong signatures of neutrino-initiated EAS that could provide efficient discrimination tools for the remaining background events.

These options are being investigated within GRAND, through simulations and experimental work. In 2017 the GRANDproto project [15] will deploy a hybrid detector composed of 35 3-arm antennas (allowing for a complete measurement of the wave polarization) and 24 scintillators, that will cross-check the EAS nature of radio-events selected from a polarization signature compatible with EAS.

4 GRAND development plan

Before considering the complete GRAND layout, several validation steps have to be considered. The first one will consist of establishing the autonomous radiodetection of very inclined EAS with high efficiency and excellent background rejection, with a dedicated setup of size $\sim 300\,\mathrm{km}^2$. This array will be too small to perform a neutrino search, but cosmic rays should be detected above 10^{18} eV. Their reconstructed properties (energy spectrum, composition) will enable us to validate this stage. The absence of events below the horizon will confirm our EAS identification strategy. A second array, 10 times larger, will allow to test the technological choices for the DAQ chain, trigger algorithm and data transfer. This will mark the start of GRAND data taking, foreseen in the mid-2020s.

5 Conclusion

The GRAND project aims at building the ultimate next-generation neutrino telescope. Preliminary simulations indicate that a sensitivity guaranteeing the detection of cosmogenic neutrinos is achievable. Work is ongoing to assess GRAND achievable scientific goals and the corresponding technical constraints. Background rejection strategies and technological options are being investigated.

Acknowledgements. The GRAND and GRANDproto projects are supported by the Institut Lagrange de Paris, the France China Particle Physics Laboratory, the Natural Science Fundation of China (Nos.11135010, 11375209), the Chinese Ministry of Science and Technology and the São Paulo Research Foundation FAPESP (grant 2015/15735-1).

References

[1] O. Martineau-Huynh, et al. (2016), Vol. 116 of *European Physical Journal Web of Conferences*, p. 03005, 1508.01919

[2] D. Fargion, International Cosmic Ray Conference **2**, 396 (1999), astro-ph/9906450

[3] X. Bertou, et al, Astroparticle Physics **17**, 183 (2002), astro-ph/0104452

[4] D. Ardouin, et al., Nuclear Instruments and Methods in Physics Research A **555**, 148 (2005), astro-ph/0504297

[5] H. Falcke, et al., Nature **435**, 313 (2005), astro-ph/0505383

[6] J. Alvarez-Muñiz, W.R. Carvalho, Jr., A. Romero-Wolf, M. Tueros, E. Zas, Phys. Rev. D **86**, 123007 (2012), 1208.0951

[7] A. Dziewonski, D. Anderson, Physics of the Earth and Planetary Interiors **25**, 297 (1981)

[8] R. Gandhi, C. Quigg, M.H. Reno, I. Sarcevic, Phys. Rev. D **58**, 093009 (1998), hep-ph/9807264

[9] S. Dutta, et al., European Physical Journal C **18**, 405 (2000), hep-ph/9905475

[10] J. Alvarez-Muñiz, W.R. Carvalho, E. Zas, Astroparticle Physics **35**, 325 (2012), 1107.1189

[11] P. Allison, et al, Phys. Rev. D **93**, 082003 (2016), 1507.08991

[12] D. Ardouin et al., Astroparticle Physics **34**, 717 (2011), 1007.4359

[13] K. Kotera, D. Allard, A.V. Olinto, J. Cos. and Astro. Phys. **10**, 13 (2010), 1009.1382

[14] A. Aab et al., Phys. Rev. D **89**, 052002 (2014), 1402.3677

[15] Q. Gou, et al. (2016), Vol. 116 of *European Physical Journal Web of Conferences*

The ExaVolt Antenna: Concept and Development Updates

Carl Pfendner[1,*] for the EVA collaboration

[1] *Dept. of Physics and CCAPP, The Ohio State University, 191 W. Woodruff Ave., Columbus, OH 43210, USA*

Abstract. A flux of ultrahigh energy neutrinos is expected both directly from sources and from interactions between ultrahigh energy cosmic rays and the cosmic microwave background. Using the cost-effective radio Cherenkov technique to search for these neutrinos, the ExaVolt Antenna (EVA) is a mission concept that aims to build on the capabilities of earlier radio-based balloon-borne neutrino detectors and increase the sensitivity to lower energies and fluxes. The novel EVA design exploits the surface of the balloon to provide a focusing reflector that aims to provide a signal gain of ~ 30 dBi (compared to 10 dBi on ANITA). This increase in gain when combined with a large instantaneous viewing angle will yield a 10-fold increase in sensitivity and will allow this balloon-borne experiment to probe the expected low neutrino fluxes even at energies greater than 10^{19} eV. This contribution will present an overview of the mission concept, recent technology developments, and the results of a hang test of a 1:20-scale model which demonstrates the effectiveness of the design.

1 Introduction

A flux of ultra-high energy (UHE) neutrinos is expected due to interactions between UHE cosmic rays and the cosmic microwave background, as predicted by Berezinsky and Zatsepin [1]. Several experiments are currently pursuing the detection of the cosmogenic UHE neutrino flux using radio detection techniques. Detectors embedded within polar ice caps either several kilometers deep [2, 3] or at the surface [4] exploit the proximity to neutrino interactions in a large target volume. The ANITA [5] balloon-borne interferometer synoptically scans a significantly larger ice volume (~ million km^3) providing high sensitivity to the high energy end of the expected neutrino spectra.

An experiment's sensitivity to UHE-induced radio emissions is determined by the combination of visible target volume and live-time. Ground-based arrays implement large numbers of high duty cycle stations (37 for ARA and ~ 1000 for ARIANNA) but the visible volume is limited by the refractive and attennuative properties of the ice. A balloon-borne detector, on the other hand, has a large visible target volume with a limited operational duty cycle. For balloon-borne detectors, the only practical improvement is to increase the antenna gain. The ExaVolt Antenna (EVA) [6] is a mission concept that builds on the success of three ANITA flights. EVA aims to implement a ~ 30 dBi gain antenna, exploiting the surface of a super-pressure balloon (SPB) to improve sensitivity to neutrino-induced radio-impulsive transients by a factor of 100 over ANITA.

*e-mail: pfendner.1@osu.edu

2 ExaVolt Antenna Design Concept

The proposed EVA design will be the largest-aperture radio telescope ever flown on a balloon payload and will improve the sensitivity to neutrino-induced impulsive radio-frequency (RF) signals by a factor 100 or more over any previous experiment. This large improvement in gain can be obtained by using a portion of the balloon surface itself as part of the radio detector with a toroidal geometry as indicated in Fig. (1), where an RF-reflective film is applied to a section of the balloon surface 10-m high about the equator. An incoming plane wave impulse enters the balloon below the lower reflector rim opposite the active focusing area that will apply for that direction. Once reaching the opposite side of the balloon reflector surface, the incoming plane wave impulse is focused onto a focal plane inside the balloon. A set of patch feed antennas will be positioned at the focal plane of the reflector to receive the focused plane wave.

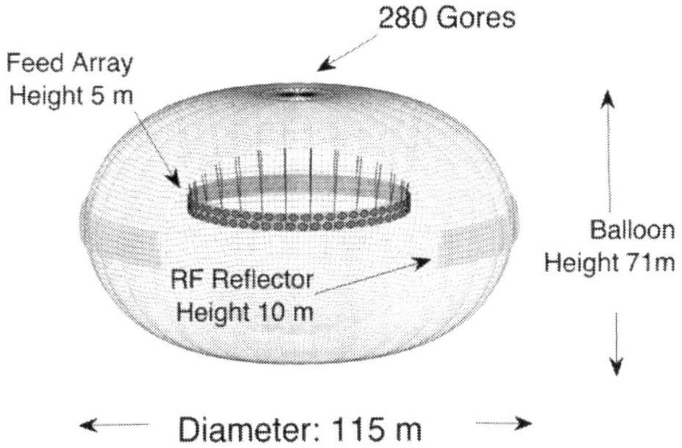

Figure 1. The proposed design of the full-scale EVA detector. Incoming plane-wave signals are reflected off the RF reflector ring on the surface of the balloon toward the interior feed panels, instrumented with receiver antennas. The curvature of the reflector focuses the signal to improve the gain by a factor of ~ 30 dBi.

3 Scale Model Hang Test

In order to test the feasibility of the concept, the EVA collaboration performed a hang test of a 5.7-m-diameter SPB constructed to our specifications by Aerostar International. The test was performed at NASA Wallops Flight Facility in September 2014. This balloon included 50-cm high aluminized-mylar reflective panels attached to the balloon at the equator by the vendor using polyethylene adhesive tape. A view of the fully-deployed balloon during the test is shown in Fig. (2).

Prior to inflation of the balloon, we inserted a scale-model of the antenna feed array membrane through the top end-plate of the balloon. In order to accommodate the insertion, the feed array was folded and was attached to internal support lines which were used to expand the array from its folded state during inflation and which provided support for the array upon final deployment. The feed array membrane was instrumented with dual polarization sinuous patch antennas over a portion of its circumference, and four of these antennas were also instrumented with microwave receivers coupled to RF-over-fiber transceivers which inserted the RF signals into optical fiber, as shown in Fig. (2).

The two primary goals of the test were to demonstrate (1) that it was possible to deploy a feed membrane antenna array within a super-pressure balloon, and (2) that the response of the balloon and patch antennas would function as predicted by antenna models when subject to an external impulsive plane-wave signal. To create this external signal, we used a 2.6 meter offset parabolic dish as an RF collimator, fed by a broadband dual-ridge horn emitting microwave impulses. We made several compromises on the scalability for the sake of the test: the number of gores was set at 28 for this

Figure 2. Left: The fully inflated hang test balloon. Photogrammetry was used to determine the geometry of the ballon and reflector after inflation. Center: One panel of the feed array. Four of these sinuous patch antennas were instrumented with electronics. Right: The hang test setup with the test signal emitter dish shown in the background.

balloon, rather than 280, to keep construction simple; and the width of the collimation dish was much smaller than the receiving width of the reflective panels due to constraints of portability of the dish. These compromises were understood and incorporated in the simulations used for comparison. The results of the microwave test and the simulations are summarized in Fig. (3). It is evident that the temporal coherence of the arriving impulse is preserved in the focused pulse. The observed pulse gain matches the model estimates within about 2 dB, a solid validation of the simulation tools and our methodology. This agreement serves as additional confirmation of the feasibility of the concept as a method to improve signal gain.

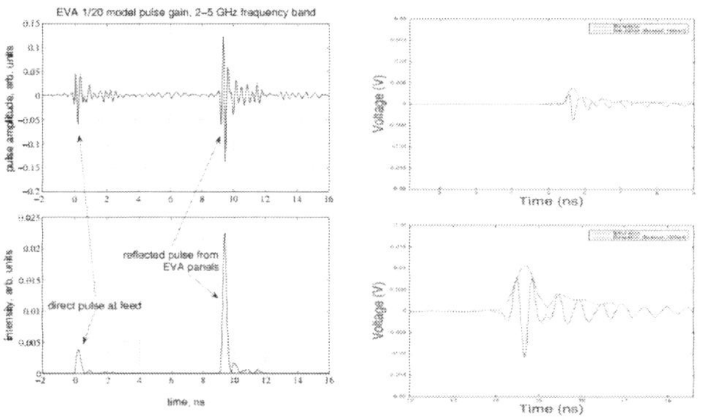

Figure 3. On the left is shown the incident and reflected pulse as measured during the hang test. The upper plot shows the recorded voltage and the lower shows the intensity. Shown on the right is the voltage for the incident (top) and reflected (bottom) pulse as predicted by the simulation program XFdtd [7] for the hang test geometry.

4 Full-scale Simulation and Prediction for the Future

In order to predict the response of the full-scale EVA detector and to investigate further possibilities of design improvements, we have created detailed antenna models for the super pressure balloon geometry using the currently available 18.7 Mcft balloon with a radius of $\sim$ 57 m. Using this model, we have investigated the response of the reflection geometry with three electromagnetic simulation packages. The packages used are as follows: Numerical Electromagnetic Code v.4 (NEC4) [8] which uses the method of moments; General Reflector Antenna Simulation Program (GRASP) [9] using physical optics; and XFdtd v.7 [7] which is based on the finite-difference time-domain method. Using each of these methods independently, we have found that simulations using low-gain feed antennas predict gains of 23-27 dBi from the reflector ring in the UHF band. Results from one of these tests is shown in Fig. (4).

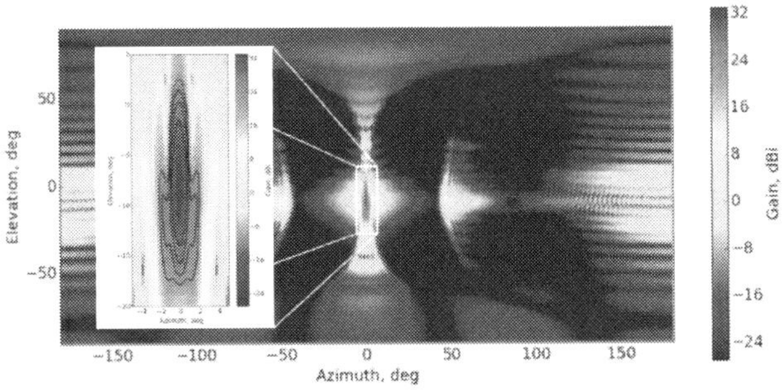

Figure 4. GRASP model power pattern for a 18.7 Mcft EVA model, using a feed with a response matched to the reflector. By using feeds matched to the antenna response of the reflector in the GRASP simualtions, the predictions of the gain can be improved up to 32 dBi as shown here.

The results of the hang test confirm both the effectiveness of the concept of using the balloon as a focusing reflector and of the agreement of the measured data with expectations from simulation. Using the expected gain in sensitivity from these results, one can predict the improvement in sensitivity to neutrino-induced RF signals and thus to the neutrino flux itself. This flux sensitivity is shown in Fig. (5) and predicts that the EVA concept will be a competitive and valuable contributor to the detection of UHE neutrinos.

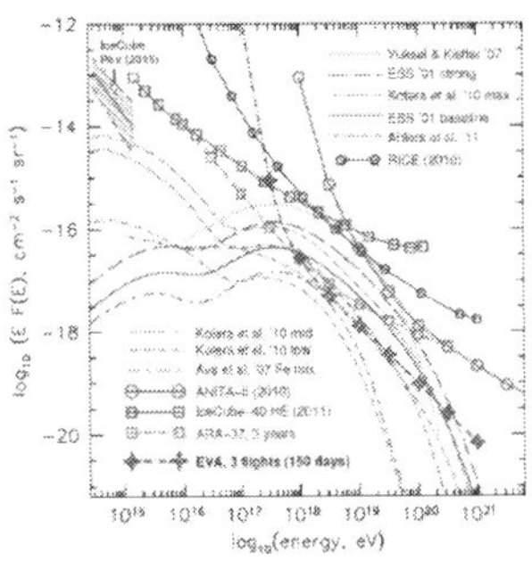

Figure 5. Estimated flux limits for a full-scale EVA detector after 150 days of flight time. Also shown are estimated sensitivity curves for other UHE neutrino detection experiments and several flux models.

References

[1] V. S. Berezinsky and G. T. Zatsepin, Phys. Lett., **B28**, 423 (1969); V. S. Berezinsky and G. T. Zatsepin, Sov. J. Nucl. Phys., **11**, 111 (1970)

[2] M. G. Aartsen *et al.*, Phys.Rev., **D88**, 112008, (2013); M. G. Aartsen *et al.*, Science, 6161, 342, 1242856 (2013).

[3] P. Allison *et al.*, Astropart.Phys., **35**, 457 (2012).

[4] S. R. Klein *at al.*, IEEE Trans. Nucl. Sci., **60** 637 (2013).

[5] P.W. Gorham *et al.*, Phys. Rev., **D82**, 022004 (2010); P.W. Gorham *et al.*, Erratum (2010).

[6] P.W. Gorham *et al.*, Astropart. Phys., **35** 242 (2011).

[7] http://www.remcom.com/xf7

[8] G. J. Burke, E. K. Miller, and A. J. Poggio, IEEE Intl Symp. National Radio Science (URSI) (2004).

[9] http://www.ticra.com/products/software/grasp

Ultimate precision in cosmic-ray radio detection — the SKA

Tim Huege[1,⋆], *Justin D.* Bray[2], *Stijn* Buitink[3], *David* Butler[4], *Richard* Dallier[5,6], *Ron D.* Ekers[7], *Torsten* Enßlin[8], *Heino* Falcke[9,10], *Andreas* Haungs[1], *Clancy W.* James[11], *Lilian* Martin[5,6], *Pragati* Mitra[3], *Katharine* Mulrey[3], *Anna* Nelles[12], *Benoît* Revenu[5], *Olaf* Scholten[13,14], *Frank G.* Schröder[1], *Steven* Tingay[15,16], *Tobias* Winchen[3], and *Anne* Zilles[4]

[1]*IKP, Karlsruher Institut für Technologie, Postfach 3640, 76021 Karlsruhe, Germany*
[2]*School of Physics & Astronomy, Univ. of Manchester, Manchester M13 9PL, United Kingdom*
[3]*Astrophysical Institute, Vrije Universiteit Brussel, Pleinlaan 2, 1050 Brussels, Belgium*
[4]*EKP, Karlsruher Institut für Technologie, Kaiserstr. 12, 76131 Karlsruhe, Germany*
[5]*Subatech, 4 rue Alfred Kastler, 44307 Nantes cedex 3, France*
[6]*Station de radioastronomie de Nançay, Observatoire de Paris, CNRS/INSU, Nançay, France*
[7]*CSIRO Astronomy & Space Science, NSW 2122, Australia*
[8]*Max-Planck-Institut für Astrophysik, Karl-Schwarzschildstr. 1, 85748 Garching, Germany*
[9]*Dept. of Astrophysics/IMAPP, Radboud Univ. Nijmegen, 6500 GL Nijmegen, The Netherlands*
[10]*Netherlands Institute for Radio Astronomy (ASTRON), 7990 AA Dwingeloo, The Netherlands*
[11]*ECAP, Univ. of Erlangen-Nuremberg, 91058 Erlangen, Germany*
[12]*Department of Physics & Astronomy, University of California, Irvine, CA 92697, USA*
[13]*Kernfysisch Versneller Instituut, Univ. of Groningen, 9747 AA Groningen, The Netherlands*
[14]*Interuniversity Institute for High-Energy, Vrije Universiteit Brussel, Pleinlaan 2, 1050 Brussels, Belgium*
[15]*Istituto di Radioastronomy, INAF, via P. Gobetti, Bologna, 4012, Italy*
[16]*International Centre for Radio Astronomy Research, Curtin University, Bentley, WA, 6102, Australia*

Abstract. As of 2023, the low-frequency part of the Square Kilometre Array will go online in Australia. It will constitute the largest and most powerful low-frequency radio-astronomical observatory to date, and will facilitate a rich science programme in astronomy and astrophysics. With modest engineering changes, it will also be able to measure cosmic rays via the radio emission from extensive air showers. The extreme antenna density and the homogeneous coverage provided by more than 60,000 antennas within an area of one km^2 will push radio detection of cosmic rays in the energy range around 10^{17} eV to ultimate precision, with superior capabilities in the reconstruction of arrival direction, energy, and an expected depth-of-shower-maximum resolution of < 10 g/cm^2.

1 Introduction

Over the past decade, radio detection of cosmic-ray air showers has made tremendous progress [1]. Today, radio antenna arrays routinely complement classical cosmic-ray observatories, and have proven to contribute valuable information with competitive resolutions achieved in the determination of the relevant air-shower parameters [2]. Arrival direction and energy of the primary particle can be measured accurately even with sparse antenna arrays where individual detector stations are spaced 100 to

⋆e-mail: tim.huege@kit.edu

200 metres apart. Determination of the depth of shower maximum (X_{max}) is also possible with such sparse arrays, with resolutions of approximately 40 g/cm^2 achieved with today's analysis approaches [3, 4]. The true power for X_{max} measurements with radio techniques, however, lies in precision measurements of the radio-emission footprints with dense antenna arrays.

In this article, we discuss the potential of ultra-precise air-shower measurements with the upcoming Square Kilometre Array (SKA)[1]. We illustrate the expected improvement over the existing Low Frequency Array (LOFAR) telescope, describe the science potential and discuss the necessary engineering changes to the SKA baseline design.

2 Air-shower detection: from LOFAR to the SKA

In low-frequency radio astronomy, LOFAR can be seen as a pathfinder towards the SKA. The same is true for cosmic-ray detection with dense antenna arrays [5]. Only with the impressive results achieved by LOFAR did the potential for air-shower detection with the SKA become apparent. In particular, the resolution in the determination of X_{max} achieved with LOFAR [6] sparked the interest in air-shower detection with the SKA.

LOFAR uses a top-down analysis approach: For each measured air-shower event nearly 100 air-shower simulations with proton- and iron-primaries are generated with the microscopic CoREAS Monte Carlo code [7]. The individual simulations are then compared with the measurements to find the air-shower parameters that yield the best agreement. It turns out that X_{max} is the one quantity that governs the agreement, and hence it can be read off directly from the best-fitting simulations. The average X_{max} resolution achieved with this analysis techique amounts to 17 g/cm^2 [6]. The best events even reach an X_{max} determination to within better than 10 g/cm^2. For reasons explained below, we expect that the SKA will yield an X_{max} measurement to within 10 g/cm^2 for each and every measured cosmic-ray shower. This is significantly better than the 20 g/cm^2 achievable with the most precise techniques available so far, namely fluorescence- and Cherenkov-light detection (see, e.g., [8]).

2.1 Increased antenna density and homogeneity

The key factor for the increased measurement precision of the SKA with respect to LOFAR lies in the layout of the antenna array. While LOFAR constitutes a fairly inhomogeneous array with significant gaps in between radio antennas (see Fig. 1 top-left), the SKA will possess a very dense, very homogeneous core of radio antennas (see Fig. 1 top-right). The radio-emission footprint has several characteristics sensitive to the depth-of-shower maximum [1], and for LOFAR only such events where these features are sampled favorably with the inhomogeneous antenna array can yield supreme X_{max} resolution. For the SKA, any "contained" event will be sampled very precisely (see Fig. 1 bottom), and will thus be reconstructable with superior precision. Initial simulation studies confirm this expectation with typical X_{max} resolutions of ≈ 6 g/cm^2 achieved with a LOFAR-style analysis [9]. In combination with experimental uncertainties, a practical resolution of ≈ 10 g/cm^2 thus seems achievable.

2.2 Increased frequency coverage

Also the increased frequency coverage of 50–350 MHz of the SKA as opposed to 30–80 MHz for LOFAR (in the most-used low-frequency observation mode) will benefit the reconstruction quality. Again, initial simulation studies [9] illustrate this improvement; a pure-simulation reconstruction for an SKA-like array with a frequency coverage from 30–80 MHz achieves an ≈ 10 g/cm^2 X_{max} resolution, while the frequency range from 50–350 MHz yields an X_{max} resolution of ≈ 6 g/cm^2.

[1]Throughout this article, we use the shorthand SKA to refer to the phase 1 implementation of the low-frequency aperture-array part of the SKA, which will measure in the frequency range from 50–350 MHz.

DOI: 10.1051/epjconf/201713502003

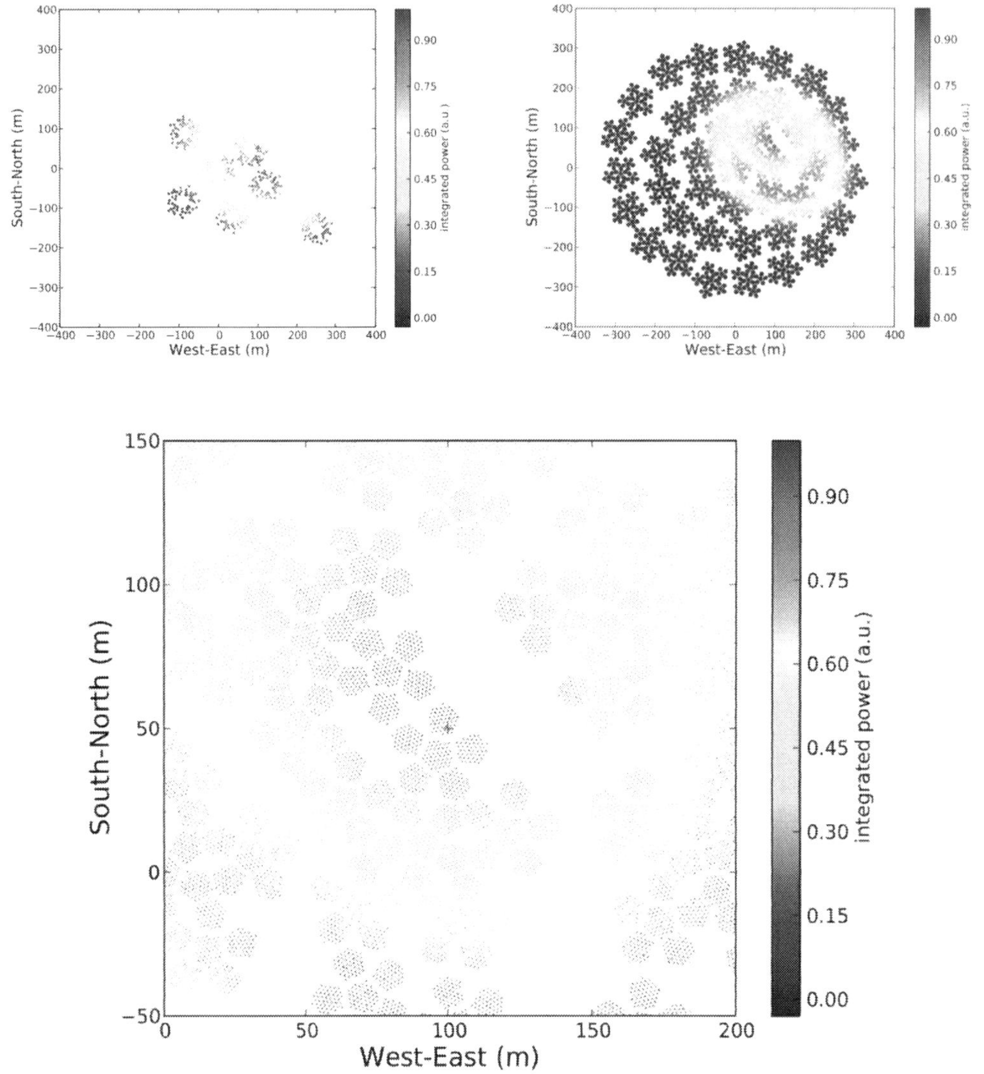

Figure 1. CoREAS-based simulation of how the air-shower radio emission would be sampled by LOFAR (top-left) and the SKA core (top-right and zoom-in at bottom). The SKA will measure the radio-emission footprint with an extremely dense and homogeneous array of antennas, yielding superior reconstruction quality on an event-to-event basis. The appearance of a clear ring-structure in the SKA simulation is related to the presence of a Cherenkov ring at higher frequencies. Plots are from the simulation study presented in [9].

2.3 Increased event statistics

While cosmic-ray detection with LOFAR has provided impressive results, it does suffer from rather limited event statistics. The reasons for this are two-fold: So far, only the innermost area of the LOFAR core can be used for air-shower detection, limiting the fiducial area to less than 0.1 km^2. Furthermore, cosmic-ray detection is not active 100% of the time, mostly because technical and or-

DOI: 10.1051/epjconf/201713502003

ganisational hurdles prevent fully commensal operation. This resulted from the fact that cosmic-ray detection capability was not a priority feature from the beginning of the design phase. With the SKA, we aim for 100% commensal operation and a fiducial area of roughly 1 km^2, increasing the event statistics by a factor of more than 100 over LOFAR. This will make the energy range from a few times 10^{16} eV up to a few times 10^{18} eV accessible for detailed measurements, with approximately 10,000 events measured per year above 10^{17} eV.

3 Science potential for air-shower detection with SKA

The science potential for cosmic-ray detection with the SKA lies in precision measurements. Several scientific goals can be addressed with SKA air-shower measurements.

3.1 High-precision mass composition measurements

The energy range accessible with the SKA, from a few times 10^{16} eV up to a few times 10^{18} eV, is suspected to harbour a transition from Galactic to extragalactic cosmic rays [10, 11]. The key metric needed to investigate this hypothesis is highly precise composition information. With supreme X_{max} resolution, the SKA will deliver very-high-quality data in this important energy range. Exploiting information beyond a per-event determination of X_{max}, we hope that even a separation between individual particle species, possibly even between proton and Helium, could become feasible.

3.2 Particle and air-shower physics

The air shower evolution, in particular X_{max}, is sensitive to hadronic interaction physics. It has been shown before that the proton-air cross section, multiplicity of secondaries in interactions, elasticity and pion charge ratios can be probed with air-shower measurements [12]. With the high-precision measurements provided by the SKA, such studies in the very forward direction and at energies beyond those accessible by the Large Hadron Collider could be carried out with minimised systematic uncertainties. If the deployed particle detector array allows a measurement of the muonic component of the air showers, even more detailed studies of the air-shower physics will become possible.

3.3 Tomography of the electromagnetic cascade

The LOFAR analysis approach employed today only uses the integrated power measured by the antennas. In addition, there are measurements of signal timing, pulse shape and polarisation that very likely carry additional information. Furthermore, there is phase information in the radio signal that is currently not taken into account at all. We expect that with near-field interferometric techniques, it should be possible to do "tomography" of the electromagnetic cascade in the air shower, and thus provide information that goes well beyond the pure determination of X_{max}.

3.4 Physics of thunderstorms and lightning

It has been shown previously that strong electric fields in the atmosphere, in particular those occurring during thunderstorms, influence the radio emission from air showers [13]. In fact, properties of the atmospheric electric fields can be probed in situ by detailed analysis of the measured radio emission and comparison with theoretical calculations [14]. The air-shower tomography development may lead as well to imaging techniques for the reconstruction of charge flows during lightnings. With the SKA, it will thus be possible to probe the physics of thunderstorms and also test possible connections between lightning initiation and cosmic-ray air showers.

4 Engineering changes to the baseline SKA-design

The SKA baseline design does not foresee the capabilities needed for air-shower detection. Here we shortly describe the required engineering changes that have been proposed to the SKA management and are currently under consideration.

4.1 Antenna buffering

Buffers are foreseen in the current SKA design to temporarily store beam-formed quantities. For air-shower detection these buffers would need to additionally store the raw signals digitised at individual antennas. With 800 MHz sampling, at least 8-bit, preferably 12-bit dynamic range, and a buffer depth of 10 ms, 1.4 TB of buffers would be needed for 60,000 dual-polarised antennas. This buffering should be carried out with 100% duty cycle, fully commensal with other observations. On a trigger, 50 μs worth of data need to be read out, which amounts to 7.2 GB of data per event.

4.2 Deadtime versus read-out time

We estimate a readout rate of 1 shower per minute at 10^{16} eV. With the above numbers, the average data rate for readout would be 120 MB/s. This is very small in comparison with the continuous data stream handled by the SKA. However, cosmic rays arrive with a Poissonian time distribution and will thus cause bursts of data that need to be read out quickly. Assuming a data rate of 2.4 GB/s, leading to 3 seconds of readout time during which the buffers are frozen, we will accumulate a tolerable deadtime of 5%.

4.3 Trigger generation

A difficult problem is the generation of an efficient and pure trigger in real time. While it would in principle be possible to generate a trigger from the radio data, we propose to install a particle-detector array for the purpose of triggering. This will also gather valuable information on the particle content of the air shower which can be included in the event reconstruction. The particle detector array should become efficient at 10^{16} eV and thus have an average detector spacing of 50–100 m. The fiducial area should extend beyond the core of antennas because non-contained events will also be usable. An attractive option is to deploy 180 scintillating particle detectors with a size of 3.6 m^2 each, available from the dismantled KASCADE array [15], between the SKA antenna stations (see Fig. 2). The detectors could be read out via the same kind of RF-over-fibre links that are used for the antenna readout. Assuming 4 channels per particle detector, this would require an additional 720 readout channels (on top of the 262,144 needed for the entire antenna array). Special care has to be taken of the radio-quietness of the particle detectors. Experience with LOFAR has shown that careful shielding ensures RFI-quietness.

4.4 Benefits for the SKA

The low-level readout capability needed for cosmic-ray measurements will greatly benefit the SKA for diagnostic purposes. Experience with LOFAR shows that many low-level problems (swapped cables, damaged low-noise amplifiers, broken antennas, time synchronisation problems, ...) could be diagnosed with the per-antenna data taken in the cosmic-ray mode. Such problems are otherwise very difficult to track down and will generally deteriorate the quality of astronomical observations.

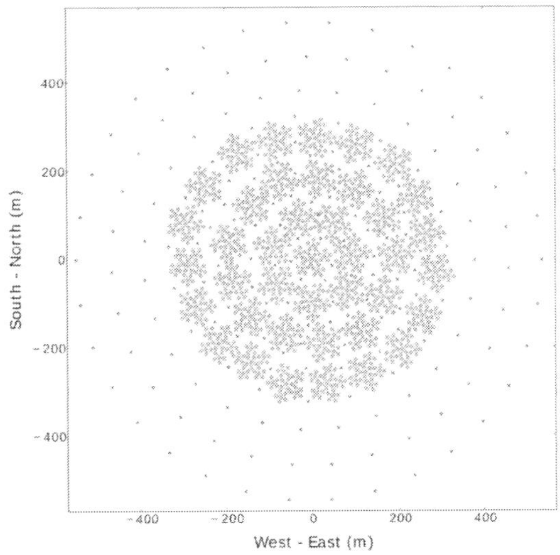

Figure 2. Possible layout of a particle detector array with 180 detectors (red squares) in between and outside the antenna stations of the SKA (gray dots).

5 Conclusion and outlook

With moderate engineering changes, the SKA can become a cosmic-ray detector that will perform extremely precise measurements of individual air showers in the energy range from a few times 10^{16} to a few times 10^{18} eV. This would enable a rich scientific harvest in return on a very limited additional investment. The SKA focus group on high-energy cosmic particles[2] has proposed the necessary engineering changes to the SKA management, where they are currently under consideration.

References

[1] T. Huege, Physics Reports **620**, 1 (2016)
[2] F.G. Schröder, Prog. Part. Nucl. Phys. (submitted), `arXiv:1607.08781`
[3] P.A. Bezyazeekov, N.M. Budnev, O.A. Gress, et al., JCAP **01**, 52 (2016)
[4] F. Gaté for the Pierre Auger Collaboration, these proceedings (2016)
[5] P. Schellart, A. Nelles, S. Buitink, et al., Astronomy & Astrophysics **560**, A98 (2013)
[6] S. Buitink, A. Corstanje, H. Falcke, et al., Nature **531**, 70 (2016)
[7] T. Huege, M. Ludwig, C.W. James, AIP Conf. Proc. pp. 128–132 (2013)
[8] A. Aab, P. Abreu, M. Aglietta, et al., Nucl. Instrum. Meth. A **798**, 172 (2015)
[9] A. Zilles, S. Buitink, T. Huege, these proceedings (2016)
[10] W.D. Apel, J.C. Arteaga-Velàzquez, K. Bekk, et al., Phys. Rev. D **87**, 081101 (2013)
[11] P. Blasi, The Astronomy and Astrophysics Review **21**, 1 (2013)
[12] R. Ulrich, R. Engel, M. Unger, Phys. Rev. D **83**, 054026 (2011)
[13] S. Buitink, W.D. Apel, T. Asch, et al., Astronomy & Astrophysics **467**, 385 (2007)
[14] P. Schellart, T.N.G. Trinh, S. Buitink, et al., Phys. Rev. Lett. **114**, 165001 (2015)
[15] T. Antoni, W.D. Apel, F. Badea et al., Nucl. Instrum. Meth. A **513**, 490 (2003)

[2] http://astronomers.skatelescope.org/home/focus-groups/high-energy-cosmic-particles/

Initial simulation study on high-precision radio measurements of the depth of shower maximum with SKA1-low

Anne Zilles[1,⋆], *Stijn* Buitink[2,3], and *Tim* Huege[4]

[1] *Institut für Experimentelle Kernphysik, Karlsruher Institut für Technologie, 76128 Karlsruhe, Germany*
[2] *Astrophysical Institute, Vrije Universiteit Brussel, Pleinlaan 2, 1050 Brussels, Belgium*
[3] *Department of Astrophysics/IMAPP, Radboud University Nijmegen, 6500 GL, Nijmegen, The Netherlands*
[4] *Institut für Kernphysik, Karlsruher Institut für Technologie, 76021 Karlsruhe, Germany.*

Abstract. As LOFAR has shown, using a dense array of radio antennas for detecting extensive air showers initiated by cosmic rays in the Earth's atmosphere makes it possible to measure the depth of shower maximum for individual showers with a statistical uncertainty less than 20 g/cm^2. This allows detailed studies of the mass composition in the energy region around 10^{17} eV where the transition from a Galactic to an Extragalactic origin could occur. Since SKA1-low will provide a much denser and very homogeneous antenna array with a large bandwidth of 50 – 350 MHz it is expected to reach an uncertainty on the X_{max} reconstruction of less than 10 g/cm^2. We present first results of a simulation study with focus on the potential to reconstruct the depth of shower maximum for individual showers to be measured with SKA1-low. In addition, possible influences of various parameters such as the numbers of antennas included in the analysis or the considered frequency bandwidth will be discussed.

1 Introduction

With its about 70.000 antennas in a dense and uniform antenna spacing on a fiducial area of one km^2 and its large bandwidth of 50 – 350 MHz, the low-frequency part of the the Square Kilometer Array (SKA1-low, first phase) will provide very precise measurements of individual cosmic ray air showers. An overview of the SKA1-low array as well as the physics motivation are described in [1]. The LOFAR experiment shows that for individual shower events initiated by cosmic rays with energies of 10^{17} – $10^{17.5}$ eV the shower depth, which is correlated to the mass of the primary cosmic ray, can be reconstructed with a mean uncertainty of 16 g/cm^2 [2], for the "best" events even with 10 g/cm^2 uncertainty. Based on results of first simulation studies, a reconstruction of the shower depth X_{max} of individual showers with a mean intrinsic uncertainty of well below 10 g/cm^2 seems feasible.

1.1 Reconstruction method of the shower depth for SKA1-low

The presented method for the reconstruction of the shower depth of individual detected showers is adapted from the strategy developed for cosmic ray detection with the LOFAR experiment [3] which is already successfully applied [2].

This simulation study on the possible uncertainty in the reconstruction of the shower depth by SKA1-low is based on sets of air-shower simulations induced by protons and iron nuclei. The air

⋆e-mail: anne.zilles@kit.edu

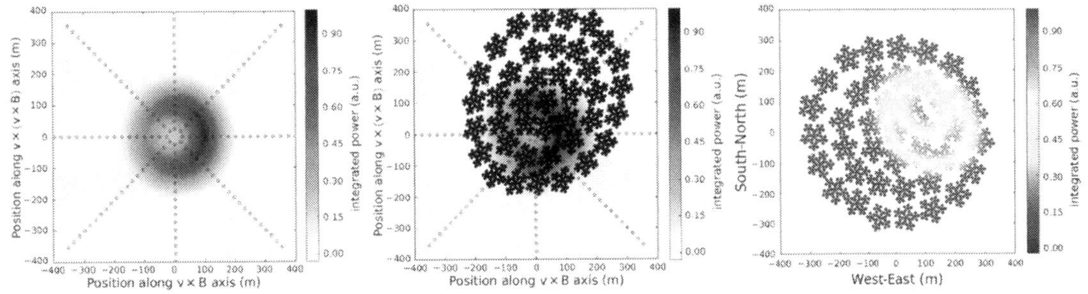

Figure 1. The left and the center plots shows the total integrated power in the shower plane defined by the axes $\mathbf{v} \times \mathbf{B}$ and $\mathbf{v} \times (\mathbf{v} \times \mathbf{B})$, while the right plot represents the integrated power on the ground with North pointing upwards. Left: Total integrated power at the simulated antenna positions (circles) and the interpolated radio footprint in the shower plane for a proton-induced air shower with an primary energy of $E = 10^{17}$ eV and a zenith angle of 36.87°, filtered to a frequency band of 50 – 350 MHz and normalised to the maximum power. Center: The SKA1-low antenna positions rotated into the shower plane. At each antenna position, a value for the total received power is interpolated. Left: The example radio footprint on ground as detected by the SKA1-low array. The different color scheme is chosen to underline to change of the coordinate system.

shower simulation was done by CORSIKA [4]. The corresponding radio signal of the shower was calculated by CoREAS [5], a plug-in of CORSIKA. As observer positions for which the radio signal was calculated 160 antenna positions in the shower plane defined by $\mathbf{v}\times\mathbf{B}$ and $\mathbf{v}\times(\mathbf{v}\times\mathbf{B})$, the so-called star-shape pattern, were chosen, as shown in Fig. (1) (left). Here, the vector $\mathbf{v}$ represents the velocity of the air shower front and $\mathbf{B}$ the Earth's magnetic field . For each position the East-West and the North-South electric field components were filtered to the SKA1-low frequency band of 50 – 350 MHz. No antenna model was included. Afterwards, the time integral of the total received power was computed for each antenna in the star-shape pattern. The footprint of the radio signal with its asymmetry due to the interference of the Askaryan and the geomagnetic effect can be reconstructed by interpolation between the rays of the star-shape pattern (see Fig. (1), center). To estimate the power which would be received by the SKA1-low antennas, the complete antenna array is rotated into the $\mathbf{v}\times\mathbf{B}$ and $\mathbf{v}\times(\mathbf{v}\times\mathbf{B})$ frame and each antenna is assigned a total power corresponding to the interpolated power at this position (see Fig. (1), right). These steps are performed for every simulation in a set which should consists in total of 50 proton-induced and 25 iron-induced air shower simulations.

Finally, each simulation acts once as "fake" data. This means that random noise is added to the interpolated total power of each antenna: 5% of the total power of each antenna as uncertainty on the antenna calibration and 1% of the maximum power in the whole array as an approximation for Galactic noise. This is just a first approximation since by this reconstruction the noise scales with the energy of the primary particle. More realistic assumptions have to be implemented in the future calculations. The remaining simulations of the set are then compared to the "fake" data on the basis of a χ^2 fit:

$$\chi^2 = \sum_{antennas} \left(\frac{P_{\mathrm{fake}} - P_{\mathrm{sim}}}{\sigma_{\mathrm{fake}}} \right)^2 \tag{1}$$

Here, P_{fake} represents the total power of a single antenna in the "fake" data set, with σ_{fake} denoting the assumed uncertainty due to noise, and P_{sim} the simulated power for the same position. This returns a χ^2 value for every comparison of the "fake" data to one of the remaining simulations which can be connected to a corresponding X_{sim} known from the simulation.

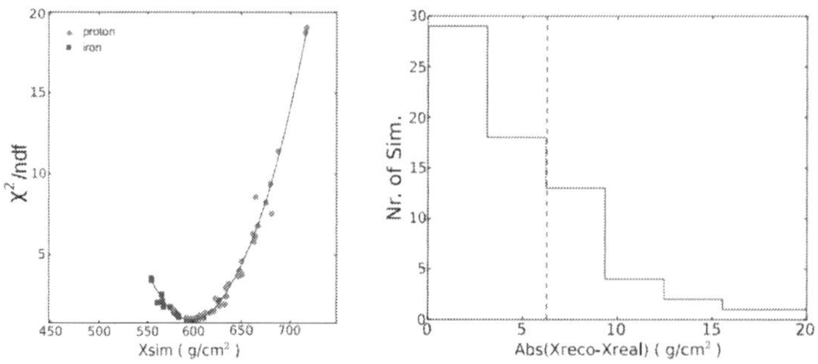

Figure 2. Left: Example χ^2 distribution for the SKA1-low array layout when comparing one "fake" event with the remaining simulations of the same set, filtered to 50 – 350 MHz. Right: Histogram for the differences of the reconstructed and the "real" shower depth: abs($X_{\mathrm{reco}} - X_{\mathrm{real}}$). The 1σ uncertainty containing 68% of the events returns a reconstruction uncertainty of 6.3 g/cm^2 for a shower of $E = 10^{17}$ eV and a zenith angle of 36.87°.

This leads to a distribution as shown in Fig. (2) (left) to which a parabola function can be fitted. This fit then returns the "reconstructed" shower depth X_{reco} as the X_{max} value for which the parabola has its minimum. The absolute difference between X_{reco} and the each "real" shower depth X_{real} of the simulated shower event which works as "fake" data is filled into a histogram (see to Fig. (2), right). In that histogram the 1σ uncertainty is defined at that value of abs($X_{\mathrm{reco}} - X_{\mathrm{real}}$) which contains 68% of entries. This returns the statistical uncertainty of the shower reconstruction shower by the method when adopting all simulations in a set are a "fake" event once.

2 First results

The results shown in Fig. (2) are based on 50 proton and 20 iron showers with a primary energy of $E = 10^{17}$ eV, an azimuth angle of 225° and a zenith angle of 36.87°. For these parameters the method returns an intrinsic reconstruction uncertainty of 6.3 g/cm^2.

Obviously, the uncertainty can depend on shower parameters such as the primary energy and the arrival direction. For a fixed zenith angle but a higher primary energy of $E = 10^{17.5}$ eV, the simulation subset of 40 p and 18 Fe returns an uncertainty of 6.0 g/cm^2, a slightly smaller uncertainty. This could be explained by a possible better resolution of structures at the Cherenkov cone since the shower evolutions starts in average deeper in the atmosphere for higher primary energies. Then, this would be a benefit for the reconstruction method because in general the structures of the footprints of the "fake" and the simulated data get compared. This could mean that the more prominent the structures in the two dimensional profiles are the better the simulation which fits best can be fitted. For lower primary energies jitter in the χ^2 distribution seems to get larger. This could be due to a flattening of the Cherenkov cone structure since the shower interacts earlier in the atmosphere. The example simulation subset of 40 p and 17 Fe for an energy of $E = 10^{16.5}$ eV leads to an intrinsic reconstruction uncertainty of 9.6 g/cm^2. This is not an effect of the signal-to-noise ratio since this scales with the primary energy.

One of the clear differences to the LOFAR experiment is the frequency band. In the LOFAR analysis the frequency band is limited to 30 – 80 MHz. In this frequency range, a bean-like structure is visible in the radio footprint instead of a clear Cherenkov ring as for the higher frequencies (see Fig. (3), left). This results in a slightly higher jitter around the parabola in the χ^2 distribution as compared to the higher frequencies and finally in an intrinsic reconstruction uncertainty of 10.6 g/cm^2.

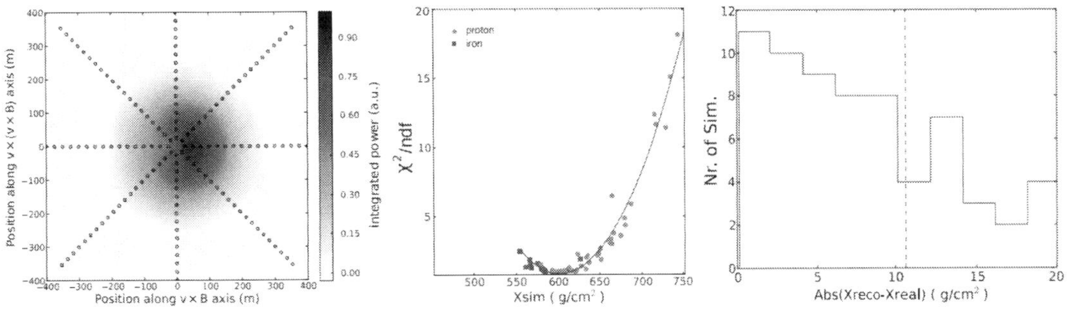

Figure 3. Left: Radio footprint in **v**×**B** and **v**×(**v**×**B**) for a proton induced air shower with a primary energy of $E = 10^{17}$ eV and a zenith angle of 36.87°, filtered to a frequency band of 30 – 80 MHz. Center: Example χ^2 distribution for the SKA1-low array layout, filtered to the LOFAR bandwidth. Right: For these parameters the uncertainty of the shower depth reconstructions is 10.6 g/cm^2 (1σ).

In addition, there are more aspects which influence the uncertainty of the shower depth reconstruction, like the numbers of antennas read out. A thinning of the antenna array, meaning just reading out a subset of the antennas, does not influence the reconstruction precision as long as the read-out are homogeneously distributed over the radio footprint. The study of the impact of the number of the read-out antennas is closely connected to and therefore limited by the number of simulated antenna positions.

The method used for the power interpolation also turned out to be a limiting factor. As visible in Fig. (1) (left), dips in the interpolated power distribution are visible between the single rays. This seems to be an artifact of the interpolation since they vanish if more antennas on rays in-between the existing ones are simulated and included for the interpolation.

3 Conclusion

The preliminary results of an initial simulation study show that a LOFAR-like approach for the reconstruction of the shower depth for single events is applicable to SKA1-low data. Since the simulation sets functioning as the basis for this study are still incomplete, the results can be interpreted as an indication of how precisely the SKA will reconstruct X_{max}. Hereby, the reconstruction profits especially from the very dense antenna array, but also from the larger bandwidth at higher frequencies in comparison to the LOFAR experiment. In this simulation study, we evaluated the intrinsic reconstruction accuracy reachable with a LOFAR-like analysis for data as it would be provided by SKA1-low. The SKA1-low array will achieve a mean intrinsic reconstruction uncertainty of less than 10 g/cm^2. The influence of experimental uncertainties (realistic background noise, uncertainty in the atmospheric properties, ...) on the reconstruction uncertainty has to be evaluated in a future study. Furthermore, it has to be studied whether the inclusion of signal timing, pulse shape and polarisation measurements would yield to additional information beyond the pure determination of X_{max}.

References

[1] T. Huege et al., *Ultimate precision in cosmic-ray radio detection - the SKA*, these preceedings.
[2] S. Buitink et al., Nature 531 (2016) 70.
[3] S. Buitink et al., Phys. Rev. D **90**, 082003 (2014).
[4] D. Heck et al., Report FZKS **6019** (1998).
[5] T. Huege, M. Ludwig and C. James, AIP Conference Proceedings **1535**, 128-132 (2012).

Cosmogenic photons strongly constrain UHECR source models

Arjen van Vliet[1],*

[1] *Department of Astrophysics/IMAPP, Radboud University, P.O. Box 9010, 6500 GL Nijmegen, The Netherlands.*

Abstract. With the newest version of our Monte Carlo code for ultra-high-energy cosmic ray (UHECR) propagation, CRPropa 3, the flux of neutrinos and photons due to interactions of UHECRs with extragalactic background light can be predicted. Together with the recently updated data for the isotropic diffuse gamma-ray background (IGRB) by Fermi LAT, it is now possible to severely constrain UHECR source models. The evolution of the UHECR sources especially plays an important role in the determination of the expected secondary photon spectrum. Pure proton UHECR models are already strongly constrained, primarily by the highest energy bins of Fermi LAT's IGRB, as long as their number density is not strongly peaked at recent times.

1 Introduction

Recently the Fermi-LAT collaboration updated their measurements on the isotropic diffuse gamma-ray background (IGRB) and extended it up to 820 GeV [1]. A possible source for part of the IGRB is secondary electromagnetic cascades initiated by interactions of ultra-high-energy cosmic rays (UHECRs) with the cosmic microwave background (CMB) or the extragalactic background light (EBL). In these same interactions secondary neutrinos can be produced. These neutrinos could possibly contribute to the astrophysical neutrino flux as measured by IceCube [2].

The UHECR energy spectrum has been measured with unprecedented statistics by the Pierre Auger [3, 4] (Auger) and Telescope Array [5] (TA) collaborations. The UHECR mass measured by these two collaborations can, however, still be interpreted in different ways. While the measurements of Auger show a depth of the shower maximum, $X_{\rm max}$, indicating an increasingly heavier mass composition for $E \gtrsim 10^{18.3}$ eV [6], TA results in the same energy range are consistent with a pure proton composition [7]. Despite these differences the $X_{\rm max}$ measurements are in good agreement with each other [8]. The predictions of different air shower simulation models, however, leave room for varying interpretations of the data. Therefore many UHECR composition models are still viable.

However, as shown e.g. in Refs. [9–11], the parameter range of possible pure proton models can be constrained when taking into account the secondary gamma-ray and neutrino production during the propagation of UHECRs from their sources to Earth. Ref. [12] even claims that the proton dip model is challenged at more than 95% C.L. by the cosmogenic neutrino flux alone.

To obtain the predicted cosmogenic gamma-ray and neutrino flux for a certain UHECR model the propagation of UHECRs through the universe, including all relevant interactions with the CMB and

*e-mail: a.vanvliet@astro.ru.nl

EBL, has to be simulated. Here the newest version of our UHECR propagation code, CRPropa version 3 [13], is used to simulate the cosmic ray, electromagnetic cascade and neutrino propagation and obtain predictions for the cosmogenic gamma-ray and neutrino fluxes. CRPropa is a full simulation framework for Monte Carlo UHECR propagation including all relevant interactions for protons as well as for heavier nuclei (photo-meson production, pair production, photodisintegration, nuclear decay and energy reduction due to the adiabatic expansion of the universe). For the electromagnetic cascade propagation the specialized code DINT [14], interfaced and shipped with CRPropa 3, is used. DINT solves the one-dimensional transport equations for electromagnetic cascades initiated by electrons, positrons or photons and includes single, double and triplet pair production, inverse-Compton scattering and synchrotron radiation.

2 Simulation setup

All the simulations done here assume a homogeneous distribution of identical sources. The cosmic rays are injected at the sources following a spectrum of

$$\frac{dN}{dE} \propto (E/E_0)^{-\alpha} \exp(-E/E_{\rm cut}) , \tag{1}$$

with E the energy of the particles, E_0 an arbitrary normalization energy, α the spectral index at injection and $E_{\rm cut}$ the cutoff energy. A minimum energy of cosmic rays at the sources of $E_{\rm min} = 0.1$ EeV is assumed. The EBL used for the cosmic ray simulations is the Gilmore 2012 model [15]. In Fig. 1 (solid lines) a reference scenario is given with pure proton injection at the sources for which $\alpha = 2.5$ and $E_{\rm cut} = 200$ EeV. In this case a co-moving source evolution up to a maximum redshift of $z_{\rm max} = 6$ is implemented. Unless stated otherwise these simulation parameters have been used for the other scenarios as well.

These parameters are not optimized to fit the cosmic ray spectrum perfectly as the systematic uncertainty of the energy spectrum measurements is much larger than the statistical uncertainty for both Auger and TA. Furthermore the assumption made here, a homogeneous distribution of identical pure proton sources with a power law injection spectrum with exponential cutoff, may not correctly depict the real situation. Additionally it is convenient to change only one parameter at a time in order to see the effect of only that parameter on the predicted spectra. Fitting a pure proton model to the TA spectrum would lead to a strong source evolution and a relatively hard injection spectrum and would be strongly constrained by both the Fermi-LAT IGRB data and the IceCube data [11, 12].

3 Source evolution dependence

The evolution in time (redshift) of the sources of UHECRs influences the expected cosmogenic gamma-ray and neutrino flux significantly. As the sources of UHECRs are unknown, a wide range of possible source evolutions could be applicable. Here the effect of the source evolution on the UHECR spectrum and the expected cosmogenic neutrino and gamma-ray fluxes is investigated by taking the reference scenario and adjusting the source evolution by a factor of $(1 + z)^m$ with z the redshift of the source and $-6 \leq m \leq 6$. A value of $m = -6$ (negative source evolution) would, for instance, correspond roughly to the source evolution of High Synchrotron Peaked (HSP) BL Lacs [9]. All the other parameters of the simulation are kept the same in order to clearly see the effect of the source evolution alone. In Fig. 1 the results are given for $m = 0$, $m = -6$ and $m = 6$. The simulated cosmic ray spectrum is normalized to the flux measured by Auger [3] at $E = 10^{18.85}$ eV and the simulated cosmogenic neutrino and gamma-ray spectra are normalized accordingly. For comparison the spectrum measured

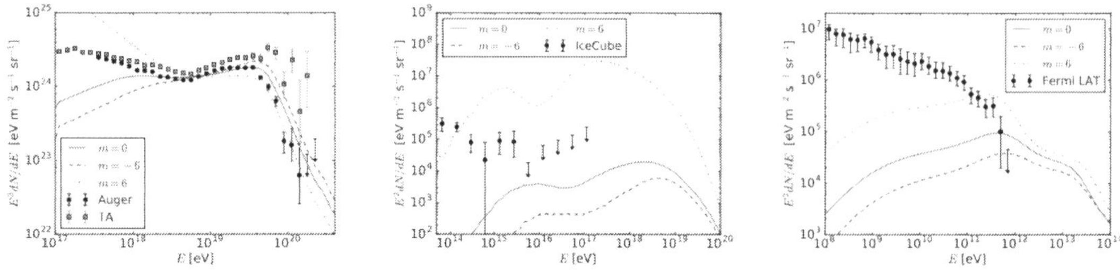

(a) Cosmic rays (b) Neutrinos (c) Photons

Figure 1. (a) Simulated cosmic ray spectra normalized to the Auger flux [3] (circles) at $E = 10^{18.85}$ eV. The spectrum measured by TA [5] (squares) is given as well. (b) Corresponding neutrino spectra compared with IceCube data [2] (circles). (c) Corresponding photon spectra compared with Fermi-LAT IGRB data [1] (circles) using Galactic foreground model A. The cosmic ray simulations were done starting a pure proton injection at $E_{\mathrm{min}} = 0.1$ EeV with an exponential cutoff at a cutoff energy of $E_{\mathrm{cut}} = 200$ EeV and an injection spectral index of $\alpha = 2.5$ using the Gilmore 2012 EBL model [15]. The simulations were done up to a maximum redshift of $z_{\mathrm{max}} = 6$ with a comoving source evolution multiplied by $(1 + z)^m$ with $m = 0$ (solid lines), $m = -6$ (dashed lines) and $m = 6$ (dashed-dotted lines). See text for further details.

by TA [5] is given as well. The simulated neutrino flux is compared with IceCube data [2] while the simulated gamma-ray flux is compared with Fermi-LAT IGRB data [1], using galactic foreground model A with foreground model uncertainties and IGRB intensity uncertainties added in quadrature.

From Fig. 1(a) can be seen that the cosmic ray spectrum is affected by the source evolution most strongly at the lower energy range ($E \lesssim 5 \times 10^{18}$ eV). This is expected as in this energy range the particles can reach us from very large distances while at higher energy, due to the different energy-loss processes, the cosmic rays can only come from relatively nearby sources. Note here that this lower energy range might also be the regime where a Galactic contribution to the cosmic ray spectrum starts playing a role, which could possibly compensate for an underprediction of the cosmic ray spectrum. Fig. 1(b) shows that a strong source evolution is clearly constrained by the astrophysical neutrino flux as measured by IceCube, but a co-moving or negative source evolution are still allowed when only the neutrino flux is considered. However, Fig. 1(c) indicates that especially the non-detection at the highest energy bin (580-820 GeV) of the IGRB measured by Fermi LAT is very constraining. Only sources with a number density strongly peaked at recent times are still allowed by this limit. Note furthermore that there are even many other sources expected to contribute to the IGRB (see e.g. Ref. [16]), which are not taken into account here.

Instead of a co-moving source evolution multiplied by $(1 + z)^m$, specific functions for certain UHECR source candidates can be tested in the same way. For this procedure the source evolution parametrizations listed in Ref. [9] for gamma-ray bursts (GRBs), high luminosity active galactic nuclei (HLAGNs), medium high luminosity AGNS (MHLAGNs), medium low luminosity AGNs (MLLAGNs) and sources following the star formation rate (SFR) are implemented. The results (with all other parameters of the simulations kept the same) are shown in Fig. 2. From Fig. 2(c) can be seen that all these source classes exceed the IGRB data measured by Fermi LAT.

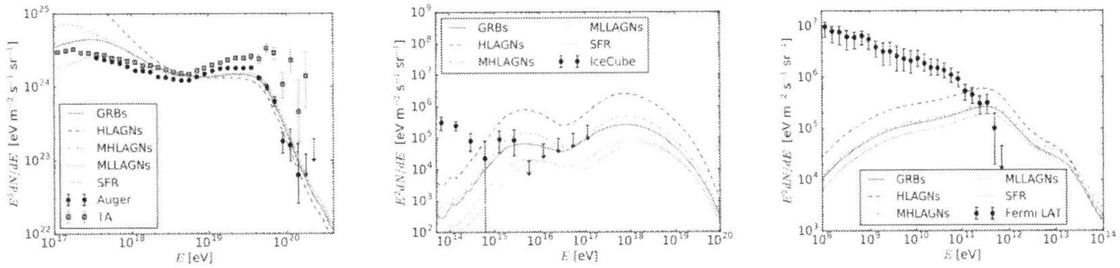

(a) Cosmic rays (b) Neutrinos (c) Photons

Figure 2. Simulated cosmic ray (a), neutrino (b) and photon (c) spectra compared with the same data as in Fig. 1. The evolution of the UHECR sources are in this case for GRBs (solid lines), HLAGNs (dashed lines), MHLAGNs (dashed-dotted lines), MLLAGNs (thick solid lines) and sources following the SFR (dotted lines). See text for further details.

4 Conclusions

Figs. 1 and 2 show that the IGRB measured by Fermi LAT is more constraining for UHECR models than the IceCube neutrino measurements. The bin with the highest energy (580-820 GeV) of the IGRB especially provides a strong constraint for pure proton UHECR models. For the scenarios investigated here, only in the case of source densities that are strongly decreasing with redshift (for instance in the case of HSP BL Lacs) is it possible to get a gamma-ray spectrum in agreement with that highest energy bin. With a few more years of data and, perhaps, an extension to even higher energies, Fermi LAT might be able to rule out all realistic UHECR pure proton models.

I want to thank Rafael Alves Batista and Jörg Hörandel for helpful discussions and the Auger PC for their suggested corrections. I acknowledge financial support from the NWO Astroparticle Physics grant WARP.

References

[1] M. Ackermann et al. (Fermi-LAT Collaboration), Astrophys. J. **799**, 86 (2015)
[2] C. Kopper, W. Giang and N. Kurahashi for the IceCube Collaboration, PoS (ICRC2015) 1081
[3] I. Valiño for the Pierre Auger Collaboration, PoS (ICRC2015) 271
[4] A. Aab et al. (Pierre Auger Collaboration), JCAP **1508**, 049 (2015)
[5] C. Jui for the Telescope Array Collaboration, PoS (ICRC2015) 035
[6] A. Porcelli for the Pierre Auger Collaboration, PoS (ICRC2015) 420
[7] T. Fujii for the Telescope Array Collaboration, PoS (ICRC2015) 320
[8] M. Unger for the Pierre Auger and the Telescope Array Collaboration, PoS (ICRC2015) 307
[9] E. Gavish and D. Eichler, Astrophys. J. **822**, 56 (2016)
[10] V. Berezinsky, A. Gazizov and O. Kalashev, arXiv:1606.09293 (2016)
[11] A. D. Supanitsky, arXiv:1607.00290 (2016)
[12] J. Heinze et al., Astrophys. J. **825**, 122 (2016)
[13] R. Alves Batista et al., JCAP **1605**, 038 (2016)
[14] S. Lee, Phys. Rev. **D 58**, 043004 (1998)
[15] R. C. Gilmore et al., Mon. Not. Roy. Astron. Soc. **422**, 3189 (2012)
[16] M. Ackermann et al. (Fermi-LAT Collaboration), Phys. Rev. Lett. **116**, 151105 (2016)

DOI: 10.1051/epjconf/201713503002

Circular polarization of radio emission from air showers in thunderstorm conditions

T.N.G.Trinh[1,★], *O. Scholten*[1,2], *A. Bonardi*[3], *S. Buitink*[4], *A. Corstanje*[3], *U. Ebert*[5,6], *J. E. Enriquez*[3], *H. Falcke*[3,7,8,9], *J.R. Hörandel*[3,7], *P. Mitra*[4], *K. Mulrey*[4], *A. Nelles*[10], *S. Thoudam*[11], *J.P. Rachen*[3], *L. Rossetto*[3], *C. Rutjes*[5], *P. Schellart*[12], *S. ter Veen*[3], and *T. Winchen*[4]

[1]*KVI-Center for Advanced Radiation Technology, University Groningen, P.O. Box 72, 9700 AB Groningen, The Netherlands*
[2]*Interuniversity Institute for High-Energy, Vrije Universiteit Brussel, Pleinlaan 2, 1050 Brussels, Belgium*
[3]*Department of Astrophysics/IMAPP, Radboud University Nijmegen, P.O. Box 9010, 6500 GL Nijmegen, The Netherlands*
[4]*Astrophysical Institute, Vrije Universiteit Brussel, Pleinlaan 2, 1050 Brussels, Belgium*
[5]*Center for Mathematics and Computer Science (CWI), PO Box 94079, 1090 GB Amsterdam, The Netherlands*
[6]*Department of Applied Physics, Eindhoven University of Technology (TU/e), PO Box 513, 5600 MB Eindhoven, The Netherlands*
[7]*NIKHEF, Science Park Amsterdam, 1098 XG Amsterdam, The Netherlands*
[8]*Netherlands Institute of Radio Astronomy (ASTRON), Postbus 2, 7990 AA Dwingeloo, The Netherlands*
[9]*Max-Planck-Institut für Radioastronomie, P.O. Box 20 24, Bonn, Germany*
[10]*Department of Physics and Astronomy, University of California Irvine, Irvine, CA 92697-4575, USA*
[11]*Department of Physics and Electrical Engineering, Linnéuniversitetet, 35195 Växjö, Sweden*
[12]*Department of Astrophysical Sciences, Princeton University, Princeton, NJ 08544, USA*

Abstract. We present measured radio emission from cosmic-ray-induced air showers under thunderstorm conditions. We observe for these events large differences in intensity, linear polarization and circular polarization from the events measured under fair-weather conditions. This can be explained by the effects of atmospheric electric fields in thunderclouds. Therefore, measuring the intensity and polarization of radio emission from cosmic ray extensive air showers during thunderstorm conditions provides a new tool to probe the atmospheric electric fields present in thunderclouds.

1 Introduction

Lightning initiation [1] and propagation [2] are driven by the electric fields in a thunderstorm. As it is very challenging to measure these fields in situ before or after the discharge and up to now essentially impossible during the discharge, there are ongoing controversies related to the physical mechanisms. A new non-intrusive method of field measurements is offered with LOFAR. It was discovered that under thunderstorm conditions, the intensity and polarizations of radio emission from

★e-mail: t.n.g.trinh@rug.nl

air showers differ considerably [3] from events in fair-weather conditions, so they can be used to probe the atmospheric electric fields present in thunderclouds.

Radio emission from air showers is measured by many experiments. Here we present the measurement with the LOw-Frequency ARay (LOFAR) radio telescope using the Low-Band Antennas (LBA). The LBA, having the frequency range of 10 - 90 MHz, are grouped into circular stations. The core of LOFAR, so-called 'Superterp', consists of 6 such stations, located in a ~ 320 m diameter region. The read-out of the LOFAR antennas for a cosmic-ray event is triggered by the LORA particle detector array at the Superterp. The measurements are done in both fair-weather and thunderstorm conditions.

2 Radio mechanism

The main mechanism for radio emission from air showers is the deflection of electrons and positrons in the showers in opposite direction by the Lorentz force exerted by the geomagnetic field [4, 5]. They form a transverse current and thus emit radiation polarized linearly along the direction of the Lorentz force, i.e. $\hat{\mathbf{e}}_{\mathbf{v}\times\mathbf{B}}$, where $\mathbf{v}$ is the velocity of the shower and $\mathbf{B}$ is the geomagnetic field.

A secondary contribution is from charge excess which is due to a build-up of negative charge in the shower front [6, 7]. This also produces a radio pulse polarized radially with respect to the shower symmetry axis. The signal observed on the ground is the superposition of the two contributions and thus depends on the viewing angle.

Under thunderstorm conditions, strong atmospheric electric fields affect the motion of shower particles and thus influence the radio emission from the showers. The electric field can be decomposed into two components

$$\mathbf{E} = \mathbf{E}_\perp + \mathbf{E}_\parallel, \tag{1}$$

where $\mathbf{E}_\perp$ and $\mathbf{E}_\parallel$ are perpendicular and parallel to the shower axis, respectively. Due to $\mathbf{E}_\perp$, there is a net force acting on particles

$$\mathbf{F} = \mathbf{q}(\mathbf{E}_\perp + \mathbf{v} \times \mathbf{B}). \tag{2}$$

which gives rise to a change in both magnitude and direction of the transverse current. Thus, the magnitude and polarization of the radiation from the tranverse current also change. Depending on the sign, the parallel component $\mathbf{E}_\parallel$ may accelerate the particles and thus increase the number of particles at a certain energy. However, this effect is hardly seen in the signals observed by LOFAR LBA. These particles are trailing far behind the shower front, and therefore increase the intensity at lower frequencies [8]. Therefore, we assume that the parallel component vanishes in this work.

The complex voltages are expressed as $\varepsilon_j = E_j + i\hat{E}_j$ where E_j is sample j of the pulse and $\hat{E}_j$ is its Hilbert transform. The Stokes parameters are given by

$$\begin{aligned}
I &= \frac{1}{n}\sum_{j=0}^{n-1}\left(|\varepsilon|^2_{j,\mathbf{v}\times\mathbf{B}} + |\varepsilon|^2_{j,\mathbf{v}\times\mathbf{v}\times\mathbf{B}}\right), \\
Q &= \frac{1}{n}\sum_{j=0}^{n-1}\left(|\varepsilon|^2_{j,\mathbf{v}\times\mathbf{B}} - |\varepsilon|^2_{j,\mathbf{v}\times\mathbf{v}\times\mathbf{B}}\right), \\
U + iV &= \frac{2}{n}\sum_{j=0}^{n-1}\left(\varepsilon_{j,\mathbf{v}\times\mathbf{B}}\varepsilon^*_{j,\mathbf{v}\times\mathbf{v}\times\mathbf{B}}\right),
\end{aligned} \tag{3}$$

In the LOFAR analysis, the summations are calculated for $n = 5$ samples of 5 ns each, centered around the peak of the pulse. Stokes I represents the intensity. The orientation of the linear polarization can be derived from Stokes Q and Stokes U. Stokes V is the amount of circular polarization.

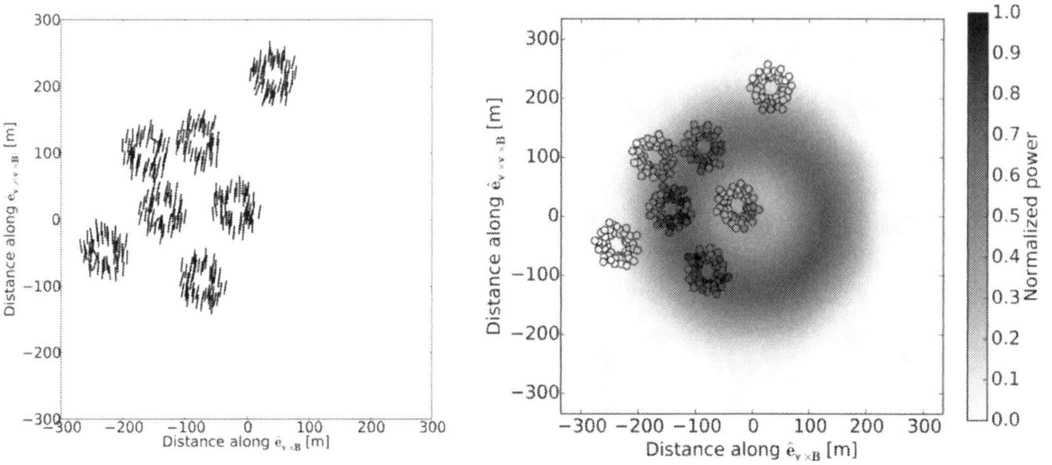

Figure 1. Left panel: Linear polarization footprint of an air shower under thunderstorm condition. Each line represents the orientation of the linear polarization for each antenna. Right panel: Intensity footprint of an air shower under thunderstorm condition. The background is the result from CoREAS simulation. Each small circle displays the intensity for each antenna. There is a good agreement between the data and the simulation.

3 Fair-weather events vs thunderstorm events.

As discussed above, the events observed under thunderstorm conditions show large differences shown in the intensity and polarizations compared to fair-weather events.

3.1 Linear polarization

The angle of linear polarization is given by $\psi = \frac{1}{2}\tan^{-1}(\frac{U}{Q})$. In fair-weather events, the polarization over all antennas is mainly along the expected $\hat{\mathbf{e}}_{\mathbf{v}\times\mathbf{B}}$ direction with some small deviation because of the contribution from charge excess. In thunderstorm events, the net force **F** changes the direction of the transverse current and thus the orientation of the linear polarization is no longer along the $\hat{\mathbf{e}}_{\mathbf{v}\times\mathbf{B}}$ direction, as it can be seen in the left panel of Fig. (1).

3.2 Intensity footprint

Unlike fair-weather events showing a bean-shaped intensity footprint due to the inteference of the transverse current and charge excess components, the ones in thunderstorm conditions show a ring-like structure as seen in the right panel of Fig. (1). The observed structure for this event can be understood as the effect of an electric field that is build up in two layers. The upper layer starts at a height of 8 km above the ground and extends to 2.9 km with a strength of $|E_U| = 50$ kV/m. At the height of 2.9 km, the electric field changes such that the net force is inversed and its strength reduces to $|E_L| = 26.5$ kV/m. Two layers are needed to introduce the destructive interference between the emission from the two layers, which results in the ring-like structure in the intensity footprint.

3.3 Circular polarization

In fair-weather events, due to the time-shift between the pulses radiated from the transverse current component and the charge-excess component, there is a small amount of circular polarization. This

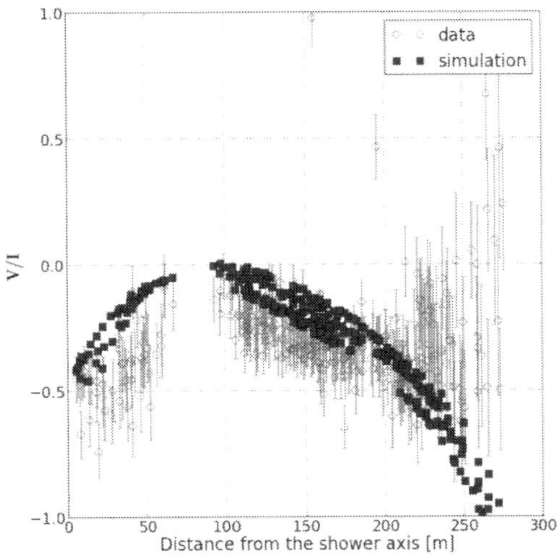

Figure 2. Circular polarization of an air shower under thunderstorm condition. The blue dots are simulated results and the red ones are data.

circular polarization vanishes at the core of the shower and depends on the azimuth angle of antennas [9]. However, in thunderstorm events we observe large amount of circular polarization near the core of the shower as shown in Fig. (2). This can only be explained by the change of the transverse current due to the change of the atmospheric electric field. Therefore, the full set of Stokes parameters helps to have a better understanding of the structure of atmospheric electric fields.

4 Conclusion

Using only the intensity footprint of the events measured under thunderstorm conditions, one can probe the atmospheric electric fields. It shows that we are sensitive to the height where the electric field changes and the relative strength between the fields in the two layers. The structure of electric fields can be recontructed better by using both the intensity and the polarization information.

References

[1] A. Dubinova et al., Phys. Rev. Lett. **115**, 015002 (2015)
[2] J.R. Dwyer, M.A. Uman, Physics Reports **534**, 147 (2014), the Physics of Lightning
[3] P. Schellart et al., Phys. Rev. Lett. **114**, 165001 (2015)
[4] F.D. Kahn, I. Lerche, Royal Society of London Proceedings Series A **289**, 206 (1966)
[5] O. Scholten, K. Werner, F. Rusydi, Astroparticle Physics **29**, 94 (2008)
[6] G. Askaryan, J. Phys. Soc. Japan **Vol: 17, Suppl. A-III** (1962)
[7] K.D. de Vries et al., Astroparticle Physics **34**, 267 (2010)
[8] T.N.G. Trinh et al., Phys. Rev. D **93**, 023003 (2016)
[9] O. Scholten et al. (2016), accepted by Phys. Rev. D (11/2016)

Lightning Imaging with LOFAR

Olaf Scholten[1,2,★], *Stijn* Buitink[3], *Roxana* Dina[1], *Qader* Dorosti Hasankiadeh[1], *Wilfred* Frieswijk[4], *Freddie* Hendriks[1], *Jeroen* Muller[1], *Danny* Sardjan[1], *Gia* Trinh[1], *Antonio* Bonardi[5], *Arthur* Corstanje[5], *Ute* Ebert[6,7], *Heino* Falcke[4,5], *Jörg* Hörandel[5], *Pragati* Mitra[3], *Katharine* Mulrey[3], *Anna* Nelles[8], *Jörg* Rachen[5], *Laura* Rossetto[5], *Casper* Rutjes[6], *Pim* Schellart[9], *Satyendra* Thoudam[10], *Sander* ter Veen[4], and *Tobias* Winchen[3]

[1]*University of Groningen, KVI Center for Advanced Radiation Technology, 9747 AA Groningen, The Netherlands*
[2]*Vrije Universiteit Brussel, Dienst ELEM, IIHE, Pleinlaan 2, 1050 Brussels, Belgium*
[3]*Astrophysical Institute, Vrije Universiteit Brussel, Pleinlaan 2, 1050 Brussels, Belgium*
[4]*Netherlands Institute for Radio Astronomy (ASTRON), 7990 AA Dwingeloo, The Netherlands*
[5]*Department of Astrophysics/IMAPP, Radboud University, P.O. Box 9010, 6500 GL Nijmegen, The Netherlands*
[6]*CWI, Centrum Wiskunde & Informatica, Amsterdam, The Netherlands*
[7]*TU/e, Eindhoven University of Technology, Eindhoven, The Netherlands*
[8]*Physics and Astronomy, University of California, Irvine, CA 92697-4575, U.S.A*
[9]*Department of Astrophysical Sciences, Princeton University, Princeton, NJ 08544, USA*
[10]*Department of Physics and Electrical Engineering, Linnéuniversitetet, 35195 Växjö, Sweden*

Abstract. We show that LOFAR can be used as a lightning mapping array with a resolution that is orders of magnitude better than existing arrays. In addition the polarization of the radiation can be used to track the direction of the stepping discharges.

1 Introduction and summary

Lightning is a familiar phenomenon, but still far from being understood [1]. There are recent suggestions for the mechanism that initiates lightning discharges [2, 3] as well as some qualitative understanding of how a lightning leader discharge propagates over distances of kilometers, but a quantitative theory is lacking. In particular, it is known that leaders with an excess of negative charge (called negative leaders) often proceed in a step-wise fashion, but the physical mechanism for this stepping is not understood. While stepping the leader emits strong radio pulses that can be used to locate its position in the sky. These radio pulses are used by Lightning Mapping Arrays (LMAs) to map the progress of leader development [4]. This makes a LMA the ideal instrument to study the physics involved in Lightning initiation and propagation.

Here we show that a new observation mode at LOFAR, a new generation radio telescope [5] build mainly for astronomy observations, allows for LMA observations with a resolution that is orders of magnitude better than presently existing LMAs, thus allowing for lightning observations with unprecedented resolution. The polarization of the signal can be used to reconstruct the direction of the

★e-mail: Scholten@KVI.nl

discharge. Combining these we can construct a 4-D vector map of the discharge process with high precision. To interpret the observations we have developed a model for calculating radio emission from a stepped leader. In Ref. [6] it was reported that the radio pulses have a fine structure on the order of 10 ns which we have confirmed in our preliminary measurements.

Recently we have developed a non-invasive method for determining the electric-field in thunderclouds [7, 8] by measuring at the core of LOFAR the radio footprint from cosmic-ray air showers. It is of much interest to combine this with the LOFAR-LMA observations.

2 State-of-the-art LMA

Nowadays there are several Lightning Mapping Arrays (LMA) [4] in operation. To determine the position in the atmosphere of the source that emitted the radio signal one measures the time of arrival (ToA) in a network of antennas on the ground. To reach sufficient accuracy they are spread over distances of typically 60 km [9]. Most arrays operate in an abandoned TV broadcasting frequency band in the lower VHF band, 60-66 MHz where the signal is digitized at a frequency of 25 MHz. The data is analyzed in blocks of 80 μseconds, called "Windows", in which the arrival time of the strongest peak is determined. The window size limits the maximum number of pulses to be analyzed to 12,500 per second. In each Window the arrival time is determined with a precision of 40 ns set by the resolution of the digitization. The relative timing of individual antennas is set using GPS. This has resulted in detailed maps of leader development that have contributed much to a better understanding of the physics of lightning, see [1].

3 LOFAR

The Low Frequency Array [5] (LOFAR) is a digital radio telescope that is primarily build for astronomy observations, consisting of many thousands of antennas. The antennas are grouped in stations, where each core station has 96 Low Band Antennas (LBA) and 48 High Band Antennas (HBA). Each LBA consists of two inverted V-shaped dipoles labeled X and Y. The LBA are sensitive to the frequency range of 10 - 90 MHz and are used for the observations discussed in this work. The sensitivity range of the HBA is 110 - 240 MHz. The signals are sampled at 200 MS/s leading to a sampling time of 5 ns. The approximately 2 km diameter LOFAR core contains 24 such stations with the highest density offered by the six stations located in a 320 m diameter region called the 'Superterp' near Exloo in the north of The Netherlands. Remote stations may lie at a distance of 1000 km from the core.

All antennas of LOFAR are equipped with ring buffers, that store the raw voltage traces of each individual antenna in the array for up to 5 s. When a trigger is received, the ring buffers are frozen and their contents are copied over the network to a central storage location. In the observation mode pertinent to this work the trigger is generated when there is lightning activity.

4 LOFAR as LMA

At each step of a lightning leader a sizable current flows in short time. By the nature of this process, strong linearly polarized radio-frequency pulses are emitted at meter wavelengths, making LOFAR an exquisitely suited instrument for measuring them with the LBA. This is corroborated by model calculations [10] showing the the angular distribution of the emitted radiation as well as the polarization can be used to determine rather accurately the direction of the discharge step. In addition it is possible to extract from the shape of the measured pulse the time profile and the length scale over which the current flows.

Data were taken with LOFAR-LBAs, in the 30-80 MHz band on June 19, 2013, while a thunderstorm was active at a distance of some 50 km from the telescope. The time trace of the LOFAR antennas on the Superterp as well as other stations up to distances of 30 km were analyzed.

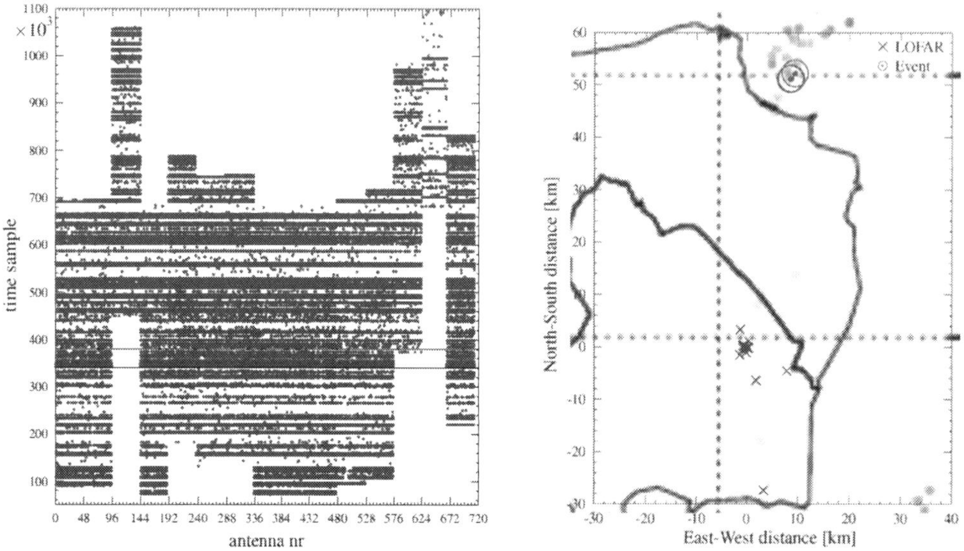

Figure 1. Left side shows pulse timing (vertical axis), indicated by dots, for all antennas (antenna number on horizontal axis) that were included in the present measurement. The traces of different stations (of 3 ms length, sampling time of 5 ns) have been time-shifted to align the strongest peak in the time-sequences. One trace could not be aligned. The pulses marked in red for three events are analyzed for their source location as shown on the right overlayed on a map of the North-East corner of The Netherlands showing the borders of the provinces of Groningen and Drenthe. The orange dots indicate lightning observations by the Dutch weather service. The positions of the LOFAR stations that are involved in the measurements are indicated by blue crosses.

As the first step in the analysis of the data the pulse-patterns seen in the different stations are roughly lined-up as shown in Fig. 1 for the different stations. A pulse is labeled as such when, after Radio-Frequency Interference (RFI) mitigation, the power exceeds a certain threshold. This lining-up is necessary as for these older data the precise timing off-sets of the stations were not available. As can be seen this works very well for all 15 stations with the exception of one for which the off-set exceeded the trace-length and no matching could be made. Each station consists of 48 antennas spread over an area with a diameter of about 30 m for which the relative timing is known accurately. For the three events marked in red in Fig. 1 the source location and emission time is found by fitting the arrival-times of the pulses for each separate antenna adjusting the station offsets, but keeping them the same for all events. The fit reproduces the measurements with an accuracy of about 1 time sample. At the right panel of Fig. 1 the source locations (green circled dots) are overlayed with a map of North-East Netherlands showing the provinces of Groningen and Drenthe. The orange and red points label the positions of observed lightning discharges in an half-hour time slot around the LOFAR observations as determined by the Dutch weather service showing a good agreement with the found source locations.

As an example the Hilbert envelopes of a short section of 200 samples of the time traces for one particular event is shown in Fig. 2 for several stations in separate plots for each of the two polarizations. The Hilbert envelope is averaged over the antennas of a station for a specific polarization. This shows that the time traces for different stations are very similar and thus that the structure is genuine, not due to interference with random noise. It also shows that the time traces for the two polarizations

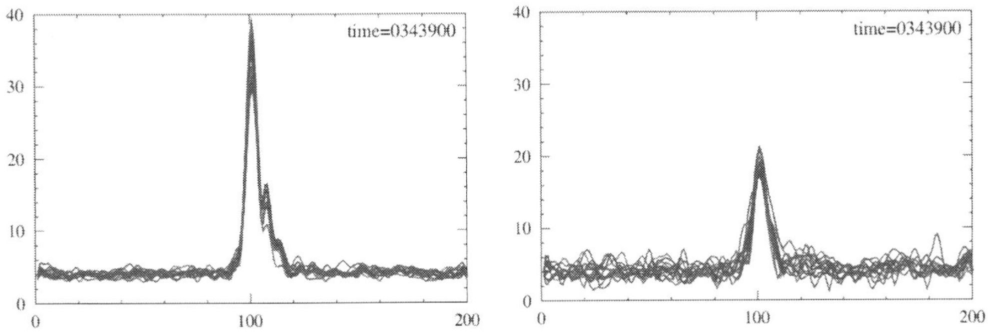

Figure 2. The average pulse shapes (Hilbert envelope), i.e. the average of the signals measured by the antennas in each station normalised to the integral in the X axis, (a different color for each station) for the two polarization directions (left and right hand sides). The total trace length shown is 200 samples of 5 ns, i.e. 1 μs.

are clearly distinct with a clear polarization-dependent structure for the pulse. Because the pulse is weaker in one polarization it is over threshold in fewer antennas giving rise to a more noisy signal. It should be noted that for some other events a much more complicated pulse structure is observed.

Work is still ongoing to determine the relative accuracy of source positions. In addition for this initial analysis [11, 12] the pulse timings were determined in single-sample precision. In the next step we refine this using pulse interferometry with which we should easily be able to obtain nanosecond timing accuracy.

References

[1] J.R. Dwyer and M.A. Uman, Phys. Rep. **534**, 147 (2014).
[2] A. Gurevich, A. Karashtin, Phys. Rev. Lett **110**, 185005 (2013).
[3] Anna Dubinova, *et al.*, Phys. Rev. Lett **115**, 015002 (2015).
[4] Some links to Lightning Mapping Array, Oklahoma LMA; SPORT; New Mexico; Ebro Valley; J.D. Hill, *et al.*, J. of Geoph. Res.: Atm **118**, 8460 (2013).
[5] M. P. van Haarlem, *et al.*, Å5562013A2; arXiv:1305.3550; LOFAR Wikipedia page.
[6] N. Murray, P. Krider, J. Willett, Atmos. Res. **76**, 455 (2005).
[7] P. Schellart *et al.*, Phys. Rev. Lett **114**, 165001 (2015), arXiv:1504.05742.
[8] G. Trinh, O. Scholten, *et al.*, Phys. Rev. D **93**, 023003 (2016), arXiv:1511.03045.
[9] R.J. Thomas *et al.*, J. Geophys. Res. **109**, D14207 (2004).
[10] Freddie Hendriks, Bachlor thesis, University of Groningen, 2016.
[11] Roxana Dina, Master thesis, University of Groningen, 2014.
[12] Danny Sardjan, Bachelor thesis, University of Groningen, 2016.

Analytic Calculation of Radio Emission from Extensive Air Showers subjected to Atmospheric Electric Fields

Olaf Scholten[1,2,⋆], *Gia* Trinh[1], *Krijn D.* de Vries[2], and *Lucas* van Sloten[1]

[1] *University of Groningen, KVI Center for Advanced Radiation Technology, 9747 AA Groningen, The Netherlands*
[2] *Vrije Universiteit Brussel, Dienst ELEM, IIHE, Pleinlaan 2, 1050 Brussels, Belgium*

Abstract. We have developed a code that semi-analytically calculates the radio footprint (intensity and polarization) of an extensive air shower subject to atmospheric electric fields. This can be used to reconstruct the height dependence of atmospheric electric field from the measured radio footprint. The various parameterizations of the spatial extent of the induced currents are based on the results of Monte-Carlo shower simulations. The calculated radio footprints agree well with microscopic CoREAS simulations.

1 Introduction and summary

Atmospheric electric fields during thunderstorm conditions can drastically modify the radio footprint of extensive air showers (EAS). This fact has been used [1, 2] to extract from the measured radio footprint the altitude dependence of the electric fields. In contrast to conventional weather-balloon measurements this offers a non-intrusive way to determine these fields that are essential for understanding lightning initiation and development [3].

The extraction of the atmospheric fields relies on a model calculation of radio emission from an EAS. Accurate codes exists that perform reliable microscopic calculations [4]. One such code that allows for setting arbitrary atmospheric electric fields is CoREAS [5] that uses results of a CORSIKA shower evolution. This uses a Monte-Carlo simulation and thus the results of two calculations with very similar but unequal electric fields may differ considerably due to random shower-to-shower fluctuations. As a result it is close to impossible to use CoREAS in a systematic optimization procedure. To resolve this problem we have developed a deterministic semi-analytic calculation that predicts the radio-footprint given a certain altitude dependence of the induced currents in the EAS. An additional advantage is that the analytic code is much less CPU intensive than the microscopic calculation (a few seconds compared to many hours). As a result it now becomes practical to solve the inverse problem from radio emission, i.e. to determine from a measured radio footprint (intensity and polarization distributions on the ground) of a shower the configuration of currents in the atmosphere that produced this.

The code is based on the macroscopic formulation of radio emission as developed in refs [6–10] and incorporated in the EVA-code [11]. An important difference between the EVA-code and the present code (called EVA-light) is that here we use a direct parametrization of the current distribution

⋆e-mail: Scholten@KVI.nl

in the EAS. Certain parameters, such as the distribution of the current density as function of distance to the shower axis, are kept fixed, while other parameters related to the altitude dependence of the currents are optimized to fit the data. In addition the new code is optimized for a fast calculation of the complete radio footprint.

Work is still ongoing to refine the parametrizations of the charge-excess and the induced-current clouds where the realistic distributions as determined in Monte-Carlo shower evolution calculations are guiding. Preliminary results show good agreement with those of microscopic CoREAS calculations.

2 Modelling Radio emission from EAS

The motion of the charged particles in the EAS are modeled as a charge cloud with a parameterized density profile moving towards Earth with the light velocity. Radio emission is subsequently calculated using the retarded vector potential where we follow the a macroscopic approach to construct the retarded potentials. Furthermore we have put some effort in optimizing the code for the calculation of a complete radio footprint.

In this paper we work with the four-current $j^\mu(t_s, x_s, y_s, h)$ where $z_s = -t_s$ (taking $c = 1$) is the distance from the shower front to the ground, called height and always measured along the shower axis, x_s, and y_s are the transverse directions, and h is the distance from the shower front to a point in the charge cloud. We use the notation where $\mu = 0$ denotes the time (charge) components and $\mu = x, y, z$ the space (current) components of a four vector. The shower front is moving with the light velocity. A particular point in the charge cloud is thus at an height of $\zeta = z_s + h$.

Following the usual notation, the vector potential for an observer at a point $(x_o, y_o, z_o = 0)$ on the shower plane, defined as the plane perpendicular to the shower axis going through the point of impact of the shower on the ground, is taken as

$$A^\mu(t_o, \vec{x_o}) = \int d^3\vec{x'} \, \frac{j^\mu(t_r, \vec{x'})}{L} \left| \frac{dt_r}{dt} \right| , \tag{1}$$

where L is the optical path-length, $L = c(t_o - t_r)$. Only for a homogeneous medium with a constant index of refraction n it is given by $L = nR = n|\vec{x_o} - \vec{x'}| = c(t_o - t_r)$. We introduce the retarded distance

$$\mathcal{D} \overset{\text{def}}{=} L\left|\frac{dt}{dt_r}\right| = nR\,(1 - n\vec{\beta}\cdot\hat{n})\Big|_{\text{ret}} = (t - t_r) - n^2\beta(h - \beta t_r) = n\sqrt{(-\beta t + h)^2 + (1 - \beta^2 n^2)d^2} \, , \tag{2}$$

for a moving point charge with velocity $c\beta$. The vector potential can now be written compactly as

$$A^\mu(t_o, \vec{x_o}) = \int d^3\vec{x'} \, \frac{j^\mu(t_r, \vec{x'})}{\mathcal{D}} \, . \tag{3}$$

The index of refraction depends on the height in the atmosphere through it density dependence as given by the Gladstone-Dale relation, $n_{\text{GD}} = 1 + n_\rho\, \rho(z)$ where ρ is the air density. For the evaluation of the retarded distance given in Eq. (2), the average value of the index of refraction over the photon trajectory will be used, assuming the straight-line approximation for the photon path [11],

$$n = (1/z) \int_0^z (1 + n_\rho \rho(\zeta))\, d\zeta \, , \tag{4}$$

which does not depend on the distance of the observer to the shower axis for vertical showers.

To evaluate the integration over z' in Eq. (3), we follow the same approach as used in Ref. [11] and replace the integral in the z-direction by an integral over $\lambda = h_c - h$ where $\mathcal{D} = 0$ at the critical height h_c. This substitution allows for an easier calculation of derivatives of the vector potential that are necessary to calculate the electric field since the derivatives vanish on the limits of the λ integration, as shown in Ref. [11].

2.1 Electric field, charge excess

The charge excess in the shower front is given by $Q(z)$ propagating with the light velocity in the $-\hat{z}$-direction thus contributing to the zero and the z component of the vector potential. Substituting the expression for the vector potential, integrating over the spatial extent of the charge cloud and converting to an integration over λ, the radial component of the electric field can be written as

$$E_r^{CX} = -\frac{\partial A^0}{\partial r_o} = -\int dx_s\, dy_s\, \frac{\partial w(r_s)/r_s}{\partial r_s}\, I_\lambda^{CX}\,; \qquad I_\lambda^{CX} = \int_0^{h_c} d\lambda\, f(h_c - \lambda)\, \frac{Q}{n\mathcal{D}}\bigg|_{\lambda=h_c-h}\,, \tag{5}$$

where $r_s^2 = (x_s^2 + y_s^2)$ is an integration point in the shower front and the position of the observer is denoted with the subscript o. The integral is separated into an integration over λ, a line-sub-shower, along a line parallel to the axis of the full shower, followed by one over the radial extent. In the numerical calculation the results line-sub-showers, $I_\lambda^{CX}(t)$, are stored for a grid of r_s and d values, where the latter is the distance of the observer to line-sub-shower.

In order to obey charge conservation a residue charge line-density $Q'(z) = (dQ/dz)$ is left behind at rest in the atmosphere which can be ignored as shown in Ref. [7].

2.2 Electric field, Transverse current

Along a similar line of reasoning as for the charge-excess radiation, the transverse current contribution is written as

$$E_x^{TC} = -\frac{\partial A^x}{\partial t^o} = \int dx_s\, dy_s \frac{w(r_s)}{r_s} I_\lambda^{TC}\,, \tag{6}$$

$$I_\lambda^{TC} = \int_0^{h_c} d\lambda\, f'(h_c - \lambda)\, \frac{J^x}{n\mathcal{D}}\bigg|_{\lambda=h_c-h} - \int_0^{h_c} d\lambda\, f(h_c - \lambda)\, \frac{J'^x}{n\mathcal{D}}\bigg|_{\lambda=h_c-h}\,. \tag{7}$$

where $J'^x = dI^x/dt_r$ and $f'(h) = df/dh$. As before the results for line-sub-showers, $I_\lambda^{TC}(t)$ are calculated once on a grid of r_s and d values to be used subsequently in the calculation of the complete footprint of the electric field.

Due to the continuity equation we have $dJ_x/dx = dQ/dt$, thus for $r_s^2 = x_s^2 + y_s^2$ we write $Q_x^D(t_r, x_s, y_s) = \left[\int_{-\infty}^{t_r} J_x(t)dt\right] \times \frac{d\,w(r_s)/r_s}{dx_s}$ and a similar term for J_y. This corresponds to a dipole charge distribution moving with the shower front. In reality these particles thermalize and will trail well behind the front and thus not contribute to coherent emission. To account for this we include a decrease proportional to $e^{-X/a^D X_0}$,

$$\vec{J}^D(t_r) = \int_{-\infty}^{t_r} \vec{J}(t)\, e^{-[X(t_r)-X(t)]/(a^D X_0)}\, dt\,, \tag{8}$$

where X depends on t through its dependence on the height in the atmosphere. In Eq. (8) X_0 is the electron mean free path and a^D a parameter that is adjusted to get best agreement with microscopic shower calculations. Following a similar notation as before the contribution to the electric field is

$$E_x^D = -\int dx_s\, dy_s\, \frac{d\,w(r_s)/r_s}{r_s\, dr_s}\, I_D^x\,; \qquad I_x^D = \int d\lambda\, f(h_c - \lambda)\, \frac{J_x^D}{n\mathcal{D}}\bigg|_{\lambda=h_c-h}\,, \tag{9}$$

DOI: 10.1051/epjconf/201713503004

where a similar expression is obtained for E_y^D.

2.3 Parameterizations

The density distribution of the four current is parameterized as

$$j^\mu(t,x,y,h) = \frac{w(r)}{r} f(h,r)\, J^\mu(t); \qquad w(r) = N_w\, r/M_0(r/M_0+1)^{-2.5} \quad ; \quad f(h,r) = N_f\, u^\alpha e^{-2u} \,, \quad (10)$$

where $r = \sqrt{x^2+y^2}$, $u = h/\alpha\,\lambda$, M_0 the Moliere radius, and h the distance behind the shower front. The normalization constants N_w and $N_f(r)$ in the functions w and f are determined by $\int w(r)\,dr = 1$ and $\int_0^\infty f(h,r)\,dh = 1\ \forall r$. In this way $J^\mu(t)$ is the charge and current at time t integrated over the whole shower front. Note that $w(r)$ corresponds to r times the NKG function for $s = 2$. The two parameters in f, the pancake thickness $\lambda(r)$ and $\alpha(X)$, depend on distance to the shower axis, r, and penetration depth, X, as

$$\lambda(r) = \max[\lambda_0, \lambda_1\,(r/r_1)]\,; \qquad \alpha(X) = 1 + 0.5 \times \sqrt{X/X_{max}}\,. \quad (11)$$

The radial dependence of the pancake thickness, $\lambda(r)$, is such that for $r = 0$ we have $\lambda(0) = \lambda_0$ increasing linearly with distance on an exponential grid where at a distance of r_1 =100 m from the core we have $\lambda(100) = \lambda_1$. With increasing α the maximum in the particle density shifts to larger distances behind the shower front.

The optimization of the parameters is still ongoing. We note that choosing $M_0 = 30$ m, $\lambda_0 = 0.15$ m, $\lambda_1 = 3$ m, reproduces the main features of the charge cloud as deduced from microscopic calculation. This also gives rise to a radio footprint that agrees well with that obtained from CoREAS simulations. We are still investigating the extent to which the distributions need to be chosen differently for charge excess and transverse current in view of the findings in Ref. [2]. We note that changing the pancake thickness with distance to the shower axis is essential to be able to reproduce the (not frequency filtered) pulse shape as obtained from the CoREAS calculations. The penetration-depth dependence of α does not have a strong influence on the calculations.

References

[1] P. Schellart *et al.*, Phys. Rev. Lett **114**, 165001 (2015), arXiv:1504.05742.
[2] G. Trinh, O. Scholten, *et al.*, Phys. Rev. D **93**, 023003 (2016), arXiv:1511.03045.
[3] J.R. Dwyer and M.A. Uman, Phys. Rep. **534**, 147 (2014).
[4] T. Huege, Phys. Rep. **620**, 1 (2016), arXiv:1601.07426.
[5] T. Huege, M. Ludwig, and C. James, AIP Conf. Proc. **1535**, 128 (2012).
[6] O. Scholten, K. Werner, and F. Rusydi, Astropart. Phys. **29**, 94 (2008), arXiv:0709.2872.
[7] Klaus Werner and Olaf Scholten, Astropart. Phys. **29**, 393-411 (2008), arXiv:0712.2517.
[8] K.D. de Vries, *et al.*, Astropart. Phys. **34**, 267 (2010), arXiv:1008.3308.
[9] K. D. de Vries, *et al.*, Phys. Rev. Lett **107**, 061101 (2011).
[10] Olaf Scholten, Krijn D. de Vries, Klaus Werner, EPJ Web of Conf. **53**, 08005 (2013), arXiv:1207.1874.
[11] K. Werner, K.D. De Vries, O. Scholten, Astropart. Phys. **37**, 5 (2012), arXiv:1201.4471.

Overview of lunar detection of ultra-high energy particles and new plans for the SKA

Clancy W. James[1,⋆], *Jaime* Alvarez-Muñiz[2], *Justin D.* Bray[3], *Stijn* Buitink[4], *Rustam D.* Dagkesamanskii[5], *Ronald D.* Ekers[6], *Heino* Falcke[7,8], *Ken* Gayley[9], *Tim* Huege[10], *Maaijke* Mevius[11], *Rob* Mutel[9], *Olaf* Scholten[11,12], *Ralph* Spencer[3], *Sander* ter Veen[7], and *Tobias* Winchen[4]

[1]*ECAP, Univ. of Erlangen-Nuremberg, 91058 Erlangen, Germany*
[2]*Depto. de Física de Partículas & Instituto Galego de Física de Altas Enerxías, Univ. de Santiago de Compostela, 15782 Santiago de Compostela, Spain*
[3]*School of Physics & Astronomy, Univ. of Manchester, Manchester M13 9PL, United Kingdom*
[4]*Astrophysical Institute, Vrije Universiteit Brussel, Pleinlaan 2, 1050 Brussels, Belgium*
[5]*LPI, Russian Academy of Sciences, Moscow Region 142290, Russia*
[6]*CSIRO Astronomy & Space Science, NSW 2122, Australia*
[7]*Dept. of Astrophysics/IMAPP, Radboud Univ. Nijmegen, 6500 GL Nijmegen, The Netherlands*
[8]*Netherlands Institute for Radio Astronomy (ASTRON), 7990 AA Dwingeloo, The Netherlands*
[9]*Dept. of Physics & Astronomy, Univ. of Iowa, IA 52242, USA*
[10]*IKP, Karlsruhe Institut für Technologie, Postfach 3640, 76021 Karlsruhe, Germany*
[11]*Kernfysisch Versneller Instituut,, Univ. of Groningen, 9747 AA Groningen, The Netherlands*
[12]*Interuniversity Institute for High-Energy, Vrije Universiteit Brussel, Pleinlaan 2, 1050 Brussels, Belgium*

Abstract. The lunar technique is a method for maximising the collection area for ultra-high-energy (UHE) cosmic ray and neutrino searches. The method uses either ground-based radio telescopes or lunar orbiters to search for Askaryan emission from particles cascading near the lunar surface. While experiments using the technique have made important advances in the detection of nanosecond-scale pulses, only at the very highest energies has the lunar technique achieved competitive limits. This is expected to change with the advent of the Square Kilometre Array (SKA), the low-frequency component of which (SKA-low) is predicted to be able to detect an unprecedented number of UHE cosmic rays.

In this contribution, the status of lunar particle detection is reviewed, with particular attention paid to outstanding theoretical questions, and the technical challenges of using a giant radio array to search for nanosecond pulses. The activities of SKA's High Energy Cosmic Particles Focus Group are described, as is a roadmap by which this group plans to incorporate this detection mode into SKA-low observations. Estimates for the sensitivity of SKA-low phases 1 and 2 to UHE particles are given, along with the achievable science goals with each stage. Prospects for near-future observations with other instruments are also described.

⋆e-mail: clancy.james@physik.uni-erlangen.de

1 Introduction

The lunar Askaryan technique began with Askaryan's predictions for coherent radio–microwave radiation arising from a build-up of a negative charge excess in particle cascades, and his identification of the lunar regolith as a suitable target material due to its relative radio transparency [1]. Its modern experimental incarnation was first proposed by [2], who noted that by using ground-based radio-telescopes, the total effective area of the visible lunar surface — approximately 20 million km^2 — could be utilised as a particle detector. The large observation distance (mean of 3.8 10^5 km) however imposes a high detection threshold, meaning the technique is ideally suited to targeting the rarest ultra-high-energy (UHE; $\gtrsim 10^{18}$ eV) cosmic rays (CR) and neutrinos. Observations of UHE CR are motivated by the as-yet unknown origin of these extreme particles, while UHE neutrinos are predicted from specific theories of topological defects and super-heavy dark matter [3–6]. It should be noted that efforts to observe the lunar regolith with a lunar orbiter are ongoing [7] — however, both the bulk of experimental efforts, and these proceedings, concentrate on the ground-based method.

Since the first search for lunar Askaryan pulses at Parkes in 1995 [8], several experiments have been performed, at Goldstone ('GLUE'; [9]), Kalyazin ('RAMHAND'; [10]), the VLA ('RESUN'; [11]); Lovell ('LaLUNA'; [12]); Westerbork ('NuMoon'; [13]); and with ATCA and Parkes ('LUNASKA'; [14–16]).

While these experiments have produced a number of limits on high-energy particle fluxes, only the diffuse flux limits from NuMoon [13], and LUNASKA limits on Centaurus A [16, 17], are competitive against other experiments, and then only at extremely high energies ($\gtrsim 10^{22}$ eV). This is due to two factors: no observation has yet exceeded 200 hr, whereas most competing particle-detection experiments operate continuously; and because detection thresholds have mostly been near 10^{21} eV.

The current experimental emphasis is therefore on performing long-duration observations to limit the UHE neutrino flux, and on increasing sensitivity to detect the known flux of UHE CR. Unlike neutrinos, most of which will penetrate too deeply into the Moon to be detected, and deposit only $\sim 20\%$ of their energy in detectable hadronic cascades, cosmic rays will interact immediately at the surface with 100% energy deposition. While it was long-thought that formation zone effects would suppress the Askaryan emission from near-surface cascades and make cosmic rays undetectable with the technique, the rapid rise and fall of the excess charge will still result in radiation regardless of the presence of a dielectric [18, 19]. Furthermore, future experiments — in particular, with the Square Kilometre Array — promise the order-of-magnitude increase in experimental sensitivity required to detect the known flux of cosmic rays, potentially allowing for arrival-direction studies.

In this overview, the main emphasis is on the planned observation strategy and simulated sensitivity of the Square Kilometre Array to UHE particles. The specific plans of the SKA's High Energy Cosmic Particles Focus Group (HECP FG[1]) for observations with the low-frequency component of SKA Phase 1 will be described — prospects for future observations with both SKA Phase 2 and other instruments will also be discussed.

2 The Square Kilometre Array

The Square Kilometre Array (SKA) is a radio telescope project consisting of three separate arrays covering different frequency ranges.[2] SKA-low (50–350 MHz) will be built in Australia, while SKA-mid dishes ($\gtrsim$ 1 GHz) and an aperture array (to cover the intervening frequencies) will be built in South Africa and nearby countries. Construction of Phase 1 of the SKA will begin shortly, with a

[1] http://astronomers.skatelescope.org/home/focus-groups/high-energy-cosmic-particles/
[2] www.skatelescope.org

target completion date of 2023; the full instrument in Phase 2 is targeted for completion in 2030. Phase 1 of SKA-low (SKA1-low) will consist of 131,072 log-periodic dipole antennas deployed in 512 stations of 256 antennas each [20], and will be deployed with a large central concentration of detectors over a $\sim$1 km^2 area.

The science goals of the SKA are diverse, and are broadly categorised into Science Working Groups. The SKA is designed primarily as an imaging telescope, but the ability to observe in other modes — pulsar searches and timing, and transient studies — is also part of standard design features. In order to enable the SKA to perform lunar Askaryan observations, members of most prior experimental collaborations have formed the SKA's HECP Focus Group, together with groups interested in performing observations of extensive air showers [21]. This group has studied the sensitivity of the SKA when using the lunar technique, with the main science case being published in the latest edition of the SKA Science book [22]. As a result of these studies, which are described further in Section 3, the group has decided to focus on SKA-low. More importantly, the group has determined the required engineering changes in order to allow Phase 1 of SKA low ('SKA1-low') to perform lunar Askaryan observations, and the target observation mode is presented in Section 5.

3 Simulating the sensitivity of the SKA

The first detailed estimates of the sensitivity of the SKA to UHE particles were made by [23], and have undergone continuous updates as the design of the instrument became more concrete. The current expectation is a sensitivity of 500 m^{2}K^{-1} for SKA1-low and perhaps 4000 m^{2}K^{-1} for Phase 2. This sensitivity is applicable to the detection of Askaryan pulses only when all telescope elements are correctly phased (added coherently in the voltage domain) in the direction of the pulse, and the signal power is contained to only one polarisation channel.[3]

This 'raw' sensitivity however tends to be limited by the sensitivity that can be brought to bear for online triggering, the details of which depend on the exact experimental setup. Such effects include self-vetoing due to strong signals exceeding noise-veto criteria, imperfect coverage of the lunar disk by the trigger beams, approximate dedispersion, artefacts of (analogue or digital) signal-processing, and non-optimal search algorithms — see [14, 24, 25, 28] for further details. Only the self-veto and lunar coverage effects are reproduced here.

In the case of SKA1-low, 16 coherent beams will be available for real-time searches (see Section 5). The best trade-off between beam width and lunar coverage is expected by beamforming with all core stations (approximately 50% of the total), here modelled using a 2D Gaussian distribution of antenna density with 1/e fall-off radius 240 m. Since the sky temperature rises rapidly at low frequencies for all pointing positions, observations in a 250 MHz band above 100 MHz are considered, although SKA1-low will be capable of observing over all 50–350 MHz simultaneously.

A preliminary optimisation of the beam pointing positions for the Moon at zenith, and assuming a frequency-independent sky plus system noise core sensitivity of 250 m^2 K^{-1}, indicates that beams placed 0.9$'$ inside the limb would be optimal (see Fig. (1), left). The event rate is evaluated assuming independently polarised triggers on each beam, for a total of 32 independent triggers.

In the case of Phase 2, 100% lunar coverage at the sensitivity of the full array (4000 m^{2}K^{-1}) is assumed. The chosen threshold of 10σ corresponds to a final false positive rate of less than once per century.

[3]Equivalently, the frequency-dependence of the polarisation can be predicted, such that a signal search competes against the effective noise of one polarisation channel only.

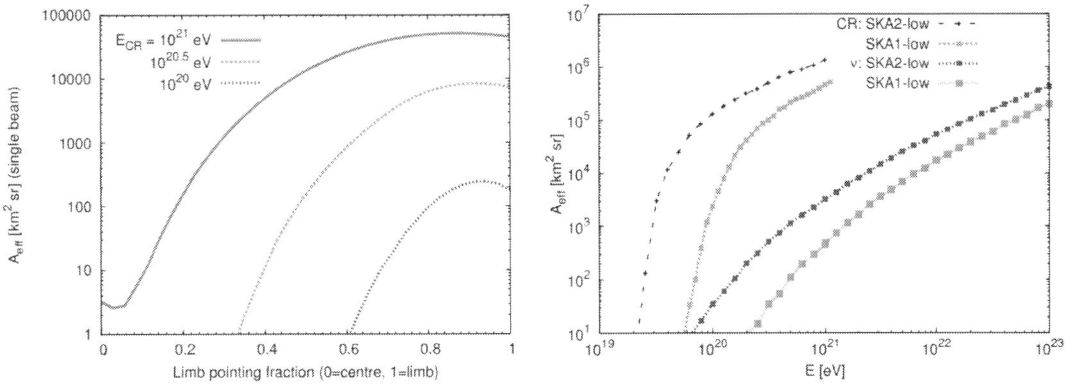

Figure 1. (left) Effective area for a single SKA1-low beam (see text) as a function of fractional lunar pointing radius, for three different cosmic ray energies $E_{\rm CR}$. (right) Effective aperture $A_{\rm eff}$ of all beams of SKA2-low and SKA1-low to cosmic rays and neutrinos as a function of their energy. [22].

4 Results for cosmic rays and neutrinos

Using these values, the resulting estimates of SKA1-low and SKA2-low sensitivity to UHE CR and neutrinos are shown in Fig. (1) (right). While SKA1-low will be sensitive to cosmic rays below 10^{20} eV, the effective aperture in this range is significantly less than what is obtained by the Pierre Auger Observatory [29]. Using an approximate parameterisation of the cosmic ray spectrum, and a lunar visibility of 29% above an elevation angle of 30° from the SKA-low site, it is estimated that SKA1-low could detect of order 1 CR yr^{-1} of observing whenever the Moon is visible (~2500 hr) [30], which may allow for a first detection. The estimated rate for Phase 2 however will be much higher: of order 100 yr^{-1}, of which half will be above the 56 EeV threshold found by the Pierre Auger Observatory to be most significant for anisotropy studies [31, 32]. The sensitivity of SKA-low in either phase to UHE ν however will not be sufficient to detect the estimated flux of cosmogenic neutrinos resulting from cosmic ray interactions with background photon fields.

The clear goal therefore of lunar Askaryan observations with SKA Phase 2 will be to perform directional studies of cosmic rays at the highest energies. Preliminary estimates indicate an angular resolution of order 5° will be achievable [30], which is sufficient to study objects such as Centaurus A. No resolution on primary composition however is anticipated, and energy resolution may be poor, so SKA observations should not be seen as a replacement for surface arrays such as the Pierre Auger Observatory or Telescope Array.

Observations with SKA1-low will, however, be able to probe beyond-the-standard-model physics at energies above 10^{20} eV. These observations could also be performed through very long ($\gg 1,000$ hr) observations with less-sensitive (and hence less-subscribed) instruments such as AuScope at 2.3 GHz [25], or LOFAR observations in UHEP mode [26]. A first detection of cosmic rays may also be possible with instruments such as FAST [27] or Parkes [25].

5 Implementation with Phase 1 of SKA-low

A sketch of the proposed SKA-lunar signal path for SKA1-low is shown in Fig. (2), based on the updated SKA1-low baseline design [20]. In standard operations, dual linear polarisations from all 256 elements in each station will be digitised, passed through a polyphase filter (PPF) to split them

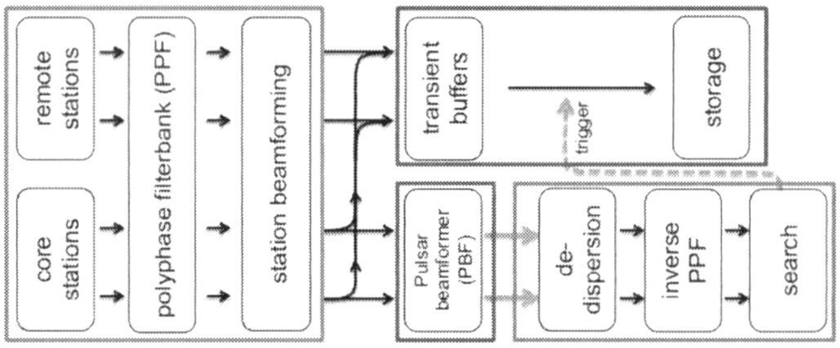

Figure 2. Sketch of the proposed signal path for SKA1-low lunar Askaryan observations, colour-coded by observation mode. Red: part of the standard imaging pipeline; blue: added functionality for pulsar and transient observations; green: specialised lunar pulse detection hardware; orange: required interfaces.

into 300 1 MHz channels, and formed into a 'station beam'. These station beams can then be passed to the correlator for imaging (not shown), the pulsar timing array beamformer (PBF) to form 'array' beams, and/or buffered to memory for transient detection. The PBF will be able to produce 16 dual polarisation array beams at 300 MHz bandwidth.. In order to search for lunar Askaryan pulses, these beams will be directed at the Moon, and their output passed to dedicated hardware.

To correct for ionospheric delays, each sub-band will undergo a bulk de-dispersion according to the best real-time STEC prediction, and then be passed to an inverse PPF. This will operate on each station beam to transform the signal back to the time-domain. Note that the bulk dedispersion, and also any noise-whitening, can be performed for free by modifying the initial inverse PPF coefficients.

In order to search for a lunar Askaryan signal, each beam will be copied to many separate streams, with each applying a different fine dedispersion to account for errors in the initial bulk dedispersion. Each of the subsequent signals will then be searched for possible lunar pulses.

The optimal search method for a coherent signal against white noise uses a matched filter. However, even if the pulse shape can be perfectly predicted, the signal phase will vary with respect to the sampling phase. Therefore, a search over multiple signal shapes must be performed. The expected trigger rate from pure noise events is expected to be in the range 0.1–1 Hz.

The final stage will be an anti-coincidence test over all 16 beams to veto RFI events. Once a trigger is generated that passes the veto, both the array beams, and the buffered station beams, will be stored offline for further analysis. Further details of the proposed implementation of these components is given in [33].

The doubled effective area available in offline processing through buffered data from non-core stations will be sufficient to discriminate these false positives from true lunar events — which necessarily implies that the final sensitivity will be trigger-limited. Returning 10 μs per trigger is expected to be sufficient to capture a fully dispersed pulse, account for arrival-time uncertainties on long baselines, and provide sufficient off-signal data to estimate the noise level. The total data volume, assuming station-beam data at 8-bit resolution, will be 8 MB per trigger.

6 Conclusion

Using the lunar Askaryan technique, the low-frequency component of the Square Kilometre Array in Phase 2 will have the ability to detect an unprecedented number of ultra-high-energy cosmic rays, in order to perform directional studies and potentially resolve the cosmic ray origin. Almost all the necessary technical development has been performed by the member groups of the High Energy

Cosmic Particles Focus Group within the SKA Organisation. Observations with LOFAR's UHEP mode, and with Phase 1 of SKA-low, will provide both critical technical development, and important limits on beyond-the-standard-model physics. There are good prospects for a first detection with SKA1-low or other instruments, and for very-long-duration observations specifically to test exotic models of extremely high-energy particles.

References

[1] G.A. Askaryan, Sov.Phys.JETP **14**, 441 (1962).
[2] R.D. Dagkesamanskii, I.M. Zheleznykh, Sov.Phys.JETP Lett. **50**, 259 (1989).
[3] P. Bhattacharjee, AIP Conf.Proc. **433**, 168 (1998).
[4] M. Kachelriess, arXiv:0810.3017 (2008).
[5] R. Aloisio, V. Berezinsky, M. Kachelriess, Phys.Rev.D **74**, 023516 (2006).
[6] R. Aloisio, S. Matarrese, A.V. Olinto, JCAP **1508**, 024 (2015).
[7] V.A. Ryabov, G.A. Gusev, V.A. Chechin, J.Phys.Conf.Series **409**, 012096 (2013).
[8] T.H. Hankins, R.D. Ekers, J.D. O'Sullivan, MNRAS **283**, 1027 (1996).
[9] P.W. Gorham et al., Phys.Rev.Lett. **93**, 041101 (2004).
[10] A.R. Beresnyak, et al., Astron.Rep. **49**, 127 (2005).
[11] T.R. Jaeger, R.L. Mutel, K.G. Gayley, Astropart.Phys. **43**, 293 (2010).
[12] R.E. Spencer et al., Proc. 10th EVN Symposium **097** (2010).
[13] S. Buitink et al., A&A **521**, A47 (2010).
[14] C.W. James et al., Phys.Rev.D **81**, 042003 (2010).
[15] J.D. Bray et al., Astropart.Phys. **65**, 22 (2015).
[16] J.D. Bray et al., Phys.Rev.D **91**, 063002 (2015).
[17] C.W. James et al, MNRAS **410**, 885 (2011).
[18] C.W. James et al., Phys.Rev.E **85**, 056606 (2011).
[19] S. ter Veen et al., Phys.Rev.D **82**, 103014 (2010).
[20] P.E. Dewdney et al, SKA-TEL-SKO-0000002 (2016).
[21] T. Huege et al., PoS(AASKA14)148 (2014).
[22] J.D. Bray et al., PoS(AASKA14)144 (2014).
[23] C.W. James, R.J. Protheroe, Astropart.Phys. **30**, 318 (2009).
[24] K. Singh et al., NIMA **664**, 171 (2012).
[25] J.D. Bray, Astropart.Phys. **77**, 1 (2016).
[26] T. Winchen, *Search for Cosmic Particles with the Moon* (these proceedings).
[27] C.W. James, J.D. Bray, R.D. Ekers, arXiv:1608.02407 (2016).
[28] J.D. Bray, *Sensitivity of lunar particle-detection experiments* (these proceedings).
[29] J. Abraham et al., Phys.Lett.B **685**, 239 (2010).
[30] C.W. James et al., PoS(ICRC2015)291 (2015).
[31] J. Abraham et al., Science **318**, 938 (2007).
[32] A. Aab et al., ApJ **804**, 15 (2015).
[33] J.D. Bray et al., PoS(ICRC2015)597 (2015).
[34] J. Bray, R.D. Ekers, P. Roberts, Exp.Astron. **36**, 155 (2013).

Sensitivity of lunar particle-detection experiments

Justin D. Bray[1,*]

[1]*JBCA, School of Physics & Astronomy, University of Manchester, Manchester M13 9PL, UK*

Abstract. The use of the Moon as a detector volume for ultra-high-energy neutrinos and cosmic rays, by searching for the Askaryan radio pulse produced when they interact in the lunar regolith, has been attempted by a range of projects over the past two decades. In this contribution, I discuss some of the technical considerations relevant to these experiments, and their consequent sensitivity to ultra-high-energy particles. I also discuss some possible future experiments, and highlight their potential.

1 Introduction

The detection of ultra-high-energy neutrinos and cosmic rays requires detectors with extremely large apertures. One technique for obtaining such apertures is to remotely monitor solar-system bodies for radio pulses from the particle cascades produced when these particles interact, using these bodies themselves as detectors [1]. This technique has been employed in a commensal experiment with the Earth-orbiting FORTE satellite, with a non-detection setting a limit on the neutrino flux [2]; the proposed dedicated SWORD satellite [3] may improve on this [4]. Jupiter has also been proposed as a target [5], although its prospects appear poor [6].

However, a target proposed before the above work [7], and the focus of most effort to date, is the Moon. A particle cascade developing in the dense medium of the lunar regolith produces a strong Askaryan radio pulse [8], and the lunar Askaryan technique employs an Earth-based radio telescope — or, potentially, a satellite in lunar orbit [9] — to search for this. Eight such experiments have been published to date [10–17]; see Ref. [18] for a summary and analysis. Below, I describe some of the limiting technical considerations for these experiments, and discuss their sensitivity to ultra-high-energy particles.

2 Anticoincidence rejection

A major difficulty for lunar Askaryan experiments is distinguishing between lunar-origin pulses and anthropogenic radio-frequency interference. One solution is to use a telescope with multiple beams directed at different parts of the moon, and to apply an anticoincidence cut, rejecting any pulse that is detected in more than one beam. A lunar-origin pulse should be detected only in the beam directed at the correct part of the moon, whereas local radio-frequency interference will generally be seen simultaneously in multiple beams. Two recent experiments that used this approach are NuMoon [14] and LUNASKA Parkes [17].

*e-mail: justin.bray@manchester.ac.uk

Table 1. Minimum and maximum detectable radio pulse amplitudes for the NuMoon and LUNASKA Parkes experiments, with and without anticoincidence cuts in their analyses [18]. Note that these values are not a complete representation of the sensitivity of these experiments.

Experiment	Anticoincidence	$\mathcal{E}_{\min}$	$\mathcal{E}_{\max}$
		(μV/m/MHz)	
NuMoon	yes	0.1365	0.1650
	no	0.1453	—
LUNASKA Parkes	yes	0.0053	0.0241
	no	∞	—

However, telescope beams overlap to some degree, and a sufficiently strong lunar-origin pulse may be detected in multiple beams, and incorrectly classified as interference [19]. This effectively sets an upper limit on the amplitude of an Askaryan pulse that can be detected by an experiment applying an anticoincidence cut. Omitting this cut, however, leads to interference contaminating the sample of recorded events and raising the noise level, raising the lower limit on the amplitude of a detectable pulse.

As seen in Table 1, the NuMoon experiment had only a narrow range between its upper and lower detection thresholds, leading to minimal sensitivity to ultra-high-energy particles; however, omitting the anticoincidence cut removes the upper limit while making little difference to the lower limit, restoring the sensitivity of the experiment. The LUNASKA Parkes experiment relied absolutely on its anticoincidence cut, having no appreciable sensitivity without it; but was sensitive to pulses across a broad range of amplitudes even with the anticoincidence cut included. In both cases, the result was unintentional, with no thorough analysis of this issue beforehand. For future experiments, a more sophisticated anticoincidence strategy will be developed, to minimise this issue [20].

3 Dedispersion

An Askaryan pulse from the moon will be subjected to a frequency-dependent dispersive delay as it passes through the ionosphere, smearing it over a timescale longer than its original length. This can be corrected by applying to the received signal a dedispersion filter that matches the dispersion characteristic of the ionosphere, defined by its slant total electron content (STEC). Some past experiments with narrow bandwidths have simply neglected dispersion (e.g. [11, 15]); others have corrected it in real time with a filter matched to the STEC (e.g. [13, 17]), or recorded the received signal with its full time resolution, allowing the correction to be applied retrospectively (e.g. [14]).

Future lunar Askaryan experiments are likely to involve large bandwidths at low frequencies ($\lesssim$ 500 MHz), for which dispersion is more severe. In this case, the uncertainty in the STEC may be sufficient that, even after dedispersion, the pulse has some significant remnant dispersion characteristic; and the larger data rates of future experiments will make it prohibitive to record the received signal with its full time resolution, so retrospective processing is impractical. The solution will likely involve a combination of parallel dedispersion at different dispersion characteristics, to cover the uncertainty range in STEC [20, 21]; and more precise measurements of STEC, either through the direct use of line-of-sight measurements of STEC to satellites of the global positioning system (GPS), or through measurements of the STEC-dependent Faraday rotation of the lunar thermal emission [22].

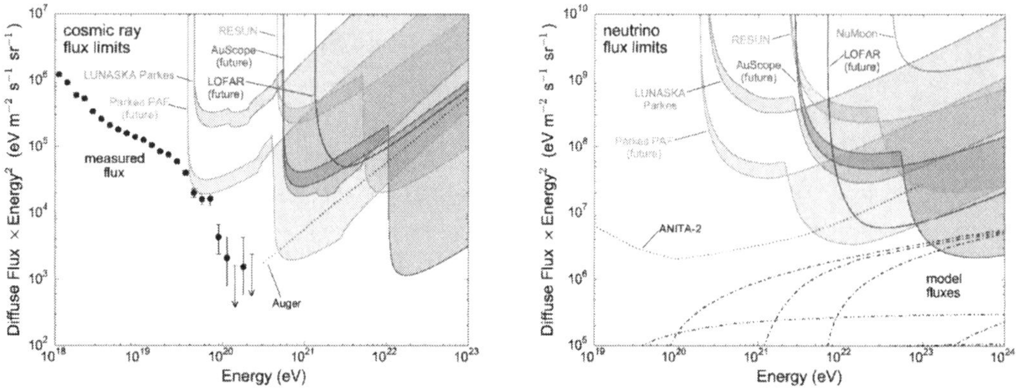

Figure 1. Limits on the fluxes of ultra-high-energy cosmic rays (left) and neutrinos (right) set by the NuMoon [14], RESUN [15] and LUNASKA Parkes [17] lunar Askaryan experiments, calculated as described in the text. Also shown are potential limits that could be set by future lunar Askaryan experiments with nominal observing times of 200 h with LOFAR [20, 25], 200 h with a phased-array feed (PAF) on the Parkes radio telescope [26], or 2900 h with AuScope [18]. Shading shows the range of uncertainty associated with models of the small-scale lunar surface roughness; the discontinuity is an artefact of the model. Filled circles show the cosmic-ray flux measured by the Pierre Auger Observatory [28], and dash-dotted lines show speculative models of the neutrino flux from exotic-physics models [29, 30]. Dotted lines show limits established by the Pierre Auger Observatory [28] and the ANITA-2 experiment [31]. All limits are 90%-confidence model-independent differential limits [2].

4 Particle sensitivity

I estimate the limits or potential limits set by some past and near-future lunar Askaryan experiments using an analytic model for the particle aperture [18, 23, 24]. To reflect the effects of small-scale surface roughness, I use the only extant model [13], which represents the aperture as a linear combination of two extreme cases: one in which these effects are ignored, and one in which they are maximised, by representing the surface as a series of small-scale facets with uncorrelated slopes. Following this model, I take the parameter $r = 0.6$, describing the proportional contribution of the first case, as a compromise based on simulations of multiple experiments [19], and scale N_S, the number of facets along the length of the shower, according to the observing frequency of each experiment. For the low-frequency experiments NuMoon and LOFAR, $N_S \sim 1$, and so no correction for small-scale surface roughness is applied.

Results are shown in Fig. (1). Uncertainty associated with the effects of small-scale surface roughness strongly affects the estimated limits for high-frequency experiments. Depending on these effects, near-future high-frequency experiments may or may not improve on existing neutrino limits at ultra-high energies. A planned low-frequency experiment with LOFAR [20, 25] will do so in any case, improving current constraints on neutrino fluxes predicted by exotic-physics models. A proposed experiment with a phased-array feed on the Parkes radio telescope [26], or on a similar large single dish, could potentially achieve the first detection of cosmic rays with this technique. A proposed experiment (not shown here) with the low-frequency component of the Square Kilometre Array [21, 27] would have a substantially greater sensitivity to both neutrinos and cosmic rays, but is a more distant prospect, as this telescope is planned for construction in 2018–2023.

5 Conclusion

The technical considerations for lunar Askaryan experiments are challenging, but are now well understood, and future experiments can be expected to address them correctly. The major theoretical problem remaining in this field is modelling the effects of small-scale lunar surface roughness, which have a dramatic effect on the particle aperture for high-frequency experiments.

Experiments practical on a near-future timescale have the potential to substantially improve on the current state of this field. A lunar Askaryan experiment with a phased-array feed on a large single dish could plausibly achieve the first detection with this technique of an ultra-high-energy cosmic ray. The planned experiment with LOFAR will constrain neutrino fluxes predicted by exotic-physics models, as well as being a technical pathfinder for a future experiment with the Square Kilometre Array.

I acknowledge support from ERC-StG 307215 (LODESTONE).

References

[1] P.W. Gorham, NASA Advanced Planning Office (2004), arXiv:astro-ph/0411510
[2] N.G. Lehtinen, P.W. Gorham, A.R. Jacobson et al., Phys. Rev. D **69**, 013008 (2004)
[3] A. Romero-Wolf, P. Gorham, K. Liewer et al., ArXiv (2013), arXiv:1302.1263
[4] P. Motloch, N. Hollon, P. Privitera, Astropart. Phys. **54**, 40 (2014)
[5] P.B. Rimmer, C.R. Stark, C. Helling, ApJ Lett. **787**, L25 (2014)
[6] J.D. Bray, A. Nelles, ApJ **825**, 129 (2016)
[7] R.D. Dagkesamanskii, I.M. Zheleznykh, Sov. Phys. JETP Lett. **50**, 259 (1989)
[8] G.A. Askaryan, Sov. Phys. JETP **14**, 441 (1962)
[9] V.A. Ryabov, V.A. Chechin, G.A. Gusev et al., Adv. Space Res. **58**, 464 (2016)
[10] T.H. Hankins, R.D. Ekers, J.D. O'Sullivan, MNRAS **283**, 1027 (1996)
[11] P.W. Gorham, C.L. Hebert, K.M. Liewer et al., Phys. Rev. Lett. **93**, 041101 (2004)
[12] A.R. Beresnyak, R.D. Dagkesamanskii, I.M. Zheleznykh et al., Astron. Rep. **49**, 127 (2005)
[13] C.W. James, R.D. Ekers, J. Alvarez-Muñiz et al., Phys. Rev. D **81**, 042003 (2010)
[14] S. Buitink, O. Scholten, J. Bacelar et al., A&A **521**, A47 (2010)
[15] T.R. Jaeger, R.L. Mutel, K.G. Gayley, Astropart. Phys. **34**, 293 (2010)
[16] R.E. Spencer, A. Macfarlane, O. Mills et al., Proc. Sci. **125**, 097 (2010)
[17] J.D. Bray, R.D. Ekers, P. Roberts et al., Astropart. Phys. **65**, 22 (2015)
[18] J.D. Bray, Astropart. Phys. **77**, 1 (2016)
[19] J.D. Bray, R.D. Ekers, P. Roberts et al., Phys. Rev. D **91**, 063002 (2015)
[20] T. Winchen, A. Bonardi, S. Buitink et al. (LOFAR Cosmic Ray KSP), these proceedings (2016)
[21] C.W. James, J. Alvarez-Muñiz, J.D. Bray et al., these proceedings (2016)
[22] R.A. McFadden, R.D. Ekers, J.D. Bray, **4**, 284 (2011)
[23] K.G. Gayley, R.L. Mutel, T.R. Jaeger, ApJ **706**, 1556 (2009)
[24] Y.S. Jeong, M.H. Reno, I. Sarcevic, Astropart. Phys. **35**, 383 (2012)
[25] K. Singh, M. Mevius, O. Scholten et al., Nucl. Instrum. Meth. A **664**, 171 (2012)
[26] J.D. Bray, R.D. Ekers, R.J. Protheroe et al., AIP Conf. Series **1535**, 21 (2013)
[27] J.D. Bray, J. Alvarez-Muñiz, S. Buitink et al., Proc. Sci. **215**, 144 (2015)
[28] J. Abraham, P. Abreu, M. Aglietta et al. (Pierre Auger Coll.), Phys. Lett. B **685**, 239 (2010)
[29] V. Berezinsky, E. Sabancilar, A. Vilenkin, Phys. Rev. D **84**, 085006 (2011)
[30] C. Lunardini, E. Sabancilar, Phys. Rev. D **86**, 085008 (2012)
[31] P.W. Gorham, P. Allison, B.M. Baughman et al. (ANITA Coll.), Phys. Rev. D **82**, 022004 (2010)

Search for Cosmic Particles with the Moon and LOFAR

T. Winchen[1,★], *A.* Bonardi[2], *S.* Buitink[1], *A.* Corstanje[2], *J. E.* Enriquez[2], *H.* Falcke[2,5,7], *J. R.* Hörandel[2], *P.* Mitra[1], *K.* Mulrey[1], *A.* Nelles[3], *J. P.* Rachen[2], *L.* Rossetto[2], *P.* Schellart[2], *O.* Scholten[4a], *S.* Thoudam[2], *T.N.G.* Trinh[4,6], and *S.* ter Veen[5](The LOFAR Cosmic Ray KSP)

[1] *Vrije Universiteit Brussel (Belgium)*
[2] *Radboud University Nijmegen (The Netherlands)*
[3] *University of California Irvine (USA)*
[4] *KVI-CART (The Netherlands)*
[5] *ASTRON (The Netherlands)*
[6] *University of Groningen (The Netherlands)*
[7] *NIKHEF (The Netherlands)*

Abstract. The low flux of the ultra-high energy cosmic rays (UHECR) at the highest energies provides a challenge to answer the long standing question about their origin and nature. A significant increase in the number of detected UHECR is expected to be achieved by employing Earth's moon as detector, and search for short radio pulses that are emitted when a particle interacts in the lunar rock. Observation of these short pulses with current and future radio telescopes also allows to search for the even lower fluxes of neutrinos with energies above 10^{22} eV, that are predicted in certain Grand-Unifying-Theories (GUTs), and e.g. models for super-heavy dark matter (SHDM). In this contribution we present the initial design for such a search with the LOFAR radio telescope.

1 Introduction

The extremely low flux of ultra-high energy cosmic rays (UHECR) makes the search for their sources a challenging task (e.g. [1]). The same challenge arises in testing specific theories of grand-unification [2] and super heavy dark matter [3, 4] by their predicted flux of neutrinos with extreme energies beyond 1 ZeV. To test these theories and increase the available UHECR data at the highest energies, detectors several orders of magnitude larger than currently available are required. One approach to achieve this necessary increase of detector area is to use Earth's Moon as detector. The Moon is 'read-out' by existing radio telescopes that record the nanosecond radio pulses that are created when a particle interacts in the lunar rock.

Several searches with radio telescopes operating in the GHz frequency regime have so far yielded only upper limits on the neutrino flux and have not been sensitive enough to observe the low UHECR flux [5–9]. The sensitivity of these searches was in particular limited by the frequency range of the used telescopes that operate in the GHz regime. While at GHz frequencies the expected pulse amplitudes reach a maximum, the pulses cannot escape the moon for most part of its surface; only

★e-mail: tobias.winchen@rwth-aachen.de

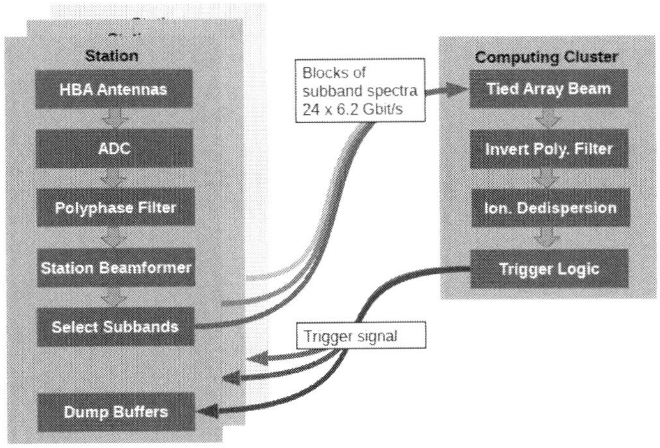

Figure 1. Overview of the online data analysis processing steps for the detection of ns-pulses with LOFAR.

pulses from particles hitting the limb are detectable at Earth. Conversely, similar searches with the LOw Frequency ARray (LOFAR) [10] can use the full visible lunar surface as sensitive area, as it operates in a frequency range that allows pulses to escape the lunar rock in more geometries and is optimal for the lunar detection of particles [11].

2 Signal Processing at LOFAR

LOFAR consists of several thousand omni-directional antennas located in more than 48 stations distributed throughout European. Twenty-four of the stations are grouped into a dense core of approximately 2 km diameter in the Netherlands. All stations are equipped with 96 low-band antennas (LBAs) with a frequency range from 10–90 MHz and at least 48 'tiles' of 16 high-band antennas (HBAs) each, with a frequency range from 110–240 MHz. By beamforming, i.e. coherent stacking the signals of the individual antennas with a corresponding delay, the sensitivity of a station is tuned towards a specific direction. A station thus assumes the role of a single dish in a classical radio telescope.

As preparation for astronomical observations, the data is preprocessed in several steps in LOFAR before it can be accessed by the user. An overview of the default processing steps together with the online-analysis steps required here is given in figure 1. After digitization, the data is first processed by a polyphase filter (PPF), which splits the data stream into 256 sub-channels of 195 kHz width. Compared to splitting data with a Fourier transformation, the PPF allows efficient sampling conversion while preventing signal-leakage between neighboring channels, thus allowing precise filtering. However, within the individual sub-channels the time resolution is reduced from 5 ns to only a few micro-seconds due to critical down-sampling. The filtered data of the individual antennas of one station is then combined into a 'station beam' before it is transmitted to a computing cluster for further processing.

While the pre-filtering of the signal is convenient for typical astronomical observations, here we require the full nano-second resolution. The thus necessary inversion of the PPF is a lossy process, but we can use the thereby approximately reconstructed signal to trigger storage of the raw voltage traces which are buffered for 5 seconds. We therefore will first combine the station beams to several tight-array beams pointing to different spots on the moon, and then construct nanosecond time traces from the combined signals. These are then corrected for ionospheric dispersion, and eventually trigger a read-out of the buffer board.

3 Polyphase Synthesis

The polyphase filter analysis is an efficient implementation of an multi channel Finite-Impulse-Response (FIR) filter with corresponding optimization of the sampling rate to split the signal $x[n]$ into well defined sub-bands $\hat{y}_i[m]$, with $i = 0 \cdots M$. The polyphase-filter mathematically corresponds to the Fourier transform $\mathcal{F}(\mathbf{y})$ of the transformed signal $\mathbf{y} = H \cdot \mathbf{x}$, with matrix H corresponding to the single-channel filter. For a K-tap filter into M-sub-channels H is a $K \cdot M \times M$ matrix. The filter principle is sketched in figure 2.

Inversion of the Fourier transformation in the polyphase analysis is possible without loss of information. The inverse of the filter matrix H^{-1}, however, does not exists as H is not square. While methods exists (e.g. [12]) to construct an approximate inverse G so that $G \cdot H = 1$, the one-step inversion produces echos of short pulses as e.g. in [13]. This limits their usability here, as we thus expect a large number of false triggers generated from RFI by this method. We here therefore investigate a more robust but computationally more complex iterative method to solve the linear-systems of equations $H \cdot \mathbf{x} = \mathbf{y}$.

As the system is under-determined, a side condition is required to constrain the solutions. The reason for the ill-determination of the inverse is here, that any $x[n]$ influences $y_i[m]$ for K values of m. Thus, for a filtered signal of finite length, $K \times M$ samples at the beginning, respectively end, of the reconstructed input signal are not completely determined. Searching for the minimum solution $|| \min x||$ can thus be used as side condition here, as this minimizes the signal in the under-determined areas at cost of a slight damping of the reconstructed signal.

An algorithm using this side condition to solve under-determined linear systems is the LSMR algorithm [14]. Compared to alternatives as e.g. the LSQR algorithm , the LSMR algorithm has also the advantage of practically monotonically conversion that allows for safe early termination, which might be advantageous for a real-time application.

In figure 3 an example of the reconstruction of a trace with a prototype implementation of this method is given. The input signal shown in fig. 3 (a) is composed of white noise and delta peaks of fixed height at random time sets. After applying a polyphase filter and its inverse to the trace, the signal is recovered almost perfectly, except for the edges of the trace as shown in fig. 3 (b). The differences between the input signal and recovered signal is exactly zero except for small time windows at intervals corresponding to the number of sub-channels as shown in fig. 3(c, d). As these

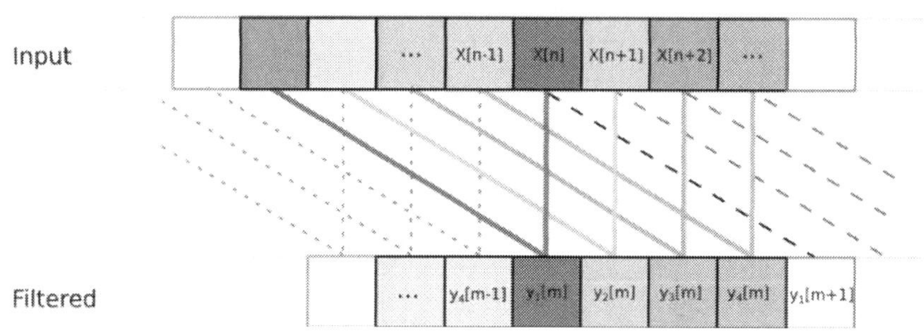

Figure 2. Sketch of the filter stage of an 2-tap 4-channel PPF before application of a 4-point FFT to $\mathbf{y}$.

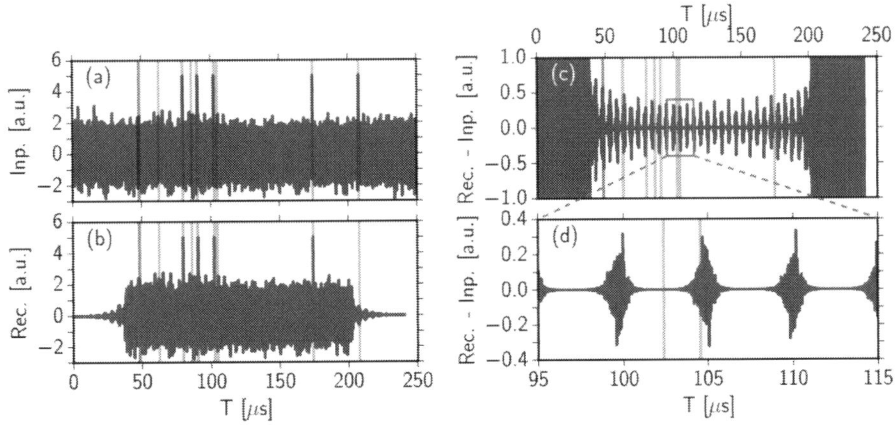

Figure 3. Accuracy of PPF-Synthesis. **(a)** Input Signal, **(b)** reconstructed signal, **(c)** difference between input and reconstructed signal, **(d)** zoom into (c). The position of input pulses are marked by vertical gray lines.

windows are known, a necessary increase of the trigger sensitivity is limited to a small time intervall. In particular, the uncertainty in the reconstruction is not related to the position of the input pulses and thus not critical for the trigger. The maximum amplitude found here corresponds to less than 30% of the root-mean squared of the white noise.

4 Conclusion

The LOFAR radio telescope is a promising instrument for the search of cosmic-particle induced radio-pulses from Earth's moon. Such a search requires triggering on preprocessed data and in particular the reconstruction of nano second resolution time traces via a polyphase synthesis. The iterative approach to polyphase synthesis investigated here is suitable for triggering the data, as it does not introduce artefacts that are likely to increase the trigger rate.

References

[1] K. Kotera, A.V. Olinto, Annual Review of Astronomy and Astrophysics **49**, 119 (2011)
[2] P. Bhattacharjee, AIP Conf. Proc. **433**, 168 (1998)
[3] R. Aloisio, V. Berezinsky, M. Kachelriess, Phys. Rev. **D74**, 023516 (2006)
[4] R. Aloisio, S. Matarrese, A.V. Olinto, JCAP **1508**, 024 (2015)
[5] S. Buitink et al., Astron. Astrophys. **521**, A47 (2010)
[6] T.H. Hankins, R.D. Ekers, J.D. O'Sullivan, MNRAS **283**, 1027 (1996)
[7] C.W. James et al., MNRAS **410**, 885 (2011)
[8] C.W. James et al., MNRAS **379**, 1037 (2007)
[9] J.D. Bray et al., Astropart. Phys. **65**, 22 (2014)
[10] M.P. van Haarlem et al., Astronomy and Astrophysics **556**, A2 (2013)
[11] O. Scholten et al., Astroparticle Physics **26**, 219 (2006)
[12] M. Unser, Proceedings of the IEEE **88**, 569 (2000)
[13] K. Singh et al., Nucl. Instrum. Meth. **A664**, 171 (2012)
[14] D. Fong, M. Saunders, SIAM Journal on Scientific Computing **33**, 2950 (2011)

TEC, Trigger and Check, preparing LOFAR for Lunar observations

Sander ter Veen[3,⋆], *Maaijke* Mevius[2], *Antonio* Bonardi[1], *Stijn* Buitink[2], *Arthur* Corstanje[1], *J. Emilio* Enriquez[1], *Heino* Falcke[1,3,4], *Jörg R.* Hörandel[1,4], *Pragati* Mitra[2], *Katey* Mulrey[2], *Anna* Nelles[5], *Jörg Paul* Rachen[1], *Laura* Rossetto[1], *Pim* Schellart[1], *Olaf* Scholten[6,7], *Satyendra* Thoudam[1], *Gia* Trinh[6,7], and *Tobias* Winchen[2]

[1] *Department of Astrophysics/IMAPP, Radboud University, P.O. Box 9010, 6500 GL Nijmegen (the Netherlands)*
[2] *Vrije Universiteit Brussel, Pleinlaan 2, 1050 Brussels (Belgium)*
[3] *Astron, Oude Hoogeveensedijk 4, 7991 PD Dwingeloo (The Netherlands)*
[4] *Nikhef, Science Park 105, 1098 XG Amsterdam (The Netherlands)*
[5] *University of California Irvine, Irvine, CA 92697, (USA)*
[6] *KVI-CART, Zernikelaan 25, 9747 AA Groningen (The Netherlands)*
[7] *University of Groningen, Zernikelaan 25, 9747 AA Groningen (The Netherlands)*

Abstract. One of the main ways to use radio to detect Ultra High Energy Neutrinos and Cosmic Rays is the Lunar Askaryan technique, that uses the Moon as a target and searches for nanosecond pulses with large radio telescopes. To use low frequency aperture arrays, such as LOFAR and the SKA, pose new challenges and possibilities in detection techniques of short radio pulses and to measure the Total Electron Content (TEC). As a prepatory work, we have used other measurements that use similar techniques, or that can answer a specific question, with the LOFAR radio telescope. This contribution reports on our work on triggering on short radio signals, post-event imaging of radio signals from buffered data and methods to determine the TEC-value.

1 Introduction

One of the ways to determine the flux and direction of ultra high energy (UHE; $> 10^{18}$ eV) cosmic particles (cosmic rays and neutrinos) is to point a large radio telescope to the Moon to search for nanosecond pulses induced by lunar showers from these particles following the Askaryan effect [1]. Previous measurements with radio telescopes such as Parkes [2–4] and WSRT [5, 6] have developed these techniques and set limits to the flux of these particles. However, it requires even larger telescopes, such as the SKA [7] or FAST, to be sensitive to the known cosmic ray flux. To set better limits and to test the techniques used for a distributed telescope such as the SKA, we are working on a method to search for ultra high energy cosmic particles with LOFAR, within the NuMoon project. The proceedings by T. Winchen gives an introduction to this subject and deals with re-obtaining a high time resolution signal to search for these pulses. This proceeding reports on two other challenges faced in this project: The influence on the ionosphere and the verification of observed pulses.

⋆e-mail: veen@astron.nl

2 LOFAR

LOFAR [8] is a low frequency radio telescope that consists of stations with fields with antennas distributed on relative distances of 10s of meters up to 100s of kilometers. The antennas of each field are combined to form one or more beams on the sky with a size of a few degrees. NuMoon observations cover 110-190 MHz. By combining the beams of the inner 24 stations, the moon can be covered with 50 beams to search for nanosecond signals induced by lunar showers from ultra high energy cosmic particles. In addition, the data from each antenna is stored in a ring buffer for 5 seconds. If a trigger is received within this time, the data can be read out to disk for further analysis.

2.1 Ionospheric corrections

The first challenge is the correction of the ionosphere. The free electrons in the ionosphere cause a dispersion in the arrival time of the signal of different frequencies, with the relation

$$\tau[ns] = k\,\mathrm{STEC}/\nu^2 = 1.34 \cdot \mathrm{STEC}/\nu^2 \quad \text{with} \quad \mathrm{STEC} = \int n_e(s)ds \tag{1}$$

with ν the observing frequency in GHz. In this STEC is the Slanted Total Electron Content following the integrated electron density along a slanted path in TEC units (TECU) of 10^{16} electrons / m^2. This dispersion causes the signal to be spread out over 10s of nanoseconds [Fig. (1)], lowering the peak amplitude. This effect needs to be corrected, to be able to detect pulses induced by cosmic particles. A second effect is Faraday rotation, in which the polarisation angle of a polarised source changes over frequency according to

$$\beta = \mathrm{RM}\lambda^2 \tag{2}$$

where λ is the wavelength in meters and the rotation measure

$$\mathrm{RM} = \frac{e^3}{2\pi m_e^2 \epsilon_0 c^3} \int_0^d n_e(s) B_{||}(s) ds, \tag{3}$$

in which $B_{||}$ is the magnetic field parallel to the line of sight, e the charge of an electron, c the speed of light in vacuum, m_e the mass of an electron and ϵ_0 is the vacuum permittivity, all in SI units. The Faraday rotation is easy to measure at low frequencies. If the magnetic field is known, this gives a measurement of the TEC value. [9] give a comparison in the measured RM variation of pulsars, as observed by LOFAR, with predicted values based on TEC and B values. This method may be used by observing pulsars in parallel or using the polarized rim of the Moon in online or offline analysis [10], and using the known B field to predict TEC values. With astronomical measurements differential[1] Faraday Rotation and differential TEC can be obtained with high precision, but the uncertainty in the differential magnetic field makes it difficult to obtain the absolute TEC value. An alternative approach is to use GPS data directly. The current maps are only available days later and have a temporal resolution of 2 hours and a spatial resolution of 2.5 x 2.5 degrees, but for the trigger real-time data is required. However, the Space Radio-Diagnostics Research Centre of the University of Warmia and Mazury and the UPC-IonSAT research group of the Technical University of Catalonia are currently developing algorithms to provide real-time high resolution TEC maps, specifically for the ILT (International LOFAR Telescope), using 119 Euref Permanent Network stations [11–13]. In the result, near real-time and high-spatial resolution maps are obtained at 15 minutes in temporal resolution and 0.5 degree in latitudinal and longitudinal spatial resolution, (Fig. (2)), with the possibility of obtaining

[1] Between two different positions 1-1000 km apart

even higher temporal resolution. The biggest challenge here is still to get to an absolute TEC value, as there is a bias due to the ambiguous 2π phase shifts, as well as the slant to vertical TEC conversion, because of the unknown height of the ionosphere (see e.g.[14]). The lower uncharged atmosphere influences the signal only by a frequency independent delay. Therefore, its expected influence on the peak height of the signal, as well as on GPS-driven TEC measurements is negligible.

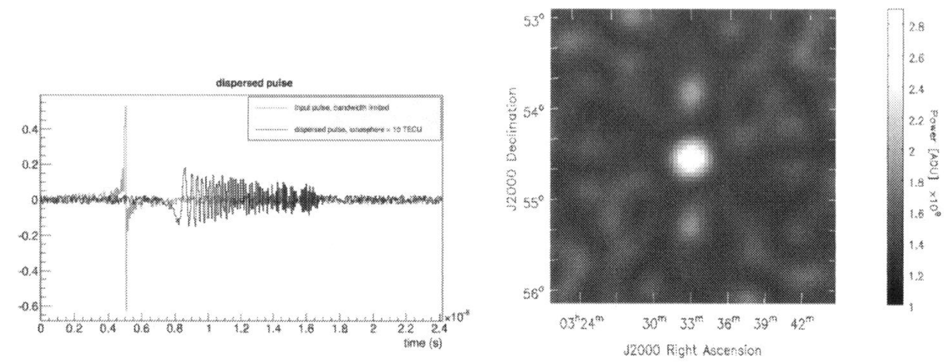

Figure 1. Left: Input pulse in red, dispersed pulse in black, with an ionospheric ionisation of 10 TECU., Right: Sky image from 6 LOFAR core stations of a dedispersed pulse of pulsar B0329+54 from TBB data obtained from triggering in real-time on beam formed data.

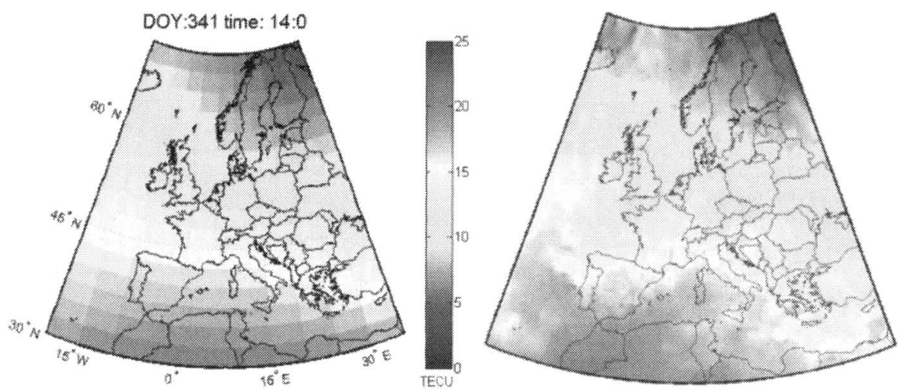

Figure 2. A comparison between the IGS global ionospheric maps [15] - on the left vs rapid ILT dedicated maps - on the right.

2.2 Pulse verification

The second challenge is the verification of pulses. For this we intend to read out the buffers from individual antennas. With this data we can check that it was not a single antenna or station creating the signal and check that the real direction is indeed coming from the Moon and not an off-beam satellite or terrestrial signal. We can further perform a better TEC correction to the pulse, and see for example if a 2x TEC value (as would be from a reflected signal) will give an even stronger signal. The technique of triggering the buffers and analysing the data was earlier performed in an astronomical context [16], to search for fast radio transients, such as pulsar pulses and the new class of fast radio bursts [17]. In this context a pulse of pulsar B0329+54 was localised [Fig. (1)], a strange signal was identified as an off-beam solar flare, and a signal was identified as coming from a single antenna.

3 Conclusion

The detection of ultra high energy cosmic particles, using the Askaryan techniques, remains challenging for telescopes such as LOFAR and SKA. But with the LOFAR observations and tests described here we are two steps closer to run successful observation in this or the next decade.

References

[1] G.A. Askaryan, Journal of the Physical Society of Japan Supplement **17**, C257 (1962)
[2] C.W. James, R.D. Ekers, J. Álvarez-Muñiz, J.D. Bray, R.A. McFadden, C.J. Phillips, R.J. Protheroe, P. Roberts, Phys. Rev. D **81**, 042003 (2010), 0911.3009
[3] J.D. Bray, R.D. Ekers, P. Roberts, J.E. Reynolds, C.W. James, C.J. Phillips, R.J. Protheroe, R.A. McFadden, M.G. Aartsen, Phys. Rev. D **91**, 063002 (2015), 1502.03313
[4] J.D. Bray, R.D. Ekers, P. Roberts, J.E. Reynolds, C.W. James, C.J. Phillips, R.J. Protheroe, R.A. McFadden, M.G. Aartsen, Astroparticle Physics **65**, 22 (2015), 1412.4418
[5] S. Buitink, O. Scholten, J. Bacelar, R. Braun, A.G. de Bruyn, H. Falcke, K. Singh, B. Stappers, R.G. Strom, R.A. Yahyaoui, Astron. & Astrophys. **521**, A47 (2010), 1004.0274
[6] S. Ter Veen, S. Buitink, H. Falcke, C.W. James, M. Mevius, O. Scholten, K. Singh, B. Stappers, K.D. de Vries, Phys. Rev. D **82**, 103014 (2010), 1010.6061
[7] J.D. Bray, J. Alvarez-Muñiz, S. Buitink, R.D. Dagkesamanskii, R.D. Ekers, H. Falcke, K.G. Gayley, T. Huege, C.W. James, M. Mevius et al., SKA Science Book Chapter, ArXiv:1408.6069 (2014), 1408.6069
[8] M.P. van Haarlem, M.W. Wise, A.W. Gunst, G. Heald, J.P. McKean, J.W.T. Hessels, A.G. de Bruyn, R. Nijboer, J. Swinbank, R. Fallows et al., Astron. & Astrophys. **556**, A2 (2013), 1305.3550
[9] C. Sotomayor-Beltran, C. Sobey, J.W.T. Hessels, G. de Bruyn, A. Noutsos, A. Alexov, J. Anderson, A. Asgekar, I.M. Avruch, R. Beck et al., Astron. & Astrophys. **552**, A58 (2013), 1303.6230
[10] R. McFadden, International Cosmic Ray Conference **4**, 284 (2011)
[11] M. Hernández-Pajares, J.M. Juan, J. Sanz, A. Aragón-Àngel, A. García-Rigo, D. Salazar, M. Escudero, Journal of Geodesy (2011)
[12] A. Krankowski, I. Shagimuratov, L. Baran, G. Yakimova, Advances in Space Research **41**, 628 (2005)
[13] A. Krankowski, I. Shagimuratov, I. Ephishov, A. Krypiak-Gregorczyk, G. Yakimova, Advances in Space Research **43** (2009)
[14] P.L. Martin, J.D. Bray, A.M.M. Scaife, MNRAS, **459**, 3525 (2016), 1604.03810
[15] M. Hernandez-Pajares, J. Juan, J. Sanz, R. Orus, A. Garcia-Rigo, J. Feltens, A. Komjathy, S. Schaer, A. Krankowski, Journal of Geodesy **83**, 263 (2009)
[16] S. ter Veen, Ph.D. thesis, Radboud University Nijmegen, http://hdl.handle.net/2066/147186 (2015)
[17] D. Thornton, B. Stappers, M. Bailes, B. Barsdell, S. Bates, N.D.R. Bhat, M. Burgay, S. Burke-Spolaor, D.J. Champion, P. Coster et al., Science **341**, 53 (2013), 1307.1628

Interpretation of the cosmic-ray air shower signal in Askaryan radio detectors

Krijn D. de Vries[1,⋆], *Stijn* Buitink[2], *Nick* van Eijndhoven[1], *Thomas* Meures[3], *Aongus* O'Murchadha[3], and *Olaf* Scholten[1,4]

[1] *Vrije Universiteit Brussel, Dienst ELEM, IIHE, Pleinlaan 2, 1050 Brussels, Belgium*
[2] *Vrije Universiteit Brussel, Astrophysical Institute, Pleinlaan 2, 1050 Brussels, Belgium*
[3] *Dept. of Physics and Wisconsin IceCube Particle Astrophysics Center, University of Wisconsin, Madison, WI 53706,USA*
[4] *University of Groningen, KVI-Center for Advanced Radiation Technology, 9747 AA Groningen, The Netherlands*

Abstract. We discuss the radio emission from a cosmic-ray air shower propagating in air before it hits an air-ice boundary after which it completes its propagation inside the ice. The in-air emission, the in-ice emission, as well as the transition radiation from the shower crossing the boundary is considered. We discuss the interpretation of the radio signal observed by an in-ice observer.

1 Introduction

In [1], we presented a calculation for the radio emission from a cosmic-ray air shower hitting an ice surface. It is shown that this signal should be observable by the currently existing Askaryan radio detectors [2–4]. The in-air emission, the in-ice emission, as well as the coherent transition radiation from the boundary were considered. In this article we focus on the interpretation of the observed signal and discuss the properties of the coherent transition radiation in view of the sudden appearance and sudden death signals. The sudden appearance signal is for example seen during beam-test experiments, for which the beam is hidden to the observer (within the observing frequency band) while inside the accelerator, after which it 'suddenly' becomes visible while exiting the accelerator. An example of the sudden death signal is the signal originating from a beam-dump process.

The importance of the cosmic-ray air shower signal in Askaryan radio detectors, if identified correctly, lies in its possibility to calibrate the currently existing Askaryan radio detectors. Furthermore, if detected, this signal immediately shows the on-site feasibility of the Askaryan radio detection technique. If misidentified, however, this signal might pose a background in the search for the signal from a high-energy neutrino-induced particle cascade in ice.

2 Macroscopic modeling

The calculations in [1] are based on a macroscopic calculation [5–7], starting from the Liénard-Wiechert potentials from classical electrodynamics. The electric field is directly obtained from the

⋆e-mail: krijndevries@gmail.com

charge and current distributions in the cascade front. The fields have to be evaluated at the negative retarded emission time, t_r, which links to the observer time, t, through the optical path length L from the emission point to the observer,

$$c(t - t_r) = L. \tag{1}$$

The optical path length inside a medium consisting out of m layers with different index of refraction n_i, can be defined by

$$L = \sum_{i=1}^{m} n_i d_i. \tag{2}$$

Here the distance d_i, traversed by the emission in layer i, is obtained by using a ray-tracing procedure based on Snell's law.

The in-air, as well as the in-ice emission originates from the time-variation of the total number of particles in the cascade, $dN(t_r)/dt_r$, in combination with relativistic boosting effects. In this article, however, we will focus on the coherent transition radiation and its link to the sudden appearance and sudden death signal. In [1], the coherent transition radiation signal was given by,

$$\vec{E}_{tr}(t,\vec{x}) = \lim_{\epsilon\to 0} \int d^2\vec{r} \frac{e\, d\, N_e(t_r)\, w(\vec{r},h)}{4\pi\epsilon_0 c} \times \left(\frac{1}{|\mathcal{D}|^2_{t_r-\epsilon}} - \frac{1}{|\mathcal{D}|^2_{t_r+\epsilon}} \right) \hat{p} \Bigg|_{h=c(t_r-t_b)} . \tag{3}$$

Following the same procedure the sudden appearance signal is given by,

$$\vec{E}_{sa}(t,\vec{x}) = -\lim_{\epsilon\to 0} \int d^2\vec{r}\, \frac{e\, d\, N_e(t_r)\, w(\vec{r},h)}{4\pi\epsilon_0 c\, |\mathcal{D}|^2_{t_r+\epsilon}} \hat{p} \Bigg|_{h=c(t_r-t_b)} . \tag{4}$$

Here d denotes the impact distance from the observer to the shower axis, where $w(\vec{r},h)$ denotes the particle distribution in the cascade front at lateral position $\vec{r}$ and a distance h behind the cascade front. The retarded distance $\mathcal{D}$ is defined by,

$$\mathcal{D} = L\frac{dt}{dt_r}. \tag{5}$$

The obtained fields agree well with the microscopic calculations presented in [8–11], and references therein.

The coherent transition radiation obtained within this formalism can be understood as the emission just above the air-ice boundary at $t_r - \epsilon$, and the emission just below the boundary at $t_r + \epsilon$. Linking Eqs. 3 and 4, the coherent transition radiation can be interpreted as the superposition of the sudden appearance signal and the sudden death signal. As will be shown in the following section, under the proper circumstances these signals will indeed be observed separately in the case of coherent transition radiation.

It should be noted that these solutions are limiting solutions. The derivation of the vector potential from Maxwell's equations has to be considered with great care, since this is done under the assumption of a continuous medium, where in our situation we consider a hard boundary. This is reflected by the fact that $t_r(t)$ is discontinuous at the boundary, and hence the retarded distance is ill defined at this point.

3 Results and interpretation

How are these results reflected in our simulation? This is shown in Fig. (1), where we show the signal observed for a perpendicular incoming air shower which penetrates the ice at an elevation of 3 km,

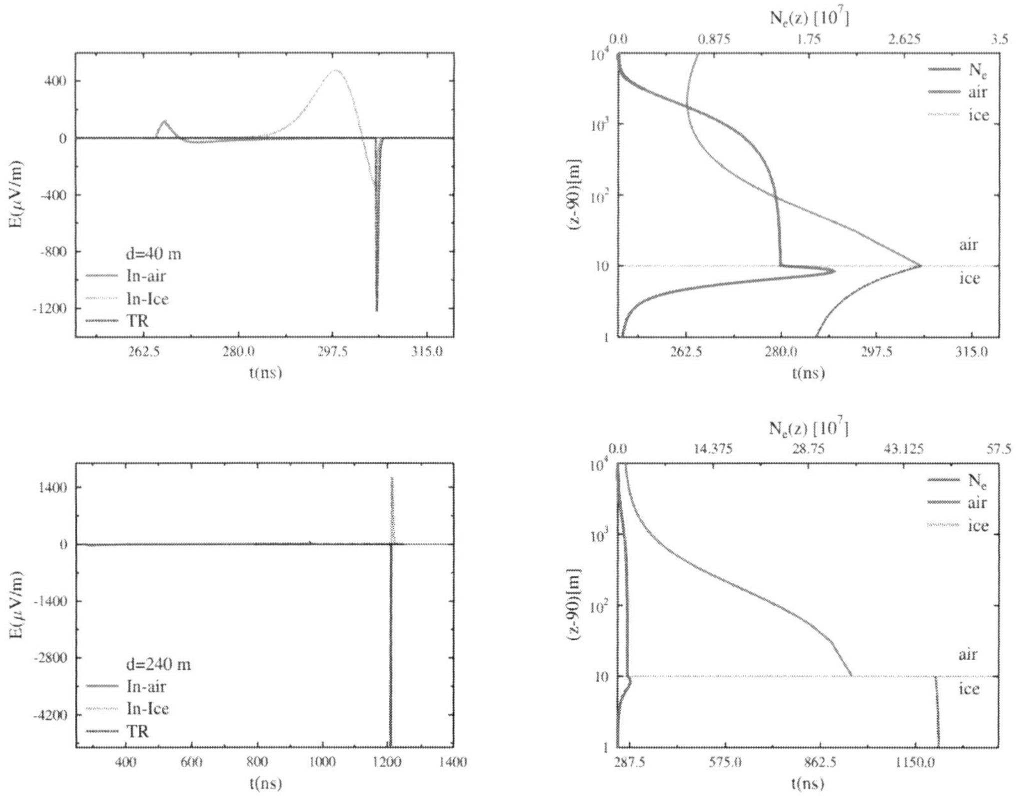

Figure 1. The electric field at different observer distances equal to, a) d = 40 m, b) d = 240 m. The figures on the right show the emission height, plotted as function of the observer time. The full red line gives the emission in air, the dotted purple line gives the transition radiation, and the dashed blue line gives the in-ice emission. For the figures on the right, the total number of particles is given by the full green line. Figure and caption taken from [1].

after which the cascade completes its propagation inside the ice. The observer is positioned in the ice, 100 m below the air-ice boundary. The detailed particle distributions are given in [1].

The top figure shows the signal observed by an observer located inside the in-ice Cherenkov cone, but outside the in-air Cherenkov cone. The interpretation of this signal can be seen in the top-right figure, where we show the emission height, linked to the emission time by $z = -ct_r$, as function of the observer time. For the in-air emission, since the observer is positioned outside the Cherenkov cone, early emission from large heights is seen before late emission, where the emission from close to the boundary is observed latest. The in-ice emission, since seen from inside the Cherenkov cone, is seen differently. Late emission is seen before early emission. It also follows that the in-ice emission is observed at the same time as the in-air emission. Lastly the emission from just above and just below the boundary superimpose to give the coherent transition radiation component.

The bottom figure shows a very interesting situation. Here the point where the cascade hits the boundary is both outside the in-air, as well as outside the in-ice Cherenkov cone. Since the in-ice Cherenkov angle also denotes the critical angle, the in-air emission only enters the ice at this angle.

Hence, the emission from just above the boundary first travels through the air, before it breaks into the ice at the critical angle. Even though this is the longer path, since the signal is not delayed by the medium for its in-air travel, it will arrive before the signal emitted from just below the boundary which traverses its full path to the observer inside the ice. It follows that in these geometries, the in-air and the in-ice emission are completely separated in time. Hence, the observer will first observe the emission from the in-air cascade, after which it appears that the cascade suddenly dies out. The dying out of the cascade is given by the sudden death signal. After some delay, from the observer point of view, the cascade suddenly appears after which the in-ice emission is seen. Hence in this situation, the emission just above the boundary is completely separated from the emission just below the boundary. Therefore, the transition radiation is not seen at a single time, but becomes separated in time. Within this framework, therefore the more fundamental description is given by the sudden appearance and sudden death signals. The observation of such a signal separated in time would confirm the interpretation of coherent transition radiation as the superposition of a the more fundamental sudden appearance and sudden death signals.

4 Summary

We discussed the interpretation of the coherent transition radiation signal, which originates from a cosmic-ray-induced air shower traversing between air and ice. It follows that within the given formalism, this can be interpreted as the superposition of two more fundamental signals given by the cascade sudden appearance and sudden death signals. This interpretation should be taken with care, since the derived potentials on which the calculations are based implicitly rely on the assumption of a continuous medium. This is reflected in the fact that the retarded distance is ill defined at the air-ice boundary. To solve for the fields, the limiting situation of the emission just above and just below the air-ice boundary is considered. The superposition of these two components gives rise to the coherent transition radiation signal. It is shown that in certain geometries, where the observer is positioned outside the in-ice Cherenkov cone for emission directly below the air-ice boundary, the coherent transition radiation is split in two signal separated in time. These signals, the sudden appearance signal and the sudden death signal, can therefore within the given formalism be interpreted as more fundamental.

References

[1] K.D. de Vries et al., Astropart. Phys. **74**, 96-104 (2016)
[2] P.W. Gorham et al., Phys. Rev. Lett. **103**, 051103 (2009)
[3] P. Allison et al., ARA Collaboration, Astropart. Phys. **35**, 457-477 (2012)
[4] A. Nelles et al., ARIANNA Collaboration, these proceedings
[5] O. Scholten, K. Werner, F. Rusydi, Astropart. Phys. **29**, 94-103 (2008)
[6] K.D. de Vries, A.M. van den Berg, O. Scholten, K. Werner, Phys. Rev. Lett. **107**, 061101 (2011)
[7] K. Werner, K.D. de Vries, O. Scholten, Astropart. Phys. **37**, 5-16 (2012)
[8] V.L. Ginzburg, V.N. Tsytovich, Transition Radiation and Transition Scattering, Adam Hilger Press, New York, 1990
[9] S. ter Veen et al., Phys. Rev. D **82**, 103014 (2010)
[10] C.W. James, H. Falcke, T. Huege, M. Ludwig, Phys. Rev. E **84**, 056602 (2011)
[11] P. Motloch, J. Alvarez-Muñiz, P. Privitera, E. Zas, Phys. Rev. D **93**, 043010 (2016)

Recent results from the ARIANNA neutrino experiment

Anna Nelles[1],* *for the* ARIANNA Collaboration

[1] *Department of Physics and Astronomy, University of California Irvine, 92697 CA, USA*

Abstract. The ARIANNA experiment is currently taking data in its pilot-phase on the Ross ice-shelf. Fully autonomous stations measure radio signals in the frequency range from 100 MHz to 1 GHz. The seven station hexagonal radio-array (HRA) was completed in December 2014, and augmented by two special purpose stations with unique configurations. In its full extent ARIANNA is targeted at detecting interactions of cosmogenic neutrinos ($> 10^{16}$eV) in the ice-shelf. Downward-pointing antennas installed at the surface will record the radio emission created by neutrino-induced showers in the ice and exploit the fact that the ice-water surface acts as a mirror for radio emission. ARIANNA stations are independent, low-powered, easy to install and equipped with real-time communication via satellite modems. We report on the current status of the HRA, as well as air shower detections that have been made over the past year. Furthermore, we will discuss the search for neutrino emission, future plans of the experiment.

1 Introduction

The ARIANNA detector is aimed at detecting neutrinos from an energy just below 10^{16} eV [1]. Whenever these neutrinos produce an electromagnetic cascade after interacting in the ice of Moore's Bay on the Ross Ice-shelf in Antarctica, the ensuing radio emission will be strong enough to be detectable. The detector profits from the fact that the water-ice interface at the bottom of the shelf-ice acts as a mirror for the radio emission and the antennas can be installed on top of the ice (see Figure 1), which has the advantage that the antenna dimensions are not restricted by borehole geometry.

2 Current status of the hexagonal-radio array (HRA)

The ARIANNA detector is currently in its pilot phase, the HRA [2]. The main array consists of 7 stations, each with four down-ward facing antennas as shown on the left in Figure 2. In addition, there is one station where two antennas have been rotated upwards as shown on the right in the same figure. While the regular stations are prototypes for the neutrino capability of ARIANNA, the altered station is a test set-up for the rejection of air shower signals. It has been found in simulations that the up-to-down ratio in amplitude is fully sufficient to veto air shower signals coming from above when looking for neutrino signals coming from below. All HRA stations are equipped with solar panels and battery systems and run autonomously during the daylight-season. The stations communicate either via long-range wifi or via satellite communication. Both allow for real-time data-transfer off Antarctica. The

*e-mail: anelles@uci.edu

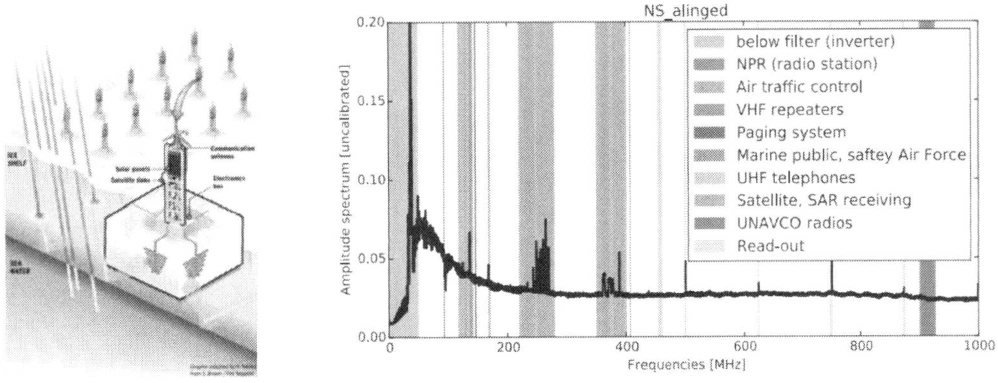

Figure 1. On the left side an artist's impression of the ARIANNA array and the detection concept is shown. On the right measurements of the background noise down to 50 MHz are shown. This data has been averaged over several measurements to reveal the narrow-band transmitters.

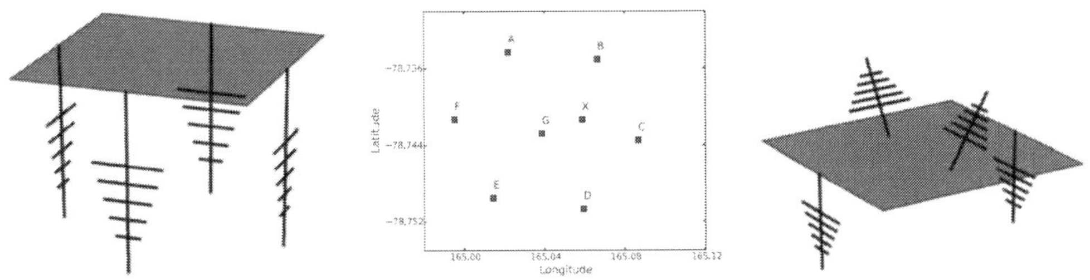

Figure 2. The HRA detector layout. At positions A - G (middle) stations with four antennas pointing down towards the water-ice interface are installed (left). At position X a station with two upward facing antennas is installed (right), which has been targeted at air shower detection. The distance between parallel antennas is 6 meters.

stations are operated with custom digitizers and amplifiers and are optimized for frequencies between 100 and 1000 MHz, with increased gain below 500 MHz. The data is sampled at 1 GHz on one board for the four antennas per stations, which delivers very high relative timing precision in the ps range. The ARIANNA Collaboration is currently developing an eight-channel alternative, which will allow for the full combination of down-and upward facing antennas in one station. The stations usually trigger on a coincidence of two or more antennas. A trigger in a station is formed, if a combination of low-and-high threshold crossings is found in a window of 5 ns. This allows for the HRA stations to run on very low threshold values of between three and four times the noise level and still maintain a very low trigger rate.

The radio emission background at the ARIANNA site is almost at the irreducible level down to 50 MHz, as shown in Figure 1 (right). The trigger rate in the station with the upward-facing antennas is mostly dominated by the Galactic background as shown in Figure 3. The trigger rates show an oscillation with a periodicity that corresponds to the period of the local sidereal time. Additional backgrounds are temporary signals from radio communication, which are easily vetoed by our station-level L1 trigger, which does not allow for a certain fraction of the power to only be in one frequency

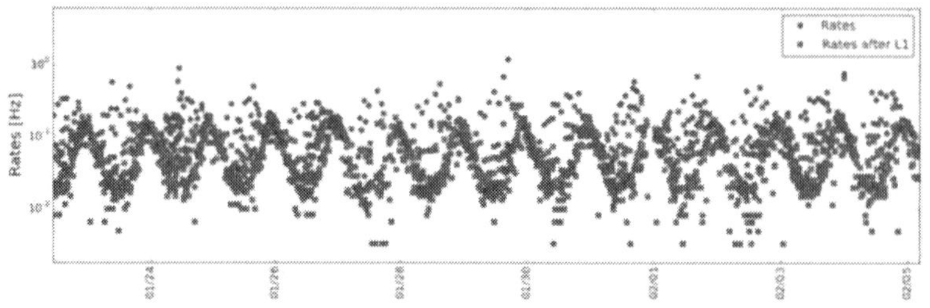

Figure 3. Trigger rates as function of time in station X triggering on the upward facing antennas. The blue circles indicate the raw trigger rates, before the cut on narrowband transmitters (L1, see text) has been applied. The green squares are the thus reduced trigger rates. The oscillation in the trigger rates has a period of 23 hours and 56 minutes. It is therefore linked to an astronomical phenomenon.

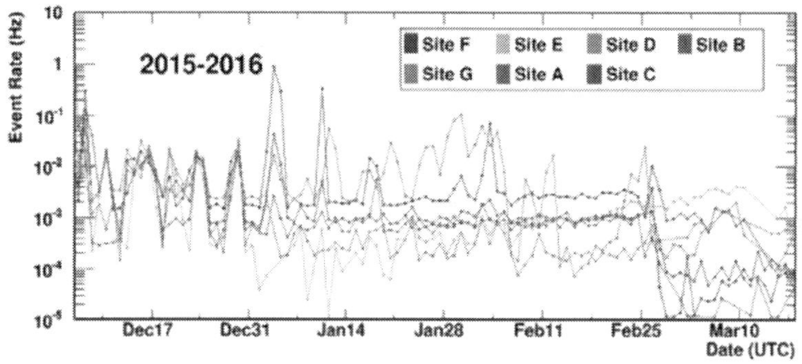

Figure 4. Trigger rates as function of time of all regular HRA stations. After December 31st a veto for narrowband transmitters (L1) has been introduced gradually. After February 25th the thresholds have been raised to lower the data rate to allow for data transfer only via satellite modem and turning off the long-distance wifi. Also this mode of operations has been stable.

bin. Still, one should note that the narrowband transmitters observed in the HRA are sparse, very intermittent and not necessarily strong in power. Moore's Bay seems to be one of the most radio-quiet and still accessible locations. The only pulsed background found at the site is correlated to heavy winds and storms and affects less than 5% of the livetime. The HRA might register discharges within the storm or from blown snow on the surface during these distinct periods. As shown in Figure 4 the stations are very sensitive to threshold settings, and further show extremely low trigger-rates. The regular stations are also affected less by the Galactic noise due to the directivity of the antennas.

3 Cosmic Ray detections

The station with the upward facing antennas has been used to detect air shower signals. Using a very simple search with an increased threshold reveals a number of very interesting candidate events. Due to the large bandwidth of the HRA system and the characteristic frequency spectrum of the air shower emission, simulations show a very distinct expectation for the shape of the air shower signals. The detailed modeling of electric field from the air shower convolved with the response of the antenna,

DOI: 10.1051/epjconf/201713505002

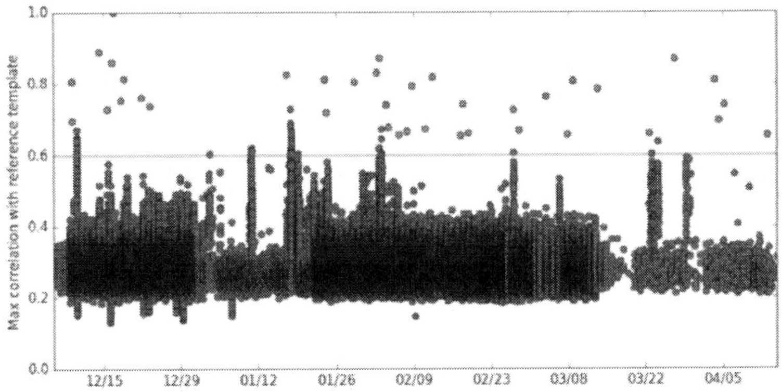

Figure 5. Best correlation per event with a reference event as function of time. One event from December was chosen as reference template. It has a correlation of one. All events not in a cluster of events that have a correlation of better than 0.6 are marked in red.

amplifier and cables predicts a time dependent waveform characterized by high to low frequency chirp, forming a well-defined pulse. The candidate events show such characteristics.

As shown earlier [1], the ARIANNA Collaboration has proposed to search for neutrino signals by the use of a correlation with a template method. We now use the interesting candidates as search templates for all other events. As shown in Figure 5 correlating all events with one of these template events reveals a distinct sub-set of events that show a high correlation, while most events show a very weak correlation. During periods of high winds, correlations of medium quality are observed, where some reach into a tentative signal space of above 0.6. All events marked in red are those, that do not occur in periods of high trigger-rates and show a good correlation. The events show a random occurrence in time and we observe roughly one per two days of livetime, which matches an energy threshold for cosmic rays of below 10^{18} eV. Consequently, these events are viable cosmic ray candidates. One example is shown in Figure 6. We will show in a forthcoming publications, how well the measured signals agree with simulated predictions and whether the number of events, in combination with livetime and efficiency, matches the known cosmic ray flux.

4 Update on neutrino analysis

The search for neutrinos follows the same approach as for cosmic rays, only that all regular stations can be used. The top panel of Figure 7 shows the probability density of the correlation parameter Chi [1] for all measured events in our data-set with a simulated neutrino signal. The bottom panel shows the distribution for the events remaining after excluding events clustered in time and those with a high single frequency contribution (L1 trigger, see above). Also shown is the expectation for Chi from the complete neutrino simulations. Most obvious is the strong peak at around 0.3 in the data, which corresponds to the correlation value of pure thermal noise, as it is also found in the forced data that is taken every 70 seconds. Some stations show a longer tail towards high values, which can be removed immediately with the cluster cut. This is explained by temporary hardware issues or increased noise at station level during storms. After cleaning, only very few events end up in the signal space. Interestingly, all of them are measured in coincidence with other stations and one has also been identified as cosmic ray in the upward-facing antennas. These are therefore the background events, which require the ARIANNA stations to have upward facing antennas for cosmic ray rejection, which

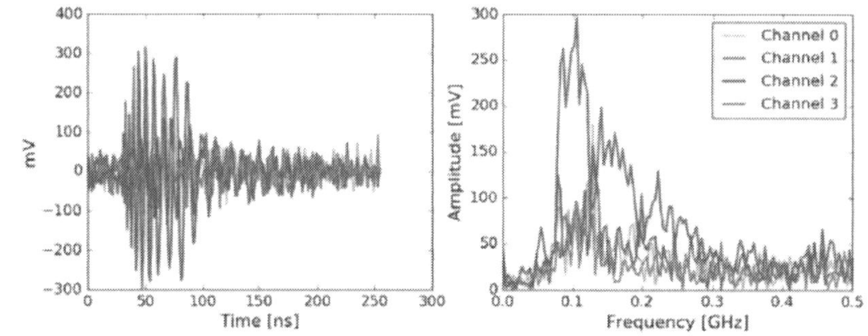

Figure 6. An example of an air shower signal as measured with the HRA. Channels 2 and 3 are upward pointing channels, while 0 and 1 point downward. All signals are strongly polarized, showing only a strong amplitude in either of the perpendicular upward facing channels. The frequency spectrum of all measured air showers is dominated by antenna response and amplifier gain, however, differences can be seen especially in the high-frequency content of the pulses.

will be the case for the future ARIANNA. For now, we increase the correlation threshold to exclude these events and arrive at the preliminary limits shown in Figure 8 for four months of data from the season 2015/16.

5 Conclusions and Outlook

The HRA pilot-array is performing according to expectations with a livetime fraction of more than 0.85. The radio environment is very quiet, the stations operate reliably on battery and solar power and first cosmic rays have been detected. The neutrino analysis only reveals cosmic rays as significant background, which can easily be suppressed using upward facing antennas. The upgrade of a station with a wind-generator and with four upward facing antennas to better be able to test reconstruction algorithms on measured air shower data is planned for the next season. ARIANNA with its 1296 stations will have a a projected sensitivity to cover many of the neutrino fluxes such as shown in [5].

References

[1] The ARIANNA Collaboration, *A First Search for Cosmogenic Neutrinos with the ARIANNA Hexagonal Radio Array*, Astroparticle Physics 70 (2015) 12-26.

[2] S. Kleinfelder for the ARIANNA Collaboration, *Design of the Second-Generation ARIANNA Ultra-High-Energy Neutrino Detector Systems*, Proceedings of the 2015 IEEE Nuclear Science Symposium, San Diego, CA, Nov. 2015.

[3] Barwick, S. W., *Progress in the Development of Radio-Cherenkov Neutrino Detectors*, Proceedings of Science, ICRC 2015, (2015), 027.

[4] The IceCube Collaboration, *Constraints on ultra-high-energy cosmic ray sources from a search for neutrinos above 10 PeV with IceCube*, arXiv:1607.05886.

[5] Kampert, K.-H. and Unger, M., *Measurements of the cosmic ray composition with air shower experiments*, Astroparticle Physics, Volume 35, Issue 10, p. 660-678, 2012.

DOI: 10.1051/epjconf/201713505002

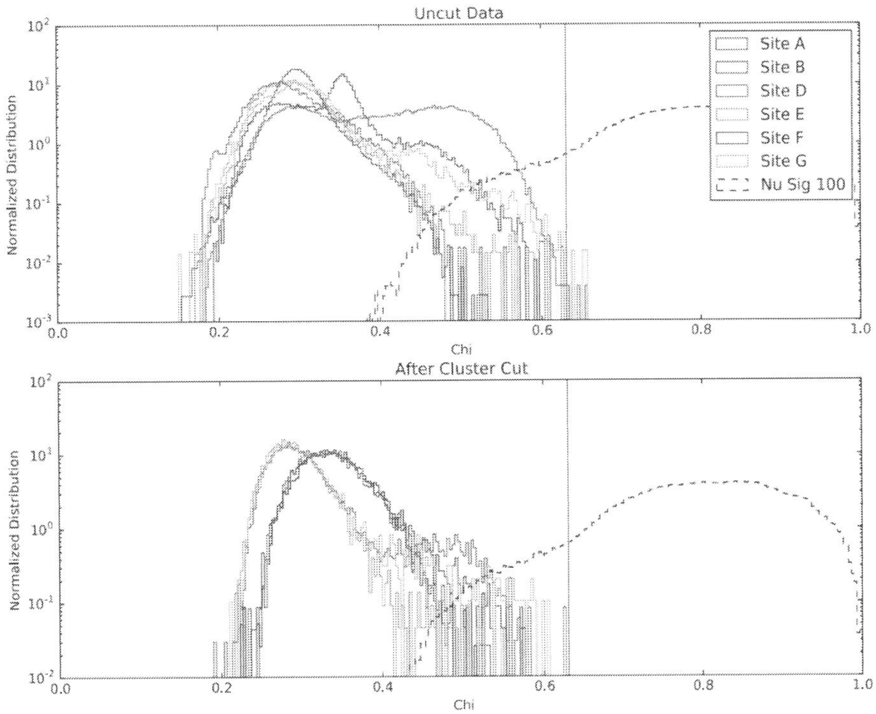

Figure 7. Normalized distribution of correlation (Chi) with a selected neutrino template. The two figures shows the data uncut and after the cluster cut, as explained in the text. The integral under the curves is normalized to one. The vertical line indicates a cut-off value of 0.63. The slightly differing average values of the curves are explained by two types of amplifiers.

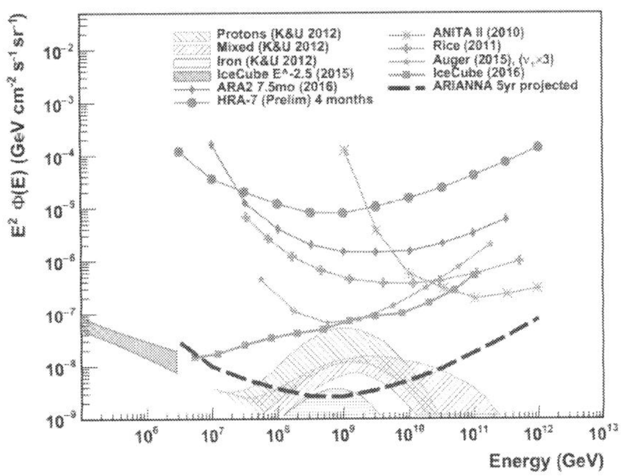

Figure 8. Current limits of the HRA and projected sensitivity of ARIANNA with 1296 stations. Also shown are models of neutrino fluxes and limits of other experiments. See [3–5] for details.

EPJ Web of Conferences **135**, 05003 (2017) DOI: 10.1051/epjconf/201713505003
ARENA 2016

Radio-interferometric Neutrino Reconstruction for the Askaryan Radio Array

Ming-Yuan Lu[1,*] for the ARA Collaboration

[1] *Wisconsin IceCube Particle Astrophysics Center, University of Wisconsin,Madison, WI 53706, USA*

Abstract. The Askaryan Radio Array (ARA) is a neutrino telescope array under phased deployment near the South Pole. The array aims to discover and determine the ultra-high energy neutrino flux via detection of the Askaryan signal from neutrino-induced showers. This novel detection channel makes ARA the most cost-effective neutrino observatory in probing the neutrino flux from 10^{17}eV – 10^{19}eV. This contribution will discuss an interferometric vertex reconstruction technique developed for ARA, taking into account the curved paths traveled by EM radiation in inhomogeneous ice. Preliminary results on the directional reconstruction of an in situ calibration pulser as well as simulated neutrino vertices will be presented.

1 Introduction

The existence of cosmic rays carrying energies up to over 10^{21}eV poses a mystery in modern astrophysics [1]. No confirmed astrophysical objects accelerate cosmic rays to such energies. However, these ultra-high-energy cosmic rays (UHECRs) are expected to engage in photopion production with cosmic microwave background photons and give rise to UHE cosmogenic neutrinos [2–4]. This neutrino flux opens a window to profound physics such as cosmic ray composition, UHECR source evolution, and so on. Detection of such neutrinos should aid physicists in solving the UHECR puzzle.

We seek to search for these neutrinos via the Askaryan emission in radio frequency (RF) transparent ice. This technique allows for a cost-effective neutrino observatory sensitive to UHE neutrinos. First predicted by G.A. Askaryan [5], the Askaryan signal arises from coherent emission due to charge asymmetry built up in a neutrino-induced shower in dense media. The angular profile of this broadband radiation is centered at the Cherenkov angle, and the total power grows quadratically with neutrino energy. In ice, this radiation peaks from 0.1 to 1GHz. Additionally, Antarctic ice near the South Pole has a measured RF attenuation length of $O(1\text{ km})$ [6], making it ideal for radio-Cherenkov based UHE neutrino experiments, such as the Askaryan Radio Array (ARA) [6]. ARA is designed to cover an area of $O(100\text{ km}^2)$, and is able to measure the timing, the frequency content, and the polarization of the said Askaryan signal. Such rich information, in principle, grants us access to neutrino vertex position, emission angle, and travel direction. Here we describe an approach to reconstruct the vertex position.

* e-mail: mlu27@icecube.wisc.edu

2 Experimental setup

An ARA station hosts four measurement strings and two calibration strings down to 200m depth. Each measurement string has four broadband antennas. Two are vertically polarized (Vpols), and the other two are horizontally polarized (Hpols). The calibration strings sit roughly 40m away from the station center (geometric center of all 16 measurement antennas). One selected calibration pulser antenna (calpulser) transmits at 1 PPS to a station at a time with a fixed polarization. We run a simple multiplicity trigger, where 3 power excursions out of 8 Vpol or Hpol channels within a pre-set duration form a trigger. Once a trigger occurs, digitization at 3.2GS/s yields a waveform of several hundred nanoseconds for each channel. A collection of such waveforms is an event. For details, see [6, 7].

3 Reconstruction method

Relative differences in signal arrival times between antennas are needed to calculate the cross correlation (CC) values for interferometry. The Antarctic ice has a depth-dependent index of refraction. This causes the rays of electromagnetic waves traversing the ice to bend downward. A raytracing code named RaySolver in the ARA simulation package AraSim [8] implements a semi-analytic algorithm to derive these rays. A significant consequence of the ray-bending effect is that some region of ice becomes invisible to a receiver. This region is called the firn shadow. Note also that for a source outside the firn shadow, more than one path to the receiver is possible. These phenomena are showcased in Fig. (1). In the next section, we present Radiospline, a software library for fast and precise signal arrival time determination taking into account ray-bending effects.

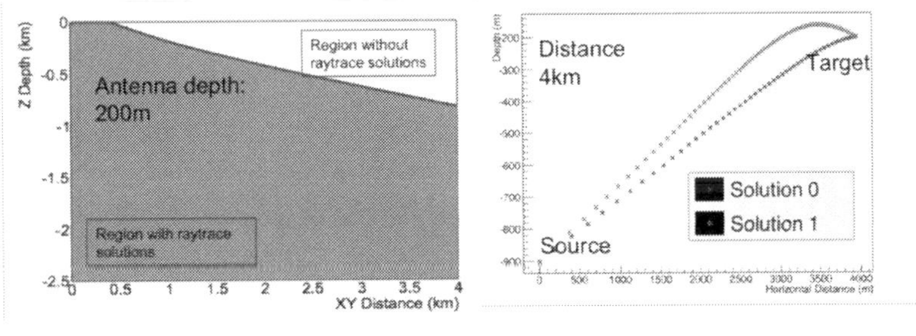

Figure 1. Left: Region where a ray path can connect from source to target is shaded. Firn shadow is blank [8]. Right: Ray paths calculated by RaySolver. The horizontal distance between the source and target is 4km. Solution 0 marks the direct ray and Solution 1 marks the indirect ray.

3.1 Radiospline and interferometry

Radiospline interpolates signal arrival times from a table of RaySolver outputs. Assuming azimuthal symmetry of ice, choosing cylindrical coordinates reduces the table to 3 dimensions: source depth, target depth, and the horizontal separation between the two. The source position moves vertically between [-2800m, 100m] and horizontally between [0m, 5000m], while the target depth varies within [-250m, -20m], all in steps of 10m. The table records arrival times from all geometric combinations. Basis spline fits are used to interpolate these time values. Basis splines are piece-wise polynomials that can be placed in linear combinations to form a fitting curve on a multi-dimensional set of data points. In practice, for a putative vertex location, Radiospline first determines if the vertex is in the firn shadow. If so, no arrival time is computed. If otherwise, source-in-ice or source-in-air tables are interpolated to give arrival times. The precision of the interpolation is $O(100\,\text{ps})$, while the computation time compared to direct use of RaySolver is reduced by a factor of >500. Note that for a source-target pair, only the direct ray is tabulated in Radiospline in this work.

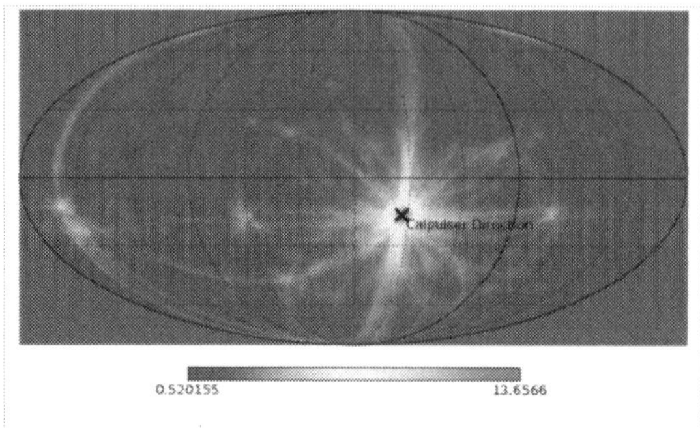

Figure 2. Reconstruction skymap of 2013 ARA3 run #673 cal event #2. Color represents the coherence value [arb. unit]. The black cross marks the known direction of the calpulser in operation at the time of event recording.

To perform interferometry, we start from a Healpix [9] skymap assumed to be surrounding our station center at a fixed radial distance. We treat each pixel center as a putative vertex position. Thus, each pixel gives a set of signal arrival times for the antennas from Radiospline. These arrival times are converted to relative time delays between channels. We then construct CC functions from the waveforms of every pair of same-polarization channels. According to the delays of each pixel, CC values are obtained and projected back to the skymap. We define the "coherence" to be the sum of all CCs projected on the same pixel, and build a skymap where each pixel holds a coherence value as proxy to the aligned power. The pixel with the maximal coherence gives the reconstructed direction. All skymaps presented here have 196,608 pixels. Mean pixel spacing is 0.4581°. One can repeat this process for a set of skymaps at distinct radial distances from the station center and attempt full vertex position reconstruction including the vertex distance. For this, 50 skymaps are generated at radial distances from 100m to 5000m in 100m steps. Note that all following reconstruction uses only Vpol channels.

4 Application to Data

4.1 Calpulser reconstruction

Here we fix the skymap radial distance to 42m, the known calpulser distance. In general we are able to reconstruct the calpulser direction to $\sim < 1°$ accuracy. An example event is shown in Fig. (2). The coherence peak is prominent and aligns with the true direction marked by a black cross. Table 1 lists the mean reconstructed zenith and azimuth for a month of filtered calpulser events versus the true values. We conclude that the reconstruction of the Vpol calpulser is successful.

Table 1. Comparison of the reconstructed zenith and azimuth means to the true values of the ARA3 calpulser.

	True	Reco Mean	Difference
Zenith [°]	106.3	106.2	-0.1
Azimuth [°]	63.8	63.5	-0.3

4.2 Simulation reconstruction

The simulated data are 10^{18}eV neutrinos, where indirect rays are excluded. An example event is shown in Fig. (3). The skymap exhibits a clear hotspot that indicates well the vertex direction. The reconstructed [zenith, azimuth] is [102.94°, 222.89°], with the true direction [102.74°, 222.69°].

A set of skymaps at different radial distances can also be constructed as described earlier. 2658 triggered AraSim events are reconstructed this way. We apply a quality cut on the maximal coherence

DOI: 10.1051/epjconf/201713505003

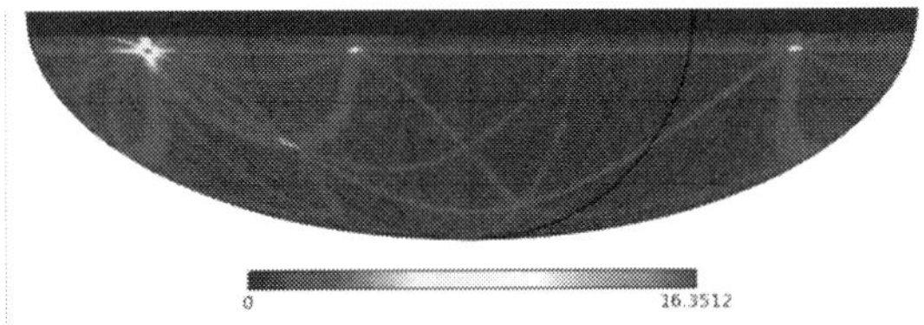

Figure 3. A 10^{18}eV simulated neutrino event reconstructed on the lower half of the skymap. Simulation-true vertex distance of 2581m was assumed in the construction of the skymap. The hotspot gives the reconstructed vertex direction.

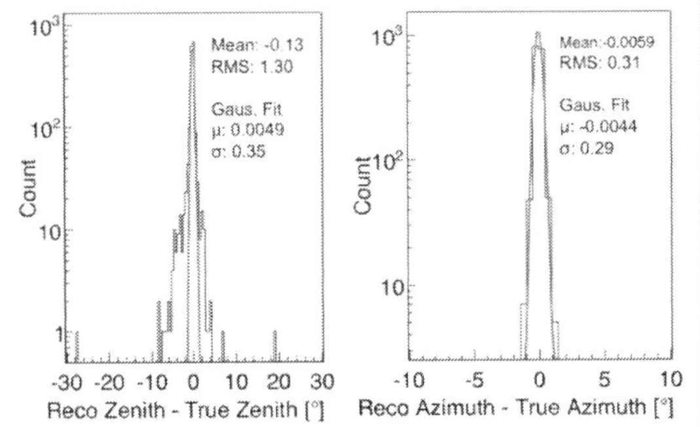

Figure 4. 10^{18}eV simulated neutrino events vertex direction reconstruction. Plots show the difference of reconstructed zenith/azimuth from true values after maximal coherence value > 5.0 cut. The fitting curves represent Gaussian fits to the histograms.

and demonstrate that in principle a zenith resolution of 1.3° and a azimuth resolution of 0.31° can be achieved (Fig. (4)). The cut efficiency is being studied. We should also mention that this reconstruction fails to resolve the true vertex distance. This, combined with ray-bending, constrains our ability to reach sub-degree zenith angle resolution. However, preliminary studies with expanded antenna spacing show success in vertex distance reconstruction up to several hundred meters.

5 Summary and future plan

An interferometric reconstruction for ARA using the Radiospline library is developed. We demonstrate the ability to reconstruct the calpulser direction with ARA3 data, and we are able to reconstruct the direction of simulated events to ~ 1° angular resolution under stated quality cuts. If implemented on a parallel-computing framework of GPUs, the interferometric techniques can potentially be used as online filters for background rejection. Moving forward, we plan to finalize the cuts and fully unblind at least two years (2013–14) of ARA2 and ARA3 data for a diffuse neutrino search. Possible ways to improve the reconstruction include updating Radiospline to describe indirect rays and applying more sophisticated methods, such as machine learning algorithms, to extract information from skymaps.

References

[1] J. Abraham et al. (Auger Collaboration), Phys. Lett. B **685**, 239 (2010)
[2] K. Greisen, Phys. Rev. Lett. **16**, 748 (1966)
[3] G. T. Zatsepin and V. A. Kuzmin, JETP Lett. **4**, 78 (1966)
[4] V. S. Beresinsky, G. T. Zatsepin, Phys. Lett. B **28**, 423 (1969)
[5] G. A. Askaryan, JETP **14**, 441 (1962)
[6] P. Allison et al. (ARA), Astropart. Phys. **35**, 457, (2012)
[7] P. Allison et al. (ARA), Phys. Rev. D **93**, 082003 (2016)
[8] P. Allison et al. (ARA), Astropart. Phys. **70**, 62 (2015)
[9] Gorski, K. M. et al., ApJ **622**, 759 (2005)

DOI: 10.1051/epjconf/201713505004

Background Rejection in the ARA Experiment

Carl Pfendner[1,★] for the ARA collaboration

[1] *Dept. of Physics and CCAPP, The Ohio State University, 191 W. Woodruff Ave., Columbus, OH 43210, USA*

Abstract. The Askaryan Radio Array (ARA) is a radio frequency observatory under construction at the South Pole that is searching for ultrahigh energy neutrinos via the Askaryan effect. Thermal fluctuations currently dominate the trigger-level background for the observatory and anthropogenic sources also introduce a significant source of noise. By taking advantage of the observatory's regular geometry and the expected coincident nature of the RF signals arriving from neutrino-induced events, this background can be filtered efficiently. This contribution will discuss techniques developed for the ARA analyses to reject these thermal signals, to reject anthropogenic backgrounds, and to search for neutrino-induced particle showers in the Antarctic ice. The results of a search for neutrinos from GRBs using the prototype station using some of these techniques will be presented.

1 Introduction

The Askaryan Radio Array (ARA) aims to measure the flux of ultra-high energy (UHE) neutrinos above 10^{17} eV. While UHE neutrinos are so far undetected, they are expected both as decay products from the GZK process [1], as first pointed out by Berezinsky and Zatsepin [2], and directly from astrophysical sources, such as Gamma Ray Bursts (GRBs).

The interaction of a UHE neutrino in dense media induces an electromagnetic shower which in turn creates impulsive radiofrequency (RF) Cherenkov emission via the Askaryan effect [3] In radio transparent media, these RF signals can then be observed by antenna arrays read out with $\sim$ GHz sampling rates.

Due to the $\sim$ 1 km radio attenuation lengths in ice [4, 5], radio arrays have the potential to view the 100s of km^3 of ice necessary to reach the sensitivity to detect $\sim$ 10 events per year from expected UHE neutrino fluxes. Next-generation detectors are under construction aiming to reach the 100s of km^3 target volume of ice. The Askaryan Radio Array (ARA) [4] is one such detector being deployed in the ice at the South Pole and the results for a search from neutrinos from Gamma Ray Bursts using data from a prototype station of this detector are presented in these proceedings. However, the radio background is considerable and, in order to perform such a search, methods to reject thermal and anthropogenic sources of noise must be developed.

★e-mail: pfendner.1@osu.edu

DOI: 10.1051/epjconf/201713505004

2 ARA Testbed Signal Algorithms

A number of background rejection methods have been previously developed for the ARA Testbed station [6]. Among these methods, two primary techniques were used to search for a diffuse flux of neutrinos: (1) requirements placed on the contents of an interferometric map (see Fig. (1)) and (2) a cut on the correlation between the peak signal strength and the cross-correlation of the aforementioned map (see Fig. (2)).

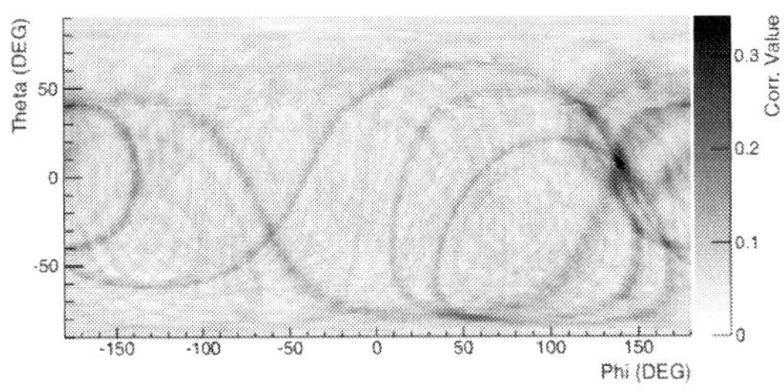

Figure 1. An interferometric map produced for a calibration pulser event. The peak of the map corresponds to the correct location of the calibration pulser in this 30-m radius map. The two criterion for the cuts based on this map are (1) that the peak is well-defined and (2) no other part of the map shows a comparable correlation strength. These are found by calculating the area of the 85% contour around the peak and the total area of comparable 85% strength on the rest of the map. The area around the peak A_{peak} must be less than a given cut value and the ratio of $A_{total}A_{peak}$ must be less than a different given value.

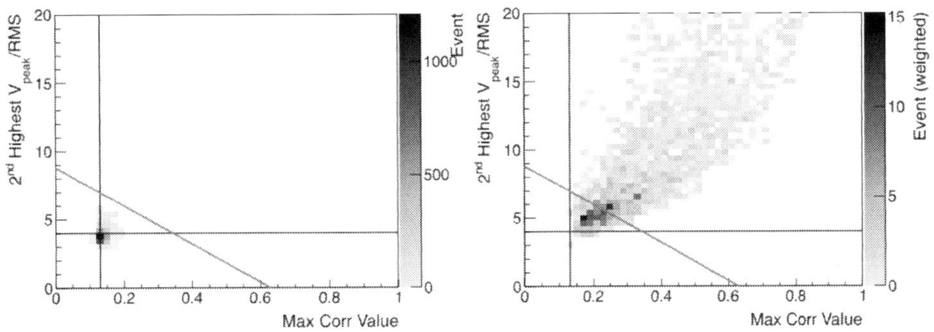

Figure 2. This pair of plots shows the application of the Peak/Correlation cut. On the left is the set of events from the 10% burn sample of the full Testbed data set remaining after all other cuts have been applied. On the right is a set of 10^{18} eV simulated events under the same conditions. The second highest V/RMS value and the peak correlation value on the interferometric map are found for each event. The cut is performed by accepting only those events above the diagonal line on the plot. The position of this line is found by optimizing for the result that gives the best expected limit.

3 ARA Testbed GRB Search

As GRBs are a potential source of UHE neutrinos, we performed a search for neutrino events correlated with the Gamma-ray Coordinates Network (GCN)[7] catalog of GRBs, as found in the GRB-web database [8], occurring during the operational period of the ARA Testbed. Of the ~ 500 GRBs in the catalog, we limited our search to the 57 GRBs that coincide with known active high-quality data-taking periods and were also found within the ARA Testbed's field of view (see Fig. (3)). Since GRBs are localized in time, we limited our signal search windows to the 10 minutes centered on each GRB event, using the hour before and the hour after the GRB as a background analysis period. We used a two-stage unblinding procedure. First 10% of the data in the background period was analyzed and the cuts optimized against these events. In the first unblinding, we apply the optimized cuts to the remaining 90% of the background analysis period and compare these results to the expectation from the 10% set. After confirmation of the expected results from the background period search, we then perform the second unblinding and apply the cuts to 100% of the events in the signal search window.

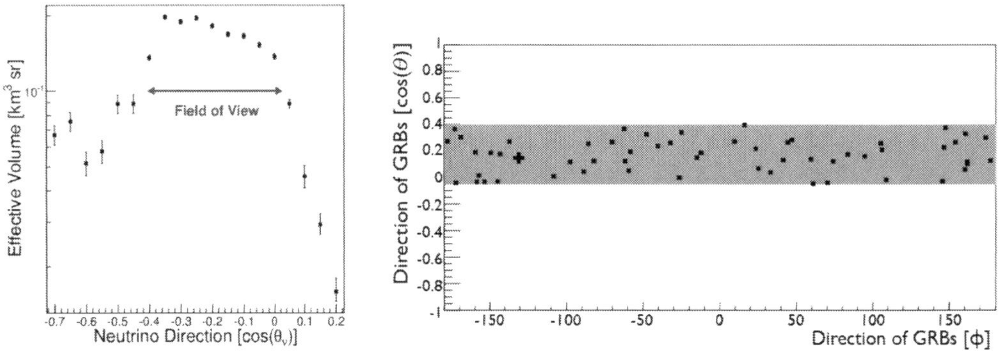

Figure 3. Left: Effective volume of the ARA Testbed as a function of the zenith angle (θ_ν) of the neutrino travel direction with a neutrino energy of $10^{7.5}$ GeV. The field of view is defined as the Full Width Half Maximum (FWHM) of the effective volume, which is $-0.4 < \cos\theta_\nu < 0.05$. This field of view covers ~ 20% of the sky. A vertically up-going neutrino has $\cos\theta_\nu = 1$. Right: The distribution map of 57 selected GRBs in Testbed local coordinates. The blue band in the map is the field-of-view cut range defined in the left figure.

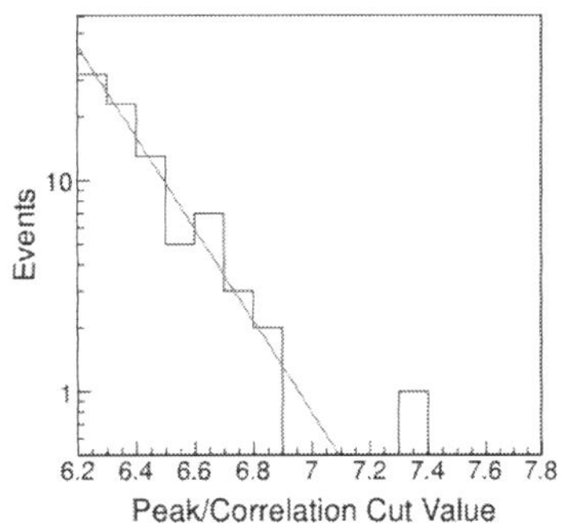

Figure 4. The differential distribution of events found in the background analysis windows of the 10% data set as a function of Peak/Correlation Cut Value after all other cuts have been applied. This distribution is fitted with an exponential function (diagonal red line) which is used to extrapolate the number of expected background events for a higher Peak/Correlation Cut Value. The optimized value is 7.6.

For each GRB, we use the observed values found in the catalog to model the expected neutrino fluence using the NeuCosmA simulation package [9]. We optimize our cuts for the best expected limit based on a simulated neutrino sample generated from the average flavor-weighted flux of all 57 GRBs examined and the data collected during 10% of the background window. Using a fit to the differential distribution of events in the background set, we find the expected number of background events as a function of cut value (see Fig. (4)). After optimization we found that all three cuts that were adjusted were loosened (see Table 1) when compared with the diffuse search performed previously. The Delay

DOI: 10.1051/epjconf/201713505004

Table 1. Comparison of cut parameter values of the analysis. See text for details.

Cut	Delay Difference Cut	Reconstruction Quality Cut		Peak/Correlation Cut
Parameter	On/Off	A_{peak}	A_{total}/A_{peak}	Peak/Corr. Cut Value
Diffuse Neutrino Search	On	< 50 deg^2	< 1.5	> 8.8
GRB Neutrino Search	Off	< 140 deg^2	< 16.4	> 7.6

Difference cut was only optimized in whether the cut was used or not; this was due to the expectation that this cut was largely redundant based on the results of the diffuse search.

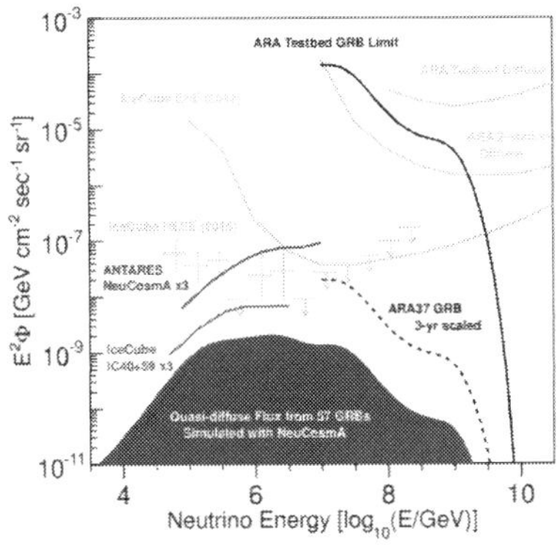

Figure 5. The ARA-Testbed quasi-diffuse all-flavor flux limit. This is the first quasi-diffuse flux limit above 10^{17} eV. Although the limit is not yet comparable to the limits placed at lower energies, applying scaling factors to the result predicts that a GRB quasi-diffuse limit from the full 37-station ARA will be competitive. Additional improvements at the trigger and analysis levels may improve these results even further.

4 Conclusions

Using interferometric cross-correlation maps and signal strength parameters, we have developed a number of effective background rejection techniques for the ARA detector. These methods are flexible enough to be adapted to perform multiple neutrino searches including a diffuse neutrino search and a search for neutrinos correlated with GRBs. Further development of these methods and the more regular geometry of deep stations will provide even greater benefits to further analyses.

References

[1] K. Greisen, Phys. Rev. Lett., 16, 748 (1966); G. T. Zatsepin and V. A. Kuzmin, JETP Lett., 4, 78 (1966)

[2] V. S. Berezinsky and G. T. Zatsepin, Phys. Lett., B28, 423 (1969); V. S. Berezinsky and G. T. Zatsepin, Sov. J. Nucl. Phys., 11, 111 (1970)

[3] G. A. Askaryan, JETP, 14, 441, (1962); G. A. Askaryan, JETP, 21, 658 (1965); P. Gorham *et al.*, Phys. Rev., 62, 8590 (2000); D. Saltzberg *et al.*, AIP Conf. Proc., 579, 225 (2001); P. W. Gorham *et al.*, Phys. Rev., D72, 023002 (2005); P. W. Gorham *et al.*, Phys. Rev. Lett., 99, 171101 (2007)

[4] Allison, P. *et al.*, Astropart.Phys., 35, 457 (2012)

[5] Barwick, S. and Besson, D. and Gorham, P. and Saltzberg, D., J. Glaciol., 51, 231 (2005)

[6] P. Allison *et al.*, Astroparticle Physics, 70, 0, 62 (2015)

[7] http://gcn.gsfc.nasa.gov

[8] http://icecube.wisc.edu/science/tools

[9] P. Baerwald, S. Hummer, and W. Winter, Phys. Rev, **D83**, 067303 (2011).

Phenomenology of transition radiation at radio frequencies from ultrahigh-energy showers

Jaime Alvarez-Muñiz[1,⋆], *Pavel* Motloch[2], *Paolo* Privitera[2], and *Enrique* Zas[1]

[1]*Departamento de Física de Partículas & Instituto Galego de Física de Altas Enerxías, Univ. de Santiago de Compostela, 15782 Santiago de Compostela, Spain*
[2]*Kavli Institute for Cosmological Physics & Dept. of Physics, Univ. of Chicago, Chicago, IL 60637, USA*

Abstract. We explore transition radiation at MHz-GHz frequencies as a possible way to detect ultrahigh-energy (UHE) particles. We have developed a general method to calculate transition radiation that extends the well-known Zas-Halzen-Stanev (ZHS) algorithm. We have applied it to the characterization of the frequency and angular properties of the electric field from high-energy showers crossing the boundary between two media. We discuss the potential of transition radiation for the detection of UHE particles in different situations of experimental interest.

1 Introduction

The origin of UHE cosmic rays (UHECRs) with energies above $\sim 1\,\text{EeV} = 10^{18}$ eV is a long-standing question in astroparticle physics [1]. CRs may interact with matter and radiation at the accelerating source, or along their path towards Earth and produce UHE neutrinos. These travel undeflected through the universe and may provide important clues on the origin of UHECRs. A promising detection technique for UHE neutrinos consists in observing the coherent Askaryan radio signal [2] at MHz-GHz frequencies from the neutrino-induced shower in a dense and radio-transparent medium [3]. In this contribution we explore transition radiation (TR) [4] at radio frequencies as a possible way to detect UHE particles. TR is emitted by charged particles when crossing the boundary between two media with different indices of refraction. There are various situations of experimental interest where TR may play a relevant role in detection, for instance Earth-skimming neutrinos inducing a shower inside the Earth's crust that emerges from ground into air, or an UHECR developing a particle shower in the atmosphere and arriving at ground [5]). We have developed a general method to calculate the radiation from a shower crossing the boundary between two media [6] that can be applied to the experimental situations described before as well as many others.

2 Calculation of the radiation

Our calculation of radio emission from particle showers crossing the boundary between two media is based on the well-known Zas-Halzen-Stanev (ZHS) algorithm [7]. A shower is modeled as a collection of charged particles whose trajectories are divided in small tracks, along which the particle travels

⋆e-mail: jaime.alvarezmuniz@gmail.com

at uniform velocity $\boldsymbol{v}$. From the position and time of the endpoints of these tracks, the frequency spectrum of the electric field from a track is derived using Maxwell's equations [7, 8]:

$$\boldsymbol{E}(\omega, \boldsymbol{x}) = \frac{e\mu_r}{2\pi\epsilon_0 c^2} i\omega \frac{e^{ikR}}{R} e^{i(\omega - \boldsymbol{k}\cdot\boldsymbol{v})t_1} \boldsymbol{v}_\perp \left[\frac{e^{i(\omega-\boldsymbol{k}\cdot\boldsymbol{v})\delta t} - 1}{i(\omega - \boldsymbol{k}\cdot\boldsymbol{v})} \right]. \tag{1}$$

where R is the distance from the track to the observer assumed to be large enough for the Fraunhofer regime to be valid. $\boldsymbol{k}$ points in the direction from the track to the observer with $k = \omega/c'$, where ω is the frequency and $c' = c/n$ the speed of light in the medium with n the refractive index of the medium. $\boldsymbol{v}_\perp = -\hat{\boldsymbol{k}} \times (\hat{\boldsymbol{k}} \times \boldsymbol{v})$ is the component of $\boldsymbol{v}$ in the plane perpendicular direction to $\hat{\boldsymbol{k}} = \boldsymbol{k}/|\boldsymbol{k}|$. t_1 and $t_2 = t_1 + \delta t$ are the absolute times at which the particle passes through the starting point and endpoint of the track, respectively. μ_r is the relative permeability of the medium. The total electric field at the observer's location is calculated by superposition of the electric field from each track, with relative phase shifts due to different spacetime positions properly taken into account. In the ZHS formula each of the endpoints of the track contributes a term to the electric field. Radiation in a particle trajectory arises from the non-cancellation of the field from adjacent tracks due to their different speeds and orientations w.r.t. the observer. This approach is completely general and it has been succesfully applied to the prediction of Askaryan radiation in particle showers [7] as well as Cherenkov radiation from individual charged particles [8].

In the extended ZHS algorithm (ZHS-TR) for the calculation of radiation from a shower crossing the boundary between a medium 1 and a medium 2, three different types of particle segments are considered (without loss of generality the observer is assumed to be located in medium 2). For tracks fully contained in the first medium, the only contribution comes from the electric field (Eq. 1) emitted into a direction such that when refracted at the boundary it reaches the observer in the second medium (the "transmitted contribution"). For tracks fully contained in the second medium, the "direct contribution" given by Eq. (1) describing the electric field radiated into the observer's direction is calculated, as well as the "reflected contribution" that comes from radiation emitted into a direction such that it is reflected off the boundary into the observer's direction. The appropriate Fresnel coefficients are used for reflection and refraction when necessary. For tracks crossing the border each track is split at the interface into two sub-tracks, as in [9]. Each sub-track is then fully contained in one medium and Eq. (1) can be applied to it. There are thus two endpoints that lie just at the interface. Since they belong to sub-tracks in different media they do not cancel exactly. This partial non-cacellation is the numerical origin of TR. Finally, the electric field produced by the entire shower is obtained from the superposition of the individual contributions of all particle tracks and sub-tracks. The ZHS-TR algorithm has been implemented in the ZHS code for an electromagnetic shower starting in a medium, separated by a planar boundary from a second medium where the observer is located, and all results presented in the following sections correspond to this geometry. If the observer is located in the first medium the roles of the two media must be reversed and two contributions (direct and reflected) are expected from medium 1 and only one (transmitted) from medium 2.

3 Radiation in showers starting in ice and crossing into air

In this section we investigate the properties of the radiation emitted by an electromagnetic shower transitioning from homogeneous ice ($n_1 = 1.78$) to homogeneous air ($n_2 = 1.0003$). The angular distribution of the electric field radiated by a 100 TeV vertical shower simulated with ZHS-TR is shown in Fig. 1. The starting point of the shower in ice was chosen so that the shower maximum occurs approximately at the ice–air boundary. A notable feature is the weak angular dependence of the emitted radiation, when compared to the strongly-beamed coherent Cherenkov radiation [7]. The fluctuations around the mean value of the field are due to the fine structure of the shower development in air. We also show in Fig. 1 the contributions to the total electric field (direct, transmitted and reflected). The direct contribution from particle tracks in air is dominant at all angles, and presents a peak at $\sim 1-2^\circ$ mainly due to Cherenkov radiation produced in air (Cherenkov angle $\sim 1.4^\circ$). The reflected contribution is negligible except for the largest angles, where it becomes comparable to the direct contribution as expected. However, the two contributions cancel each other for an observer looking orthogonally to the shower axis, since a π phase shift is gained upon reflection. The transmitted contribution from particles in ice is sub-dominant for a vertical shower because the coherent Cherenkov radiation from the Askaryan effect in ice undergoes total internal reflection at the ice–air boundary plane (the Cherenkov angle is larger than the critical angle for $n_1=1.78$) and cannot reach the observer. This component can become significant for inclined showers, see Figs. 11-14 in [6], when coherent Cherenkov radiation is refracted from ice into air. The frequency spectrum of the radiation is also shown in Fig. 1. For this

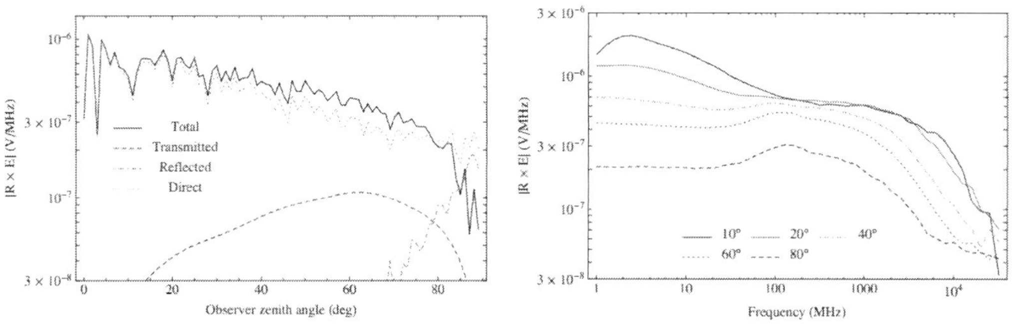

Figure 1. (Left) Angular distribution of the electric field at 1 GHz radiated by a 100 TeV vertical shower crossing from ice to air: total field (black), direct (green short dashes), reflected (red dot-dashes) and transmitted (blue dashes) contributions. The observer is located in air at different angles w.r.t. shower axis. (Right) Frequency spectrum of the electric field radiated by a 100 TeV vertical shower crossing from ice to air, at 10° (black solid), 20° (red solid), 40° (green dot dashes), 60° (blue short dashes), and 80° (black long dashes) zenith angle of observation (top to bottom). Each curve is an average of a hundred showers. Figs. taken from [6].

geometry TR is coherent up to $\sim$ 1 GHz over a wide range of angles, the reason being that showers developing in ice are rather compact in size (Molière radius $\sim$ 10 cm) at the boundary. We have also found that the electric field scales linearly with the charge excess (electrons minus positrons) flowing through the boundary. As a result, its magnitude depends on the stage of evolution of the shower when crossing the boundary [6].

4 Radiation in showers starting in air and crossing into ice

We now consider electromagnetic showers transitioning from air into ice. The angular distribution of the electric field radiated by a 10 TeV vertical shower simulated with ZHS-TR is shown in Fig. 2. The observer is located in ice at different angles w.r.t. shower axis and at two azimuthal angles (see caption of Fig. 2). The central peak around 0° corresponds to the Cherenkov radiation in air refracted into ice that dominates the emission in that angular range, while the two peaks at $\sim \pm 56^\circ$ are due to the direct emission of Askaryan radiation in ice peaking at the Cherenkov angle. The frequency spectrum of the radiation is shown in Fig. 2. Showers developing first in air are rather spatially extended in the lateral dimension (Molière radius in air $\sim$ 1 km) and for that reason the electric field increases linearly with frequency up to a few MHz above which full coherence is lost for observers around $\sim 0^\circ$

DOI: 10.1051/epjconf/201713505005

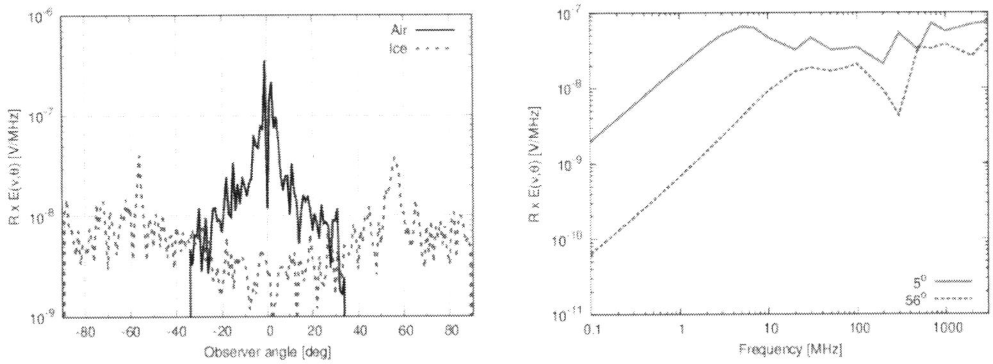

Figure 2. (Left) Angular distribution of the electric field at 1 GHz radiated by a 10 TeV vertical shower crossing from air to ice: total field from the shower in air transmitted into ice (black) and direct+reflected field from the shower in ice (blue dashed). The observer is located in ice at different angles w.r.t. shower axis and at azimuthal angles 0° (positive values of observer angle) and +180° (negative values). (Right) Frequency spectrum of the electric field radiated by a 10 TeV vertical shower crossing from air to ice, at 5° (red solid) and 56° (green dashed) observation angles.

where the transmitted radiation dominates the emission (see left panel of Fig. 2). Observers at angles around the Cherenkov angle in ice ~ 56° mainly see the direct radiation from the shower in ice, whose spectrum is known to increase linearly with frequency up to a few hundreds of MHz [7], above which full coherence is lost.

5 Conclusions

We have developed a general method based on the ZHS formalism to calculate the electric field radiated by the charged particles in a high-energy shower developing in two different media. Transition radiation naturally arises from this procedure, in addition to the coherent Askaryan radiation produced by the shower particles within each medium. Our so-called ZHS-TR algorithm has general applicability and could be used in various situations of experimental interest. The ZHS-TR algorithm could be also implemented in other shower Monte Carlo codes such as ZHAireS [10] that can handle showers induced by protons, heavier nuclei and neutrinos in the atmosphere or in dense media.

References

[1] A.A. Watson, Rep. Prog. Phys. **77**, 036901 (2014).
[2] G.A. Askar'yan, Soviet Physics JETP **14**, 441 (1962); Soviet Phys. JETP **21**, 658 (1965).
[3] A. Connolly and A. Vieregg, arXiv:1607.08232v1 [astro-ph.HE].
[4] V. Ginzburg and I. Frank, J. Phys. (USSR) **9**, 353 (1945).
[5] K.D. de Vries *et al.* Astropart. Phys. **74**, 96 (2016).
[6] P. Motloch, J. Alvarez-Muñiz, P. Privitera, E. Zas Phys. Rev. D **93**, 043010 (2016).
[7] E. Zas, F. Halzen, T. Stanev, Phys. Rev. D **45**, 362 (1992).
[8] D. García-Fernández *et al.*, Phys. Rev. D **87**, 023003 (2013).
[9] C.W. James, H. Falcke, T. Huege, M. Ludwig, Phys. Rev. E **84**, 056602 (2011).
[10] J. Alvarez-Muñiz, W.R. Carvalho Jr., M. Tueros, E. Zas, Astropart. Phys. **35**, 287 (2012); J. Alvarez-Muñiz, W.R. Carvalho, E. Zas, Astropart. Phys. **35**, 325 (2012).

Modeling the radar scatter off of high-energy neutrino-induced particle cascades in ice

Krijn D. de Vries[1,⋆], *Nick* van Eijndhoven[1], *Aongus* O'Murchadha[2], *Simona* Toscano[1], and *Olaf* Scholten[1,3]

[1] *Vrije Universiteit Brussel, Dienst ELEM, IIHE, Pleinlaan 2, 1050 Brussels, Belgium*
[2] *UW Wisconsin, Madison, Wisconsin*
[3] *University of Groningen, KVI-Center for Advanced Radiation Technology, 9747 AA Groningen, The Netherlands*

Abstract. We discuss the radar detection method as a probe for high-energy neutrino induced particle cascades in ice. In a previous work we showed that the radar detection techniqe is a promising method to probe the high-energy cosmic neutrino flux above PeV energies. This was done by considering a simplified cascade geometry and scattering model. In this article we discuss the scattering in more detail. We provide a model for the radar cross-section based on the induced plasma properties, and discuss the angular dependence of the scatter.

1 Introduction

In [1], we discussed the radar detection technique as a probe for high-energy neutrino-induced particle cascades in ice. It was shown that if one is able to scatter efficiently, this technique should allow for the cost-efficient detection of neutrino-induced particle cascades above a few PeV. It follows that the radar detection method is a promising technique to cover the currently existing sensitivity gap between a few PeV where the IceCube neutrino observatory runs out of statistics [2], and a few EeV where the currently operating Askaryan radio detectors start to gain sensitivity [3, 4].

The main scatter occurs off of the over-dense ionization plasma left behind when a neutrino induced cascade traverses through the medium. To model the effective scattering cross-section, in [1], several approximations were made. The average free charge density within a tube of 2 cm radius was considered to estimate the plasma frequency. The scattering cross-section itself was consequently obtained from an emperically derived equation [5, 6]. Furthermore, skin-effects were ignored where the scatter was assumed to be isotropic.

In this article, we consider a more detailed model for the radar cross-section. Assuming radial symmetry, a two-dimensional model of the ionization plasma is given. Having the detailed free charge distribution, the scattering efficiency for layers of equal density will be considered by taking into account for the skin effect. Furthermore, the scattering itself is discussed where it is argued that instead of isotropic scattering a slit-like interference pattern can be expected.

⋆e-mail: krijndevries@gmail.com

2 The plasma

Following [1], the starting point for the plasma model is the high-energy particle cascade which develops inside the ice. The electromagnetic cascade development as a function of depth $X(\mathrm{g/cm^2})$ can be parameterized following [7, 8],

$$N(X) = \frac{0.31\ \exp[(X/X_0)(1-1.5\ln s)]}{\sqrt{\ln(E/E_{\mathrm{crit}})}}, \tag{1}$$

where the radial particle distribution can be parameterized following [9],

$$w(r) = \frac{\Gamma(4.5-s)}{\Gamma(s)\Gamma(4.5-2s)}\left(\frac{r}{r_0}\right)^{s-1}\left(\frac{r}{r_0}+1\right)^{s-4.5}. \tag{2}$$

Here $X_0 = 36.08\ \mathrm{g/cm^2}$ denotes the electron interaction depth, $E_{\mathrm{crit}} = 0.0786$ GeV gives the critical energy for electrons in ice, and E denotes the energy of the primary cascade-inducing particle. Additionally the shower age is given by $s(X)$, where $s = 1$ denotes the shower age at shower maximum. The radial distance from the shower axis is given by r, where $r_0 = 7$ cm for ice. For more detailes we would like to refer to Appendix A of [1].

The energy deposit from the high-energy cascade by ionization of the medium is large. Typically 90% of the cascade energy is finally deposited by ionizing the medium. The ionization profile is modeled considering an ionization loss of 2 $\mathrm{MeV/g/cm^2}$ for a minimum ionizing particle. The ionization loss scales logarithmically with the energy of the ionizing particle and is therefore considered to be constant for the high-energy particles in the cascade front. Numerically it can easily be checked that the total energy loss due to ionization given by, $E^{ion}_{loss} = \int N(X)E_{ion}\,dX$, leads to an energy deposit of roughly 90%. The ionization charge density is consequently obtained by assuming an ionization energy for ice equal to $E^{ice}_{ion} = 20$ eV.

3 Over-dense scattering

For over-dense scattering to occur, we need to fulfill the condition that the detection frequency is below the plasma frequency $\nu_d < \nu_p$. The plasma frequency on itself directly depends on the free charge density and mass, $\nu_p = 8980\sqrt{\frac{m_e n_e}{m_p}}$ Hz. Here m_e denotes the electron mass, m_p denotes the effective mass of the plasma constituent, and n_e gives the free charge density. Following [10, 11], two types of plasma can be expected. A rather short lived free electron plasma, and a longer lived free proton plasma. In Fig. (1), we show the two dimensional plasma density profile for a 1 PeV particle cascade. The over-dense region for the a proton-like plasma probed at $\nu_d = 50$ MHz is indicated by the (white) contour line. It follows that already at 1 PeV energies an over-dense tube with O(cm) radius and a longitudinal extension of O(m) can be expected for a proton-like plasma with O(μs) lifetime. The longitudinal extension of the plasma is given by either the plasma lifetime, $c\tau_p$, or the length of the cascade itself.

Having the density profile, we can now calculate the scattering cross-section. To illustrate the calculation, in this article we consider a simplfied geometry for which the incoming wave is normal, and the polarization is aligned, to the casacade direction. The scattering efficiency is accordingly obtained by considering (over-dense) layers of approximate equal density. Defining the skin depth $\delta_p = c/2\nu_p$, to be the depth in which the intensity falls by e^{-1}, the fraction of the signal absorbed and re-scattered in a single layer i is given by,

$$f^i_{scatt} = (1 - f^{i-1}_{scatt})(1 - e^{-(\Delta x)^i/\delta^i_p}). \tag{3}$$

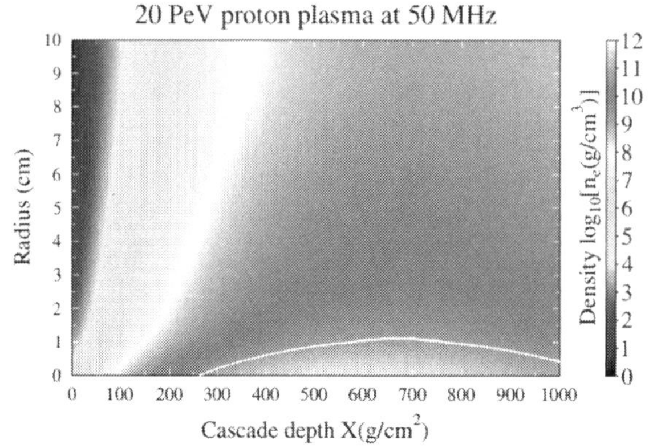

Figure 1. The two-dimensional plasma density profile for a 1 PeV neutrino-induced particle cascade in ice. The over-dense proton-like plasma region probed at ν_d = 50 MHz is indicated by the (white) contour.

Here the factor $(1 - f^{i-1}_{scatt})$ takes into account the part of the signal which has already scattered, $(\Delta x)_i$ denotes the thickness of layer i, and δ^i_p denotes the plasma frequency in the considered layer. The total fraction of the incoming wave intensity which is scattered can now be obtained by,

$$f_{scatt} = \sum_i f^i_{scatt} = \sum_i (1 - f^{i-1}_{scatt})(1 - e^{-(\Delta x)^i/\delta^i_p}). \quad (4)$$

For an incoming wave with normal incidence and its polarization aligned with the plasma, the scattering cross section is given by a folding with the geometrical area A_i of layer i. This is a sigar-like surface, for which its area is estimated by, $A_i = L_i r^{max}_i/2$. The effective over-dense scattering cross section for a wave of normal incidence with its polarization aligned to the cascade and taking into account for skin-effects is thus given by,

$$\sigma^{norm}_{od} = \sum_i f^i_{scatt} A_i = \sum_i A_i(1 - f^{i-1}_{scatt})(1 - e^{-(\Delta x)^i/\delta^i_p}). \quad (5)$$

For the general case, a geometry factor has to be included to take into account the angle with respect to normal incidence and the polarization angle. Furthermore, it should be noted that there are still several unknowns that should be included into the effective radar cross section. So-far we did not consider the free charge collision rate, as well as the detailed plasma lifetime. These quantities are hard to model. Therefore, to determine the radar cross section the scattering efficiency has to be obtained experimentally. Initial experimental work has been performed during a beam-test experiment early 2015, where a first hint for a radar scatter was observed [12]. In the near future, a follow-up experiment is planned to confirm this hint.

4 Diffraction

Where in [1], the scattering was assumed to occur isotropically, in reality this only holds in the limit where the wavelength of the incoming wave is much larger than the scattering surface, $\lambda_{sc} >> A_{scatt}$. The other limiting situation is given when $\lambda_{sc} << A_{scatt}$, for which specular reflection is expected. For efficient scattering, the detection frequency has to be below the plasma frequency, but still large enough to make multiple oscillations within the plasma lifetime which is expected to be in the nanosecond to microsecond range for the ionization plasma considered in this article. This immediately fixes the possible detection regime to MHz-GHz frequencies.

For scattering off the plasma, we have to consider two length scales. The radial extension r is small compared to the detection wavelength up to GHz frequencies, where the longitudinal extension L is of the order of the detection wavelength. Hence, in this situation $\lambda_{sc} \approx A_{scatt}$, and we expect a diffraction pattern which can be approximated by Fraunhofer single slit diffraction. The intensity of the reflected signal therefore becomes an angular dependence given by,

$$f_{diff} = \frac{I(\alpha)}{< I(\alpha) >}, \tag{6}$$

with,

$$I(\alpha) = \text{sinc}^2(\beta), \tag{7}$$

where $\beta = (\pi L/\lambda)\sin\alpha$.

5 Outlook

In this article we consider a more realistic scattering model for the radar reflection off the ionization plasma due to a high-energy neutrino-induced particle cascade in ice. We describe a method to determine the over-dense radar cross-section for the ionization plasma. This method takes into account the skin-effect and is based on the two-dimensional plasma density profile. Furthermore, the emission pattern for a scattered signal is discussed. It follows that this pattern can be approximated by a single slit diffraction pattern. These considerations can be generalized to general scattering geometries, which should allow us to perform a detailed calculation for the radar scattering. The final goal of such a modeling effort is to calculate the sensitivity for a possible future high-energy neutrino detector based on the radar detection technique.

References

[1] K.D. de Vries, K. Hanson, T. Meures, Astropart. Phys. **60**, 25-31 (2015)
[2] IceCube Collaboration, Science **342**, 1242856 (2013)
[3] P. Allison et al., ARA Collaboration, Astropart. Phys. **35**, 457-477 (2012)
[4] A. Nelles et al., ARIANNA Collaboration, these proceedings
[5] P.W. Gorham, Astropart. Phys. **15**, 177 (2001)
[6] J.W. Crispin Jr., A.L. Maffett, Proc. Roy. Soc. A **53**, 833 (1965)
[7] K. Kamata, J. Nishimura, Suppl. Progr. Theoret. Phys. **6**, 93 (1958)
[8] K. Greisen, in: J.G. Wilson (Ed.), Prog. Cosmic Ray Phys., vol. III, North Holland, Amsterdam, 1965, p. 1
[9] K. Werner, K.D. de Vries, O. Scholten, Astroparticle Physics **37**, 5-16 (2012)
[10] J.B. Verberne et al., Nature **272**, 343-344 (1978)
[11] M. Kunst, J.M. Warman, Nature **288**, 465-467 (1980)
[12] K.D. de Vries et al., PoS **ICRC2015**, 1168 (2015)

Phased arrays: A strategy to lower the energy threshold for neutrinos

Stephanie Wissel[1,*], *Jessica* Avva[2,3], *Keith* Bechtol[4,3], *Tyler* Chesebro[5], *Linda* Cremonesi[6], *Anusha* Gupta[6], *Andrew* Ludwig[7,3], *Wesley* Messino[8], *Christian* Miki[9], *Ryan* Nichol[6], *Eric* Oberla[3], *Andrew* Romero-Wolf[10], *David* Saltzberg[5], *Chandler* Schlupf[5], *Nora* Shipp[11,3], *Gary* Varner[9], and *Abigail* Vieregg[7,12,3]

[1] *Physics Dept., California Polytechnic State Univ., San Luis Obispo CA 93407, USA*
[2] *Dept. of Physics, Univ. of California Berkeley, Berkeley, CA 94720, USA*
[3] *Kavli Instit. for Cosmological Physics, Univ. of Chicago, Chicago, Illinois 60637, USA*
[4] *Wisconsin IceCube Particle Astrophysics Center, Univ. of Wisconsin-Madison, Madison, WI 53703, USA*
[5] *Dept. of Physics and Astronomy, Univ. of California Los Angeles, Los Angeles, CA 90095, USA*
[6] *Dept. of Physics and Astronomy, Univ. College London, London, United Kingdom*
[7] *Dept. of Physics, Univ. of Chicago, Chicago, IL 60637, USA*
[8] *Electrical Engineering Dept., California Polytechnic State Univ., San Luis Obispo, CA 93407, USA*
[9] *Dept. of Physics and Astronomy, Univ. of Hawaii, Manoa, HI 96822, USA*
[10] *Jet Propulsion Laboratory, California Instit. of Technology, Pasadena, CA 91109, USA*
[11] *Dept. of Astronomy and Astrophysics, Univ. of Chicago, Chicago, IL 60637, USA*
[12] *Department of Physics, Enrico Fermi Institute, Univ. of Chicago, Chicago, Illinois 60637, USA*

Abstract. In-ice radio arrays are optimized for detecting the highest energy, cosmogenic neutrinos expected to be produced though cosmic ray interactions with background photons. However, there are two expected populations of high energy neutrinos: the astrophysical flux observed by IceCube ($\sim$1 PeV) and the cosmogenic flux ($\sim 10^{17}$ eV or 100 PeV). Typical radio arrays employ a noise-riding trigger, which limits their minimum energy threshold based on the background noise temperature of the ice. Phased radio arrays could lower the energy threshold by combining the signals from several channels before triggering, thereby improving the signal-to-noise at the trigger level. Reducing the energy threshold would allow radio experiments to more efficiently overlap with optical Cherenkov neutrino telescopes as well as for more efficient searches for cosmogenic neutrinos. We discuss the proposed technique and prototypical phased arrays deployed in an anechoic chamber and at Greenland's Summit Station.

1 Introduction: The Phased Array Concept

In-ice radio neutrino experiments are optimized to detect cosmogenic neutrinos [1] produced through the interaction of ultra-high-energy cosmic rays with the cosmic backgrounds at energies above 50 PeV. However, the recent IceCube observations of astrophysical neutrinos with shower energies of a

*e-mail: swissel@calpoly.edu

few PeV [2–4] demonstrate that radio experiments with lower energy thresholds (10-PeV scale) are desirable.

Typical experiments [5, 6] rely on a coincidence trigger referenced to the local thermal noise. The temperature of the surrounding ice determines the minimum signal strength that generates a trigger. Coherently summing the signals from several antennas in a phased array involves delaying and summing their signals by the geometric delay between antennas from a given incoming direction, a technique known as beam-forming or the phased array technique. Recent studies [7] point out that beamforming can reduce the receiver noise, and therefore, the trigger and energy threshold of radio neutrino detectors. The received voltage from coherently summed, in-phase plane-wave like signals, such as those expected from the Askaryan effect, would add linearly, whereas the voltages from incoherent background noise would add in quadrature. With this technique, one could use low gain antennas, such as dipoles or bicones, to achieve an improvement in signal-to-noise ratio (SNR) that scales with the square-root of the number of antennas employed in the phased array.

2 Beamforming in an Anechoic Chamber

In order to validate the phased array technique, we performed several tests in an anechoic chamber using two to three antennas with identical amplifier chains. Several different antennas were used, including folded dipole antennas from Telewave, Inc. (ANT275D and ANT400D), custom broadband (220-500 MHz) dipoles, and broadband bicones and quad-slot antennas developed for the ARA experiment. Signals from the antennas were amplified by 83 dB using dual-stage amplifiers (MITEQ AFS4-00100200-10-15P-4 and Mini-Circuits ZKL-1R5 separated by a 3-dB attenuator) that were powered by pairs of bias tees. The signals were filtered using an Mini-circuits NHP-200 and NLP-450 for all antennas except for the broadband dipoles, which were filtered with a Mini-Circuits NLP-600. The individual antenna voltages as well as the summed beam-formed signal were recorded on a Tektronix MSO5204B oscilloscope. The system noise ($\sim$ 75 K) was measured by recording waveforms from the same amplifier chains, but with terminated inputs. For impulsive beamforming tests (Section 2.2 and Section 2.3), we also split the signals before the oscilloscope and combined them into the third channel on the oscilloscope.

2.1 Thermal Noise Correlations

The expected improvement in signal-to-noise ratio going from individual antennas to a phased array of antennas is predicated on the assumption that thermal noise in adjacent antennas is uncorrelated. For completely overlapping antennas, the 250 K thermal noise radiating from the ice should be perfectly correlated, whereas antennas separated by a wavelength or more should receive completely uncorrelated noise. Smaller separation distances between antennas reduces the length of the array as well as the overall time delays required to phase up the entire array. Optimally spaced antennas are packed as tightly as possible, without introducing a significant correlated noise component.

To determine the minimum allowable spacing between antennas, we recorded noise traces on the oscilloscope for several different spacings between antennas ranging from as close as physically possible to the maximum separation distance allowed by the dimensions of the anechoic chamber (1.5 m). For all separation distances, the antenna phase centers of the antennas were aligned parallel to their nulls. The setup was comparable to deploying vertically polarized dipole-like antennas as a phased array in a borehole.

Measurements of thermal noise were compared to a simulation of thermal noise for different antenna spacings following the phasor theory of thermal noise from Goodman [8] combined with

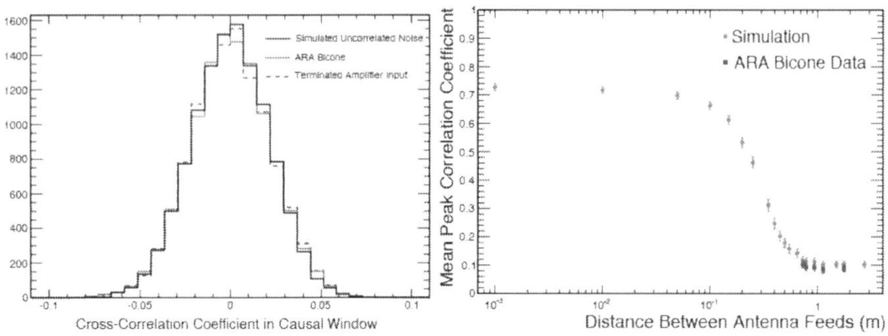

Figure 1. Left: cross-correlation coefficients between two antenna channels spaced 0.73 m apart over a causal time window ±2.4 ns around zero time delay for each of ~ 500 recorded events compared to results from terminated amplifiers and the simulation. Right: Mean (points) and standard deviation (error bars) peak cross-correlation coefficient for 500 simulated or measured events in a ±6.0 ns time window, which is the causal window for the largest feed-to-feed spacing for which we made measurements (~ 2 m).

uncorrelated system noise of 75 K.The left-hand panel of Fig. (1) shows the cross-correlation coefficient in a causal time window normalized by the number of samples and standard deviation of each waveform for the ARA bicones spaced as close as possible, terminated amplifier inputs, and the noise simulation. The measured distribution is consistent with uncorrelated noise, as it matches the distributions for both simulated uncorrelated noise and the uncorrelated system noise of the amplifiers. We observed no significant correlations for measurements at larger separation distances and using other antennas.

Significant noise correlation is only apparent in the simulations at phase-center separation distances that are not physically allowable. The righthand panel of Fig. (1) shows the peak correlation coefficient in a causal window averaged over 500 simulated events and events received on ARA bicones. At small spacings, the correlation coefficient plateaus at 0.75, corresponding to a fully correlated thermal noise contribution of 300 K combined with system noise of temperature of 75 K. For physically allowable spacings (> 0.73m), the simulation overestimates the peak correlation coefficient. We attribute this to imperfect modeling of the antenna and filter phase responses, which would further decorrelate the signals.

2.2 Analog beam-forming

The second assumption in the phased array technique is that plane-wave signals add linearly in voltage. We tested this assumption by transmitting broadband, impulsive signals from 4 m away from the center of a two-antenna phased array of broadband dipoles. The two channels had slightly different impulse responses, gains, and system temperatures due to differences in the antennas and first-stage amplifiers. Due to the differences in the two channels, the beam formed from these two channels should not show the expected $\sqrt{2}$ improvement in SNR expected from identical signal chains.

An analog beam corresponding to an angle broadside to the antennas was formed using a 3 dB combiner. The SNR averaged over 500 events of the two antennas is 2.6σ and 2.7σ, while the hardware and idealized beam have SNRs of 3.1σ and 3.3σ. The hardware beam matches the idealized beam in SNR, but the signal suffers 12% loss due to the additional components in the signal chain.

2.3 Digital beam-forming

With a digital phased array, different time delays in firmware or in software can be used to form many beams. Even though beam-forming tends to reduce the overall beam width of an individual beam, the full angular coverage can be recovered by forming multiple beams. The number of independent beams and the individual beam widths are determined by the antenna spacing and the sample rate.

We studied the trigger rate and trigger efficiency of a digital phased array and compared it to a simulation that includes both bandlimited impulses and uncorrelated noise. Using the smallest antennas, the Telewave ANT400D folded dipoles, we mounted three antennas spaced 55 cm apart on a steel ground mast. An ARA bicone was used as a transmitter along with an Avtech pulser and an attenuator of varying loss. We recorded signals on each antenna at 5 GSa/s for 500 events at several transmitter attenuations. The signals were then decimated to 1.67 GSa/s to mimic the Nyquist sampling rate for a system with a bandwidth of ~ 800 MHz. The measured traces were tested against several thresholds to study the rates and efficiency of a digital interferometric trigger.

A trigger is formed when the time-windowed power on each coherently summed beam exceeds a given threshold. The power is calculated for each sample and averaged over the time window. Calculating the power in a wide window (16 samples calculated every 8 samples) reduces the trigger rates for the dispersed pulses in this system, but if the signals were deconvolved first, a shorter sample window may be more efficient. The thresholds–extracted from the simulation due to the limited amount of data recorded in the anechoic chamber–at singles rates of 10, 100, 1000 Hz, are 2.34σ, 2.21σ, and 2.07σ, respectively.

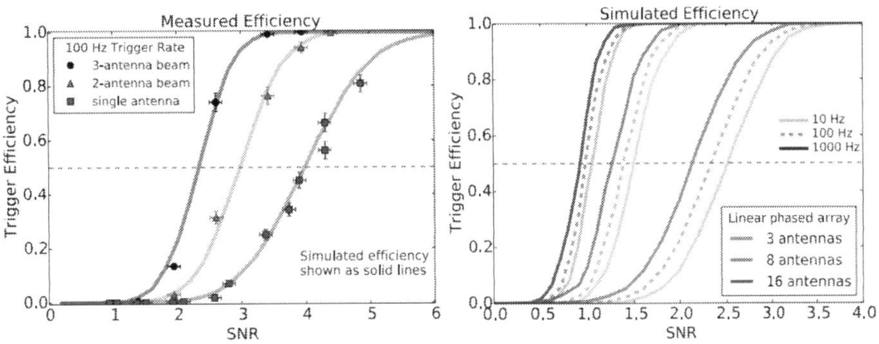

Figure 2. Left: Trigger efficiency vs. SNR in the anechoic chamber measurements (points) compared to a simulation of uncorrelated noise added to the system impulse response (lines). Right: Simulated trigger efficiency for larger phased arrays.

The trigger efficiency for a given input SNR ($\frac{V_{pk2pk}}{2\sigma}$ on a single antenna) is shown in Fig. (2). Adding additional antennas reduces the 50% point of the trigger efficiency curves from four for a single antenna down to 2.21σ at a fixed trigger rate of 100 Hz. The expected trigger efficiency extrapolated from the simulation of a 16-antenna phased array reduces the trigger threshold to about one.

3 In-Ice Prototype in Greenland

In moving from the idealized environment of antennas in air in an anechoic chamber to a phased array deployed in the ice, several factors are of concern. The ice properties and radio backgrounds local to the site may be problematic. The length and size of the antenna array may also present a challenge for deployment. To address some of these concerns, we conducted two tests in in the ice at Summit Station in Greenland in both June 2013 [9] and June 2015 [10].

With the lower trigger threshold expected from a phased array, the total visible ice volume at low energies will increase. However, this will have a minimal effect if the propagation loss of the electric field is high. The average field attenuation length of the ice at Summit Station is $\langle L_\alpha \rangle = 1022^{+230}_{-253}$ m. The measured attenuation length was extrapolated to a frequency of 300 MHz from 75 MHz for comparison with other proposed radio neutrino experiments and averaged over the top 1500 m of ice relevant for neutrino telescopes [9]. The total loss in the electric field strength when combines the losses due to attenuation in the medium with the path loss is shown for three proposed sites in Fig. (3). At likely distances for neutrino events – a few hundred meters – the electric field loss at Summit Station is comparable to the South Pole [12] and lower than at the Ross Ice Shelf [13].

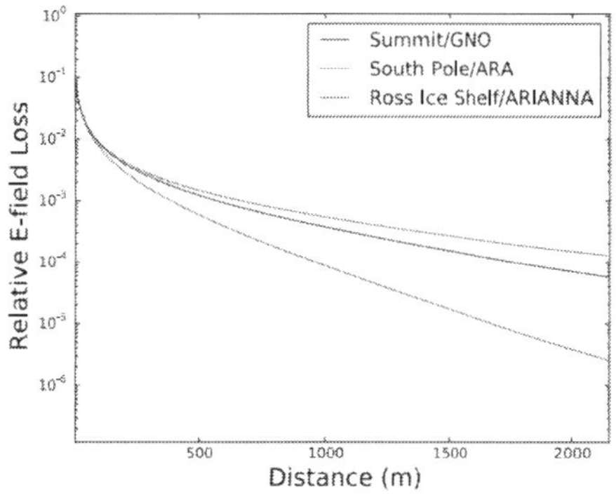

Figure 3. Electric field loss at 300 MHz due to the attenuation of the ice and path loss for potential neutrino detector sites.

During the June 2015 test, a prototypical phased array system, which is described in more detail in Ref. [10], was deployed to characterize the radio backgrounds. The system included 8 broadband dipoles, arranged vertically and separated by 76 cm feed-to-feed. The bottom of the array was 96 m below the surface of the ice. RF signals from the antennas were carried along 100-m lengths of coaxial Times Microwave LMR-240 along with DC power for the dual stage amplifier chains described in Section 2 mounted between the antennas in the borehole. Signals from each antenna were split into a beamformer chain and a digitizing chain. The beamformer chain further split each antenna signal eight ways. Eight beams at fixed angles from the horizon were formed using the split antenna channels and fixed delay lines of Times Microwave LMR-200. Traces from both the antenna and beam channels were digitized on four DRS4 evaluation boards that were triggered by a Tektronix MSO5204 oscilloscope. The global trigger provided by the oscilloscope was formed from either automatic (unbiased) or threshold crossing triggers on either an antenna or beam channel.

The improved trigger efficiency found in Section 2.3 assumes that the background noise is uncorrelated thermal noise. The noise averaged over 200 power spectra of 500-sample waveforms observed a downhole antenna, corrected for the overall gain in the system is shown in on the left in Fig. (4). The systematic uncertainty on the absolute scale reflects the uncertainties in the temperature of the cables and amplifiers in the borehole. The observed power is consistent with the expectation from thermal noise in the band between 160 and 440 MHz, the 3 dB points of the filters used in the system.

If the noise in adjacent antennas is uncorrelated, then the RMS voltages in a beamformed channel would be the quadrature sum of the RMS voltages in the individual channels after correcting for the additional loss in the beamformer (-19.9 dB due to the 8-way splitters and 8-way combiners). The RMS voltages of a beam of unbiased triggered events recorded over the course of one night are shown in the righthand panel of Fig. (4). The RMS voltages lie within 10% difference of the expectation for

DOI: 10.1051/epjconf/201713505007

an incoherent signal, and are inconsistent a coherent sum of the signals. In agreement with our results from the anechoic chamber measurements, the noise observed on the downhole beams is incoherent and uncorrelated.

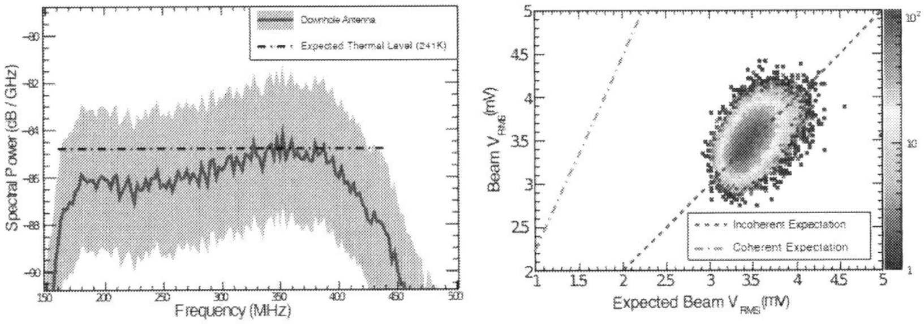

Figure 4. Left: Average power spectral density of thermal-noise triggers of a downhole antenna (blue solid line) and ±3 dB systematic uncertainty (green band) compared with thermal noise (dashed line). Right: Measured V_{rms} in a beam compared with the expected RMS voltage for the sum of five borehole antennas.

4 Conclusions

The phased array technique relies on two basic assumptions: that a plane-wave like signal adds coherently and the background noise in the antennas adds incoherently. We observed the expected increase in SNR on beams formed in an anechoic chamber. Measurements in an anechoic chamber and at a borehole in Summit Station demonstrate that closely packed antennas aligned along their nulls receive uncorrelated thermal noise.

Triggering with a phased array of low gain antennas shows promise for lowering the energy threshold of in-ice radio neutrino experiments. Using data from the anechoic chamber to inform a simulation that includes realistic parameters for an FPGA, we observe a substantial reduction in trigger threshold when compared to a single-antenna threshold. Such a trigger will be implemented at the South Pole on the ARA experiment, potentially lowering ARA's energy threshold to provide overlap in sensitivity with IceCube.

References

[1] V.S. Beresinsky, G.T. Zatsepin, Physics Letters B **28**, 423 (1969)
[2] IceCube Collaboration, M.G. Aartsen et al., Phys. Rev. Lett. **113**, 101101 (2014)
[3] IceCube Collaboration, M.G. Aartsen et al., Phys. Rev. Lett. **111**, 021103 (2013)
[4] IceCube Collaboration, M.G. Aartsen et al., Astrophys. J. **809**, 98 (2015)
[5] ARA Collaboration, P. Allison et al., Phys. Rev. **D93**, 082003 (2016)
[6] S.W. Barwick et al., Astroparticle Physics **70**, 12 (2015)
[7] A.G. Vieregg, K. Bechtol, A. Romero-Wolf, J. Cosm. and Astropart. Phys. **2**, 005 (2016)
[8] J.W. Goodman, *Statistical Optics, 2nd Ed.* (John Wiley and Sons, Inc., 2015)
[9] J. Avva et al., J. Glaciol. **61**, 1005 (2015)
[10] S. Wissel et al., Proc. of Science **ICRC2015**, 1150 (2015)
[11] R.B. Alley, B.R. Koci, Annals of Glaciology **10** (1988)
[12] T. Barrella, S. Barwick, D. Saltzberg, J. Glaciol. **57**, 61 (2011)
[13] J.C. Hanson et al., J. Glaciol. **61** (2015)

EPJ Web of Conferences **135**, 06001 (2017) DOI: 10.1051/epjconf/201713506001
ARENA 2016

Acoustic detection of high energy neutrinos in sea water: status and prospects

Robert Lahmann[1,*]

[1] *Friedrich-Alexander-Universität Erlangen-Nürnberg, Erlangen Centre for Astroparticle Physics (ECAP), Erwin-Rommel-Straße 1, 91058 Erlangen, Germany*

Abstract. The acoustic neutrino detection technique is a promising approach for future large-scale detectors with the aim of measuring the small expected flux of neutrinos at energies in the EeV-range and above. The technique is based on the thermo-acoustic model, which implies that the energy deposition by a particle cascade – resulting from a neutrino interaction in a medium with suitable thermal and acoustic properties – leads to a local heating and a subsequent characteristic pressure pulse that propagates in the surrounding medium. Current or recent test setups for acoustic neutrino detection have either been add-ons to optical neutrino telescopes or have been using acoustic arrays built for other purposes, typically for military use. While these arrays have been too small to derive competitive limits on neutrino fluxes, they allowed for detailed studies of the experimental technique. With the advent of the research infrastructure KM3NeT in the Mediterranean Sea, new possibilities will arise for acoustic neutrino detection. In this article, results from the "first generation" of acoustic arrays will be summarized and implications for the future of acoustic neutrino detection will be discussed.

1 Introduction

Measuring acoustic pressure pulses in huge underwater acoustic arrays is a promising approach for the detection of ultra-high-energy (UHE, $E_\nu \gtrsim 10^9$ GeV) neutrinos. These are expected to be produced in interactions of cosmic rays with the cosmic microwave background [1]. The pressure signals are produced by the particle showers that evolve when neutrinos interact with nuclei in water. The resulting energy deposition of the hadronic cascade within a cylindrical volume of a few centimetres in radius and several metres in length leads to a local heating of the medium, see e.g. [2] and references therein.

According to the so-called thermo-acoustic model [3], the energy deposition of particles traversing liquids leads to a local heating of the medium which can be regarded as instantaneous with respect to the hydrodynamic time scales. Due to its temperature change, the medium expands or contracts according to its bulk volume expansion coefficient. The accelerated motion of the heated medium generates a pulse whose temporal signature is bipolar with a spectrum peaking around 10 kHz, propagating through the medium. Coherent superposition of the elementary sound waves, produced over the cylindrical volume of the energy deposition, leads to a propagation within a flat disk-like volume, often referred to as "pancake", in the direction perpendicular to the axis of the particle shower. The

*e-mail: robert.lahmann@fau.de

DOI: 10.1051/epjconf/201713506001

main advantage of using sound for the detection of neutrino interactions, as opposed to Cherenkov light, lies in the much longer attenuation length of the former type of radiation – several kilometres for sound compared to several tens of meters for light in the respective frequency ranges of interest in sea water.

In addition to water and ice, which is the medium of acoustic detection test experiments presently or recently conducted, acoustic detection in salt domes [4, 5] and in permafrost [6] has been discussed. In this article, the detection of neutrinos in sea water will be discussed, with an emphasis on the achievements so far and the lessons learned for the future. Discussions on neutrino detection in fresh water and ice can be found in [7] and [8], respectively.

2 Sound in water and acoustic neutrino detection

Two processes must be understood to asses the potential of acoustic neutrino detection: The conversion of the energy released by a neutrino interaction into a detectable acoustic signal and the attenuation of the resulting sound wave in the medium.

Acoustic signal:

For a simplified analytic derivation it can be assumed that the shower energy in the plane perpendicular to the direction of the incoming neutrino is described by a Gaussian distribution with standard deviation σ_p. If the total energy deposited in the hadronic shower is denoted by E_0, then p_{max}, i.e. half the peak-to-peak amplitude of the bipolar pulse – measured as a function of time at a fixed distance in the far field – is given by

$$p_{\mathrm{max}} \propto \gamma_{\mathrm{G}} \frac{E_0}{\sigma_p^2} \tag{1}$$

where the dimensionless quantity $\gamma_G \equiv v_s^2 \alpha / c_p$ is the Grüneisen parameter, which depends on the speed of sound v_s, the bulk volume expansion coefficient α and the specific heat capacity at constant pressure c_p. For a derivation, see [9] or e.g. [10] and references therein. In Fig. (1), the Grüneisen parameter is shown for several bodies of water. For the Mediterranean Sea, its value is quite favourable for acoustic neutrino detection, basically due to its relatively high temperature of ~13°C even at great depth. For Lake Baikal, while otherwise well suited for acoustic neutrino detection, the Grüneisen parameter is small since in the deep zone of the lake, the water temperature is only 1.5 – 2°C higher than the maximum density at the respective depth [11]. Reflected by the Grüneisen parameter is the temperature change vs. depth in the top layers of the respective bodies of water. For greater depth, where the temperature gradient becomes small, the Grüneisen parameter basically follows the profile of the speed of sound, which increases almost linearly with depth due to the increasing pressure. The temperature profile for the Mediterranean Sea and tropical ocean waters is shown in Fig. (2).

Attenuation:

Attenuation in distilled water is caused by viscous absorption and is proportional to the square of the frequency of the sound wave. In sea water, in addition ionic relaxation of chemical compounds solved in the water contributes. The latter effect is due to the dependence of the association $\rightleftharpoons$ dissociation process of the chemical compounds on temperature and pressure [12].

The frequency-dependent attenuation of sound in water is shown in Fig. (3). Although NaCl is the principal constituent of salt in sea water and magnesium sulphate $MgSO_4$ only amounts for a fraction of 4.7 % by weight of the total dissolved salts in sea water, the latter dominates the absorption process for frequencies above a few kHz, up to about 100 kHz [13], i.e. in the region relevant for acoustic

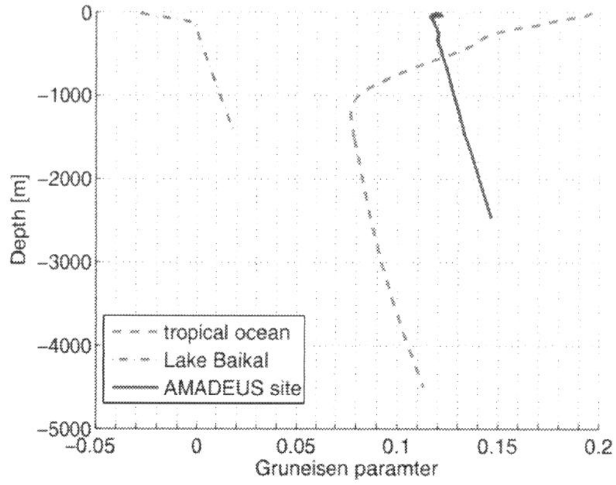

Figure 1. Grüneisen parameter for Lake Baikal (fresh water), Mediterranean Sea at the ANTARES site, and tropical ocean waters (sea water). Figure from [10].

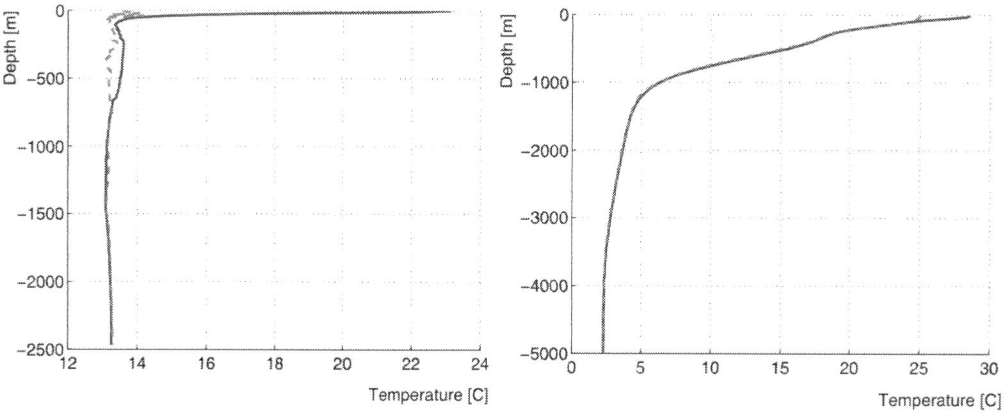

Figure 2. Temperature as a function of the depth below the sea surface, for different seasons at two different sites; *left*: Measurement from the ANTARES site in August 2007 (solid blue line) and in March 2010 (dashed red line); *right*: Temperature profile in the tropical ocean, 24°30'N and 72°30'W, for the average of the three summer months (solid blue line) and the three winter months (dashed red line). Figure from [10].

detection. The absorption length at the ANTARES site is about 5 km at 10 kHz, compared to about 100 km for distilled water. Clearly, if conditions provide for Grüneisen parameters of equivalent size, fresh water is superior to sea water for acoustic neutrino detection as it allows to survey a larger volume with the same number of sensors[1].

3 Results of the "first generation" of acoustic neutrino test setups

Current or recent test setups for acoustic neutrino detection have either been add-ons to optical neutrino telescopes or have been using acoustic arrays built for other purposes, typically for military use.

[1]The actual layout and instrumentation density of an acoustic neutrino telescope also depends on further factors, such as the noise background and the "pancake" shape of the three-dimensional pressure field.

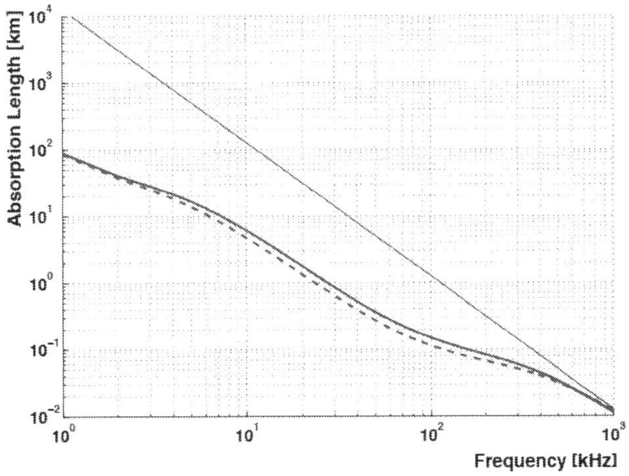

Figure 3. Absorption length according to the parametrisation given in [14] for conditions at the ANTARES site with temperature $T = 13.2°$ C, salinity of 3.8 % and a pH-value of 8. The solid thick curve is for a depth of 2000 m, the broken thick curve for a depth of 100 m. The straight thin line is the absorption length for pure water at a depth of 2000 m. Figure from [10].

The acoustic sensors are based on the piezoelectric effect [15] and $O(10)$ of these sensors have been installed in the respective setups. An overview of these "first generation" experiments in sea water – and for completeness also in fresh water and ice – that are currently taking data or have done so until recently is given in Table 1. Details on the experiments can be found in the references given in the table. At the writing of this article, only SPATS and the Lake Baikal experiment were still taking data[2].

Limits on the flux of UHE neutrinos have been derived by the SPATS [16], SAUND [17] and ACoRNE [18] experiments. While the size of the experiments is too small to derive competitive limit, it could be shown that the tools and techniques for setting such limits for acoustic neutrino detection are in place.

Table 1. Overview of existing and recent acoustic detection test sites.

Experiment	Location	Medium	Sensor Channels	Host Experiment
SPATS [19, 20]	South Pole	Ice	80	IceCube
Lake Baikal [7]	Lake Baikal	Fresh Water	4	Baikal Neutrino Telescope
OvDE [21]	Mediterranean Sea (Sicily)	Sea Water	4	NEMO
AMADEUS [22]	Mediterranean Sea (Toulon)	Sea Water	36	ANTARES
ACoRNE [23]	North Sea (Scotland)	Sea Water	8	Rona military array
SAUND [24]	Tongue of the Ocean (Bahamas)	Sea Water	7/49$^{(\star)}$	AUTEC military array

$^{(\star)}$ The number of hydrophones was increased from 7 in SAUND-I to 49 in SAUND-II

[2]For the military arrays, this means that data are not used for investigations of neutrino detection methods any more

To assess the potential for acoustic neutrino detection in a given natural body of water, ambient noise and transient background at the site have to be investigated. As a next step beyond setups of the "first generation", the KM3NeT infrastructure, currently under construction in the Mediterranean Sea, will allow for investigations of acoustic neutrino detection in a volume of several cubic kilometres of sea water [25]. The KM3NeT neutrino telescope [26] will comprise a huge array of acoustic sensors for position calibration that can be used for further studies of neutrino detection. Concepts employing fibre-based hydrophones as an alternative to piezo-based sensors for a potential future extension to KM3NeT are discussed elsewhere [27].

Studies of the conditions in the Mediterranean Sea, as performed by the OνDE and AMADEUS experiments, provide input for simulations for KM3NeT and beyond. The ambient noise is broadband and is mainly caused by agitation of the sea surface [28], i.e. by rain, wind, breaking waves, spray, and cavitations. It is predominantly the ambient noise that determines the energy threshold for neutrino detection. The ambient noise was measured to be low and stable by the OνDE [21] and AMADEUS [29, 30] experiments, and was found to be generally favourable for the operation of an acoustic neutrino detector.

Transient noise signals have short duration and an amplitude exceeding the ambient noise level. When of bipolar shape, these signals can mimic pulses from neutrino interactions. Sources of such signals can be anthropogenic, e.g. shipping traffic, or marine fauna. In particular dolphins emit short signals with a spectrum similar to that of acoustic emissions from neutrino interactions.

The transient background, as measured at the ANTARES site, has a relatively high rate. It is essential to reduce this background completely in order to measure the low expected rate of UHE neutrinos. A strong background reduction is already achieved with methods developed with AMADEUS [31]. Machine learning algorithms are used to identify bipolar pulses. Furthermore, events that are spatially and temporally clustered are discarded, as neutrinos are unlikely to produce several signals from one position within a short period of time. For further reduction it is essential to use the pancake shape of the signal as a selection criterion. This is not possible with AMADEUS due to its small size – such investigations can be pursued with the acoustic array of the KM3NeT detector [32].

In order to reduce the recorded data volume, the AMADEUS system employs an online pulse-shape-recognition trigger which is sensitive to the bipolar pulse expected from neutrino interactions [22]. This trigger selects events with a wide range of shapes. Improving the online selection may lead to a reduction of the signal-to-noise ratio of the data available for offline analysis, which would be an important step towards reducing the energy threshold for neutrino detection and thereby improving the sensitivity of a potential future acoustic neutrino telescope.

4 Conclusions and outlook

Acoustic arrays of the "first generation", either existing military acoustic arrays or additions to Cherenkov neutrino telescopes, have been used to investigate neutrino detection methods. Their sizes are far too small to yield competitive limits on the flux of UHE neutrinos but they allow for the investigation of experimental techniques for a future acoustic neutrino detector and for the investigation of background conditions for feasibility studies.

In the Mediterranean Sea, the KM3NeT neutrino telescope is currently under construction. It will comprise a system for acoustic position calibration of its optical sensors which can also be used for investigations towards acoustic neutrino detection. This would be an intermediate step towards an even bigger acoustic detector for UHE neutrinos. New concepts, in particular fibre-based hydrophones, can be used in the long run to further increase the instrumented volume of KM3NeT beyond the volume instrumented for optical detection, see [27] and references therein.

Acknowledgements

The author wishes to thank the organisers of the ARENA workshop for the invitation to give a presentation. The AMADEUS project is supported by the German government (Bundesministerium für Bildung und Forschung, BMBF) through grants 05A08WE1 and 05A11WE1.

References

[1] V.S. Berezinsky and G.T. Zatsepin, PL **B 28**, 423 (1969)
[2] R. Lahmann et al., Astropart. Phys. **65**, 69 (2015), arXiv:1501.01494 [astro-ph.IM]
[3] G.A. Askariyan et al., Nucl. Inst. and Meth. **164**, 267 (1979)
[4] G. Manthei, J. Eisenblätter, and T. Spies, Int. J. Mod. Phys. **A21S1**, 30 (2006)
[5] P.B. Price, J. of Geophys. Res **111**, B02201 (2006), arXiv:astro-ph/0506648v1
[6] R. Nahnhauer, A.A. Rostovtsev, and D. Tosi, Nucl. Inst. and Meth. **A 587**, 29 (2008)
[7] N. Budnev, *Acoustic detection of high energy neutrinos in fresh water*, in *Proc. of ARENA 2016 workshop* (2016), to be published in EPJ Web of Conf.
[8] T. Karg, AIP Conf. Proc. **1535**, 162 (2013), arXiv:1210.7974 [astro-ph.IM]
[9] J.G. Learned, Phys. Rev. **D 19**, 3293 (1979)
[10] R. Lahmann, *Ultra-High-Energy Neutrinos and Their Acoustic Detection in the Sea*, Habilitation thesis (2011)
[11] V. Aynutdinov et al., Nucl. Inst. and Meth. **A 662**, S210 (2012), doi:10.1016/j.nima.2010.11.153
[12] L. Liebermann, Phys. Rev. **76**, 1520 (1949)
[13] R.J. Urick, *Principles of Underwater Sound* (Peninsula publishing, 1983), ISBN 0-932146-62-7
[14] M.A. Ainslie and J.G. McColm, J. Acoust. Soc. Am. **103**, 1671 (1998)
[15] G. Anton et al., Astropart. Phys. **26**, 301 (2006)
[16] R. Abbasi et al. (IceCube Coll.), Astropart. Phys. **35**, 312 (2012)
[17] N. Kurahashi, J. Vandenbroucke, and G. Gratta, Phys. Rev. **D 82**, 073006 (2010)
[18] S. Bevan, Nucl. Inst. and Meth. **A 604**, 143 (2009)
[19] T. Karg, for the IceCube Coll., Nucl. Inst. and Meth. **A 662**, S36 (2012)
[20] Y. Abdou et al. (IceCube Coll.), Nucl. Inst. and Meth. **A 683**, 78 (2012)
[21] S. Aiello et al.(NEMO Coll.), L. Cosentino et al. (2008), arXiv:0804.2913 [astro-ph]
[22] J.A. Aguilar et al. (ANTARES Coll.), Nucl. Inst. and Meth. **A 626-627**, 128 (2011)
[23] S. Danaher for the ACoRNE Coll. , J. Phys. Conf. Ser. **81**, 012011 (2007)
[24] N.G. Lehtinen et al., Astropart. Phys. **17**, 279 (2002), arXiv:astro-ph/0104033v1
[25] F. Simeone, *Acoustic detection of UHE neutrinos in the Mediterranean sea: status and perspective*, in *Proc. of ARENA 2016 workshop* (2016), to be published in EPJ Web of Conf.
[26] S. Adrián-Marínez et al.(KM3NeT Coll.), J. Phys. G: Nucl. Part. Phys. **43**, 084001 (2016)
[27] E.J. Buis, *A large fiber sensor network for an acoustic neutrino telescope*, in *Proc. of ARENA 2016 workshop* (2016), to be published in EPJ Web of Conf.
[28] R.J. Urick, *Ambient Noise in the Sea* (Peninsula publishing, 1986), ISBN 0-932146-13-9
[29] D. Kießling, Master's thesis, Univ. Erlangen-Nürnberg (2013), ECAP-2013-041
[30] R. Lahmann for the ANTARES Coll. (2011), arXiv:1104.3041 [astro-ph.IM]
[31] M. Neff, Ph.D. thesis, Univ. Erlangen-Nürnberg (2013), ECAP-2013-023
[32] D. Kießling, *Signal classification and event reconstr. for acoustic neutrino detection in sea water with KM3NeT*, in *Proc. of ARENA 2016 workshop* (2016), to be published in EPJ Web of Conf.

EPJ Web of Conferences **135**, 06002 (2017) DOI: 10.1051/epjconf/201713506002
ARENA 2016

Acoustic detection of UHE neutrinos in the Mediterranean sea: status and perspective

Francesco Simeone[1,★] and *Antonio* Capone[1,2]

[1] *INFN Roma1*
[2] *University "Sapienza", Roma*

Abstract. In recent years the astro-particle community is involved in the realization of experimental apparatuses for the detection of high energy neutrinos originated in cosmic sources or produced in the interaction of Cosmic Rays with the Cosmic Microwave Background. For neutrino energies in the TeV-PeV range, optical Cherenkov detectors, that have been so far positively exploited by Baikal[1], IceCube[2] and ANTARES[3], are considered optimal. For higher energies, three different experimental techniques are under study: the detection of radio pulses produced by showers induced by a neutrino interaction, the detection of air showers initiated by neutrinos interacting with rocks or deep Earth's atmosphere and the detection of acoustic waves produced by deposition of energy following the interaction of neutrinos in an acoustically transparent medium. The potential of the acoustic detection technique, first proposed by Askaryan[4], to build very large neutrino detectors is appealing, thanks to the optimal properties of media such as water or ice as sound propagator. Though the studies on this technique are still in an early stage, acoustic positioning systems used to locate the optical modules in underwater Cherenkov neutrino detectors, give the possibility to study the ambient noise and provide important information for the future analysis of acoustic data.

1 Introduction

The interest in the study of neutrinos above 10^{17}eV, as a tool to investigate several open questions in the fields of high energy astro-particle physics and astrophysics, has grown steadily in recent years. Neutrinos created in cosmic sources or produced in the interaction of Ultra High Energy (UHE) protons with the cosmic microwave background via the Greisen-Zatsepin-Kuzmin mechanism[5], could provide complementary information to UHE charged cosmic ray and high energy gamma ray measurements. Moreover neutrinos are the only particle usable to probe the UHE non-thermal universe at distances above tens of megaparsec; all other particles are subject to interactions or decay, thus their use is limited to the local universe. Neutrino flux at these energies ($> 10^{17}$eV), derived from cosmic ray[6] and gamma ray[7] measurements, is very small and the expected event rate is of the order of $0.1\text{km}^{-2}\text{yr}^{-1}$; detection of UHE neutrinos with a reasonable statistics may need a detector of at least 100km^3 size. The acoustic detection technique could be the basic ingredient to instrument such a huge volume with a reasonable number of sensors thanks to the large attenuation length(~km) of the generated sound signal in water.

★e-mail: francesco.simeone@roma1.infn.it

 DOI: 10.1051/epjconf/201713506002

2 The Thermo-Acoustic Model

The acoustic detection technique of neutrino induced cascades in water is based on the thermo-acoustic effect[4]; the cascade energy is deposited in a narrow region of the medium, it induces a local heating and results in a rapid expansion of the water. The expansion originates a pressure wave that propagates perpendicularly to the cascade direction, whose amplitude depends on the energy density deposited in the medium. Neutrinos interact with the nucleons of the medium through neutral or charge current interactions; in the NC interaction only the hadronic shower at the interaction point can be detected, in the CC interaction the lepton, created at the interaction point, releases its energy according to its nature; While for ν_e and in part for ν_τ, the lepton energy is deposited as a shower close to the interaction vertex, for ν_μ the muon energy is deposited along its long track. In a wide range of energy useful for the acoustic detection, the ratio of the CC cross section[8] over the total is $\approx$70% (see [9]). As pointed out in [9] the mean inelasticity, so the fraction of the incident neutrino energy deposited in the induced hadronic shower and shown in Fig. (1), is weakly dependent on energy and its value is about 20%. From these considerations it might seems that the golden events are the electromagnetic showers produced in CC interactions, which are the most abundant and retain about 80% of the incident neutrino energy. Unfortunately this is not the case at very high energy; it was first noted by Landau, Pomeranchuk and Migdal[10] that above a threshold of about 10^{17}eV, the cross sections for bremsstrahlung and pair production decrease as $\sqrt{E}$. As a consequence the cascade becomes longer and sub showers will develop along the main shower as the particle energy drops below the LMP threshold; this prevents the energy density to increase linearly with the neutrino energy: hence the charged lepton contribute little to the production of the acoustic signal. The longitudinal profile of one cascade above the LPM threshold[11] is shown in Fig. (1)

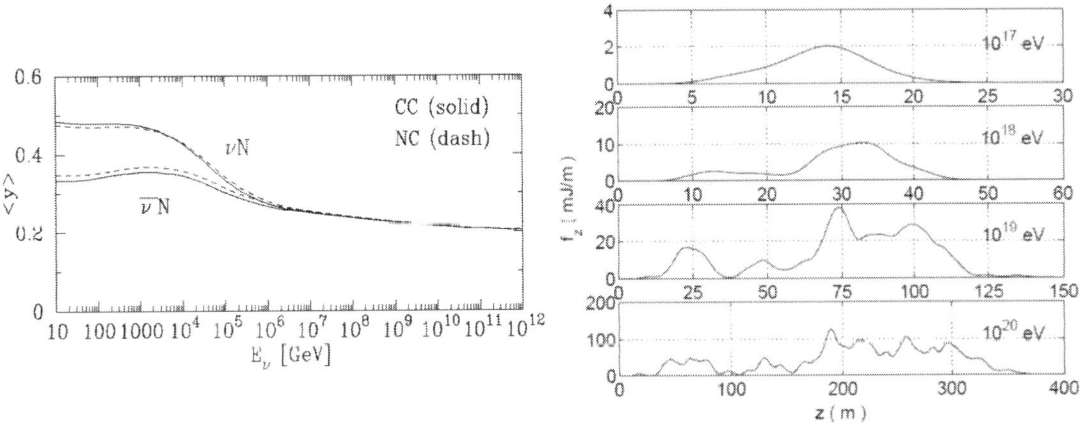

Figure 1. On left side the mean inelasticity as function of the neutrino energy and on the right the energy density profile of an electromagnetic cascade at different energy

Simulations of the acoustic signal generated by an hadronic shower in water has been performed by many authors, but uncertainty in the particle interaction and in the energy loss process at these energy as well as large fluctuation between showers, limit somehow the absolute signal prediction. To authors knowledge one of the most accurate simulation, performed in recent years, that takes into account the frequency attenuation of the medium, has been carried out by the ACORNE Collaboration[12]; their results(see Fig. (2)) shows a peak to peak amplitude of about 150mPa at 1Km from the interaction point for a neutrino energy of about $5{\cdot}10^{11}$GeV. Their results show also an extremely well collimated acoustic pulse whose amplitude decreases of about one order of magnitude in some degrees off the perpendicular plane passing through the maximum, the so called acoustic "pancake". It is worth noting that the characteristic collimation of the acoustic bipolar pulse is the fundamental property of the signal that will allow to make astronomy: sampling the "pancake" profile it's possible to reconstruct the shower direction and thus the neutrino one. Moreover this property is also the key to reject the background with high efficiency since the other sources of noise, like surface noise or biological one, even if could mimic the pressure signal, have different geometrical shape.

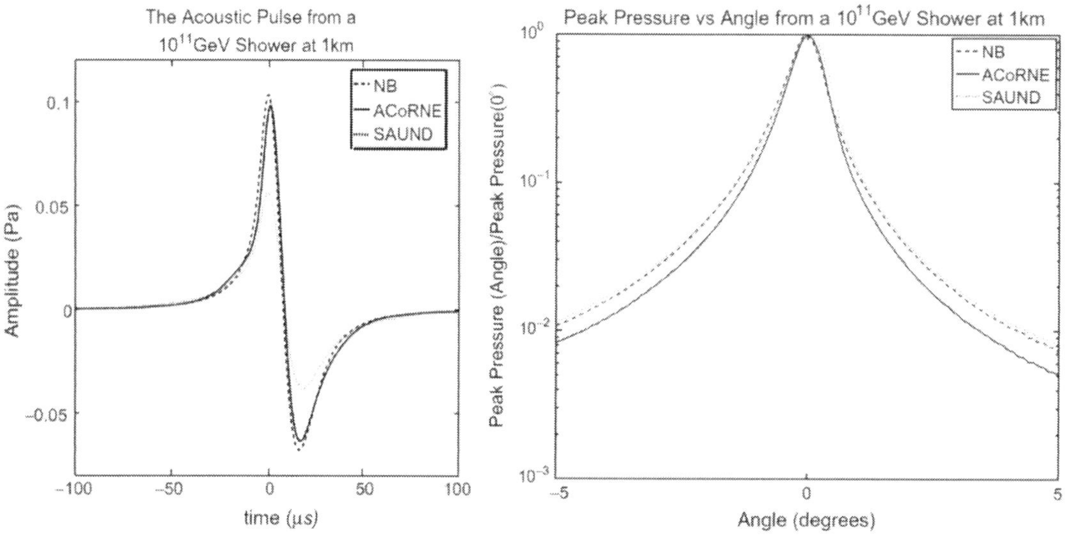

Figure 2. Acoustic pulse(left) and its angular dependency(right) for an hadronic shower of 10^{11}GeV evaluated in a plane perpendicular to the shower axis passing to the shower maximum(taken from [12]).

3 Acoustic infrastructures in Mediterranean sea

In the framework of the activities of different underwater Cherenkov neutrino telescopes, many studies have been carried out on the acoustic properties of the medium, on the sensors technology development and on their calibration. The AMADEUS[13] group (ANTARES Modules for Acoustic Detection Under the Sea) has deployed a system, designed principally as acoustic positioning system of the ANTARES neutrino telescope, made of both commercial and self-made piezoelectric sensor as well as self-made hydrophones. The AMADEUS system is fully integrated into the ANTARES acquisition system. It has been exploited to monitor the deep sea noise in the site[14], to identify and trace different sources of underwater noise, of anthropic and biological origin, and to performed useful comparisons between different technological solutions for underwater acoustic sensors tested in real environmental conditions. In 2005 the NEMO[15] Collaboration deployed OνDE (Ocean Noise Detection Experiment)[16] at the NEMO test site, 2000m depth, 25km east of the Sicily coast, composed of four commercial hydrophones arranged on a pyramidal-shaped. All data, sampled at 96 kHz and 24 bits, were transmitted in real-time and recorded for 5 minutes every hour. OνDE performed, for the first time, the underwater noise study in deep Mediterranean sea; data acquired allowed to characterize the deep sea noise as function of time and of weather conditions as well as in presence of ships

DOI: 10.1051/epjconf/201713506002

and biological sources (see Fig. (3)). The study of acoustic signals of biological origin had important drawbacks in bioacoustic. The NEMO collaboration deployed in 2013 a prototype of a Cherenkov detection unit (the NEMO phase II tower [17]) that included an acoustic positioning system. This positioning system was developed and realized by the SMO Collaboration[18]: it was composed by 18 acoustic sensors continuously sampled at 192kHz/24 bit and transmitted to shore for real time analysis. In such NEMO-SMO detector three different acoustic sensors were employed: 14 SMID TR-401 hydrophones, two free flooded ring hydrophones, model Sensor Technology Ltd SX-30, and two custom piezo-sensors. The SMO SMID hydrophones sensitivity (see Fig. (3)-right), as function of the pressure, changes of about 1dB and are the first devices calibrated at target pressure using a novel procedure developed in collaboration with the NATO Undersea Research Center of La Spezia (Italy). These sensors were also time calibrated taking into account both acquisition electronic and ceramic latency [19].

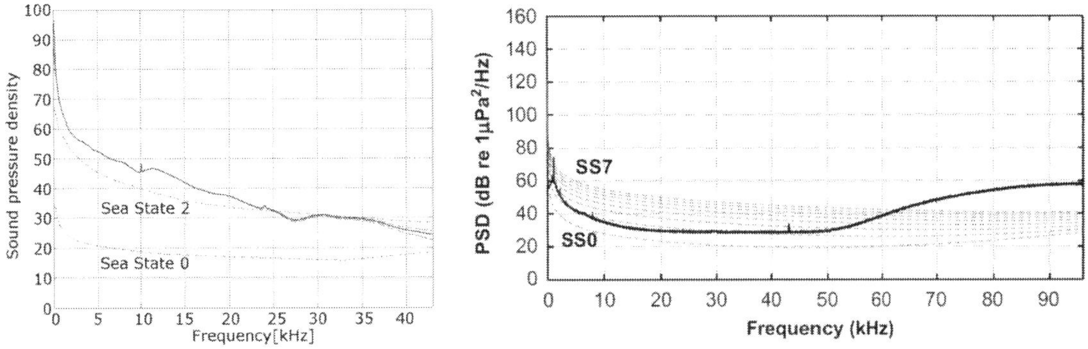

Figure 3. Noise power spectral density(dB re $\mu Pa^2/Hz$) measure by Ovde at 2000m depth (Left). Power spectral density of the intrinsic noise of the NEMO-SMO SMID hydrophone (right).

ANTARES, NEMO, NESTOR [20] and other European Collaborations merged their effort to build an European large scale Cherenkov Neutrino Telescope forming the KM3NeT Collaboration [21]. The work performed in the recent years has been focused in the development of the digital optical module (DOM [22]) and the Detection Unit (DU) layout optimization for different fields of neutrino physics (see Fig. (4)). The aim of the KM3NeT Collaboration is to deploy two different setup, based on the same technology, in the Mediterranean sea:

- ORCA[23] a dense Megaton detector with 20m horizontal spacing between adjacent DU and a vertical inter DOM space of about 6m, mainly focused on the low energy(E>10GeV) neutrino detection and neutrino mass hierarchy studies.
- ARCA[23] a neutrino detector with 100m horizontal spacing between adjacent DU and a vertical inter DOM space of about 36m, optimized for high energy neutrino detection and astroparticle studies.

Both setup will share the same basic technology and will use the DOM as building block of their DU; this means that their acoustic positioning system[24] could be use as pilot experiment for acoustic neutrino detection.

4 Future perspective

In view of the future availability of huge hydrophones array, like the one expected to be used as acoustic positioning system of neutrino Cherenkov detectors, simulation studies have been carried

Figure 4. Scheme of the KM3NeT detection unit(left). Picture of the digital optical module(right) with detail of the internal piezoelectric sensor.

out by the AMADEUS [25] and ACORNE [26] groups, to evaluate the sensitivity of the acoustic technique as a function of sensors density and detection thresholds. The AMADEUS group simulated a randomly sparse sensor array with a threshold of 5mPa and evaluated the effective volume as a function of the sensor density (see Fig. (5)); they found an optimal density of about 200 sensors per km^3. The ACORNE group has taken into account the realistic water properties, principally refraction and evaluated two different configurations (see Fig. (5)): a small dense array of 1km^3 containing 1000 hydrophones with a threshold of 35mPa and a huge 1500km^3 sparse array of 100 sensors per km^3 with a threshold of 5mPa. ACORNE results show that GZK neutrinos could be detected only with this second configuration.

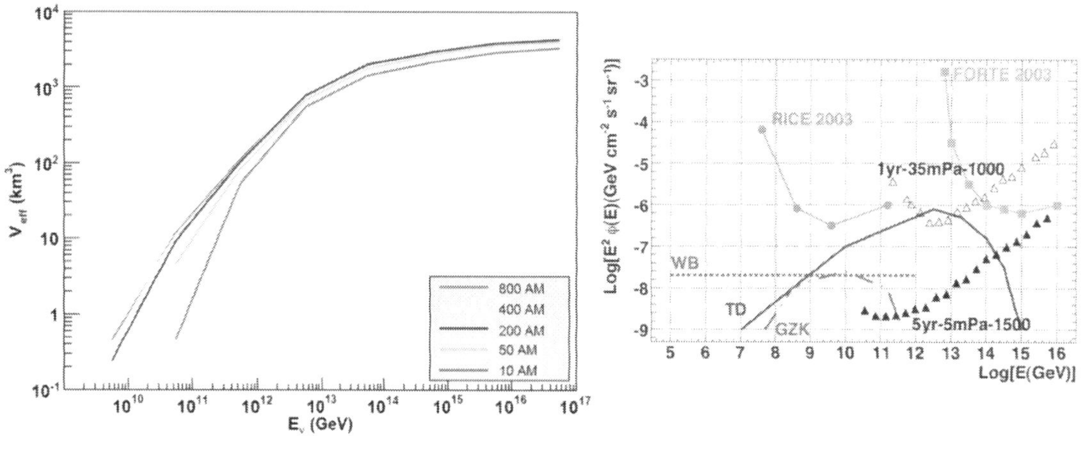

Figure 5. Left: effective volume as a function of the neutrino energy for different sensor density; the simulated instrumented volume is 1km^3. Right: the predicted limits on neutrino flux for different detector configurations, for details refer to [26]

5 Conclusions

In last years many activities focused on signal simulation, sensor development and calibration have been carried out. Fast code for acoustic neutrino signal generation and propagation are at present available. Commercial hydrophones sensitivity, in the frequency range of interest for the acoustic neutrino detection, reaches the limit of underwater noise background in relatively quite sites and procedures to calibrate sensors as a function of pressure, have been developed. In next years KM3NeT Collaboration will deploy a km^3 scale optical Cherenkov detector in the Mediterranean sea. Its acoustic positioning system will be the biggest underwater acoustic array suited to detect UHE neutrino interactions and will allow to validate and improve the background rejection algorithms as well as the reconstruction techniques. The acoustic detection of neutrinos in the energy range 10^{18}eV - 10^{21}eV will require further improvements in signal detection: it will be needed to lower the energy threshold for neutrino detection, to define a specific detector layout and to instrument a much bigger active volume.

References

[1] The Baikal homepage (official), *http://baikalweb.jinr.ru/*
[2] The IceCube homepage (official), *http://icecube.wisc.edu/*
[3] The ANTARES homepage (official), *http://antares.in2p3.fr/*
[4] G.A. Askaryan et al, Nuclear Instruments and Methods, **164**, 267 (1979)
[5] G.T. Zatsepin and V.A. Kuz'min, JETP Letters **4**, 78 (1966)
[6] L.A. Anchordoqui et al., Physics Review Letter D **76**, 123008 (2007)
[7] M. Ahlers et al., Astroparticle Physics **34**, 106 (2010)
[8] R. Gandhi et al., Physics Review Letter D **58**, 093009 (1998)
[9] R. Gandhi et al., Astroparticle Physics **5**, 81 (1996)
[10] A. B. Migdal, Phys. Rev. **103**, 1811 (1956)
[11] V. Niess et al., Astroparticle Physics **26**, 243 (2006)
[12] S.Bevan et al., Astroparticle Physics **28**, 366 (2007)
[13] J. A. Aguilar et al., Instrumentation and Methods A **626-627**, 128 (2011)
[14] F. Simeone et al., Proceedings of the 31st ICRC conference **arXiv:0908.0862v2** (2009)
[15] G. Riccobene et al., Nuclear Instruments and Methods A, **604** 604 (2009) 149
[16] G. Riccobene et al., e-Print, **arXiv:0804.2913**, (2008)
[17] T. Chiarusi et al., Journal of Instrumentation **9**, C03045 (2014)
[18] F. Simeone et al., Proceedings of the MASS Conference **DOI:10.1109/MASS.2011.107** , (2011)
[19] S. Viola et al., Nuclear Instruments and Methods in Physics Research A **725**, 207 (2013)
[20] J. Carr et al., New Journal of Physics **6 Issue 1**, 112 (2004)
[21] The KM3NeT homepage (official), *http://www.km3net.org/*
[22] S. Adrián-Martínez et al., Eur. Phys. J. C **74**, 3056 (2014)
[23] S Adrián-Martínez et al, Journal of Physics G, **43 no.8**, 084001 (2016)
[24] S. Viola et al., Proceedings of Science **ICRC2015** 1169 (2016)
[25] Doctoral thesis, T. Karg, **arXiv:astro-ph/0608312**, (2006)
[26] Doctoral thesis, J. Perkin, **arXiv:0801.0991**, (2008)

Development of an acoustic sensor for the future IceCube-Gen2 detector for neutrino detection and position calibration

Stefan Wickmann[1,*], *Dmitry* Eliseev[1], *Dirk* Heinen[1], *Peter* Linder[1], *Martin* Rongen[1], *Franziska* Scholz[1], *Lars Steffen* Weinstock[1], *Christopher* Wiebusch[1], and *Simon* Zierke[1]

[1] *III. Physikalisches Institut B, RWTH Aachen University, D-52056 Aachen, Germany*

Abstract. For the planned high-energy extension of the IceCube Neutrino Observatory in the glacial ice at the South Pole the spacing of detector modules will be increased with respect to IceCube. Because of these larger distances the quality of the geometry calibration based on pulsed light sources is expected to deteriorate. To counter this an independent acoustic geometry calibration system based on trilateration is introduced. Such an acoustic positioning system (APS) has already been developed for the Enceladus Explorer Project (EnEx), initiated by the DLR Space Administration. In order to integrate such APS-sensors into the IceCube detector the power consumption needs to be minimized. In addition, the frequency response of the front end electronics is optimized for positioning as well as the acoustic detection of neutrinos. The new design of the acoustic sensor and results of test measurements with an IceCube detector module will be presented.

1 Introduction

The IceCube Neutrino Observatory [1], located at the geographical South Pole, instruments about one cubic kilometer of Antarctic ice to detect neutrinos. The detector consists of 86 strings, each with 60 digital optical modules (DOM) forming a photomultiplier (PMT) array with a spacing of 125 m between the strings. Most of the DOMs are equally spaced (17 m) among the strings at a depth between 1450 m and 2450 m beneath the ice surface. Geometry calibration of the detector is performed with deployment data of the ice drilling and an optical calibration system, with which an accuracy of less than 1 m can be achieved.

IceCube-Gen2 is a planned extension of IceCube [2]. It will allow for the detection of high-energetic neutrinos with an increased detection rate and precision. To accomplish this an extension of the detector volume is planned by adding more strings to the existing IceCube array. The spacing between these additional strings will therefore be increased to 240 m - 300 m, depending on the planned configurations of IceCube-Gen2.

Because of the larger spacing between the strings the quality of the optical geometry calibration is expected to deteriorate due to the extinction length (20 m - 200 m) [3] of optical signals in the Antarctic ice. A possible solution could be the use of an independent geometry calibration system based on acoustic signals due to their larger attenuation length (~300 m, [4]).

*e-mail: wickmann@physik.rwth-aachen.de

The idea is to integrate acoustic sensors into the IceCube-Gen2 DOMs to use them for acoustic positioning independent of optical properties. In addition, such sensors could also be used for acoustic neutrino detection based on the thermo-acoustic effect [5] [4].

The required steps to integrate such a system into IceCube-Gen2 would be to test the concept of acoustic positioning with a DOM, to develop a suitable acoustic sensor for the required frequency ranges of the acoustic positioning and the acoustic neutrino detection with the desired low power consumption and finally to integrate the sensor into the IceCube-Gen2 DOMs. Acoustic emitters could be deployed as separate modules and do not need to be integrated into the DOMs.

Such a sensor for in-ice acoustic positioning has already been developed by the Enceladus Explorer Project (EnEx) [6]. Within this project navigation technology for a future space mission has been developed. A main aspect of the project has been the development of an in-ice acoustic positioning system for a maneuverable melting probe. The system measures the propagation time of acoustic signals between six ultrasonic emitters at the ice surface and four acoustic sensors located in the melting head of the probe. The probes position is then determined by trilateration with an accuracy of better than 1 m which is comparable to the accuracy required for IceCube's geometry calibration with optical signals.

The acoustic sensor developed in the EnEx-project is a piezo-based sensor, optimized for ultrasonic signals (18 kHz) [7]. It consists of a piezo disc (Ø16 mm × 3 mm) and two stacked sensor front end boards. One board amplifies, filters and digitizes the received signals and the second board converts the single-ended digital signals to LVDS signals. The two front end boards and the piezo disc are encapsulated in a brass housing (Ø20 mm).

The sensor front end contains two channels. After passing different filters the signal is digitized for each channel with a sampling rate of up to 3 MS/s at a resolution of 12 bits. The power consumption of one sensor front end with two channels is about 150 mW (30 mA at 5 V).

2 Prototype Development

To meet the requirements of IceCube-Gen2 the EnEx acoustic sensor front end was modified in respect of the desired frequency ranges and power consumption [8]. The new prototype front end utilizes two channels as well: channel one with a frequency range from 16 kHz to 20 kHz, optimized with regard to noise suppression of the acoustic positioning and channel two with a frequency range from 10 kHz to 100 kHz, optimized with regard to sensitivity of the acoustic neutrino detection. To substantially reduce the power consumption the LVDS conversion was removed which also leads to a reduced pin count of the sensor front end.

The power consumption of the new prototype front end was reduced to typically 50 mW (max. 53 mW) in idle mode (no digitization), which is about 33% compared to the original EnEx front end, and typically 75 mW (max. 78 mW) when digitizing both channels. Further improvement could be achieved by implementing a one channel operating mode, where only one channel is digitized and the other is in idle mode which will lead to an estimated power consumption of about 35 mW. By using enhanced electronic components optimized for low power application further reduction is possible.

A basic FPGA-firmware module was developed which controls and reads-out the prototype board. The Firmware was designed as a QSys component for use with the Altera QSys build system which will likely be used in IceCube-Gen2.

3 Test Measurements

The acoustic positioning concept for IceCube-Gen2 and its performance was evaluated in water (swimming pool) [8]. For this purpose, three EnEx acoustic sensors were integrated into an Ice-

Cube DOM[1] (Acoustic DOM) along with a data acquisition system (see fig. 1). The sensors were mounted into the Acoustic DOM using an aluminum construction that was glued to the glass sphere with vacuum cups. The sensor itself was pressed towards the glass sphere with metal springs (see fig. 1).

The Acoustic DOM was mounted under a floating platform and held under water with weights. On top of the floating platform two targets were mounted for reference measurements with a laser range-finder (LRF): one to measure the position of the Acoustic DOM and another to determine its heading. In addition six EnEx acoustic emitters were distributed in the swimming pool at different depths to reconstruct the position of the Acoustic DOM via acoustic measurements.

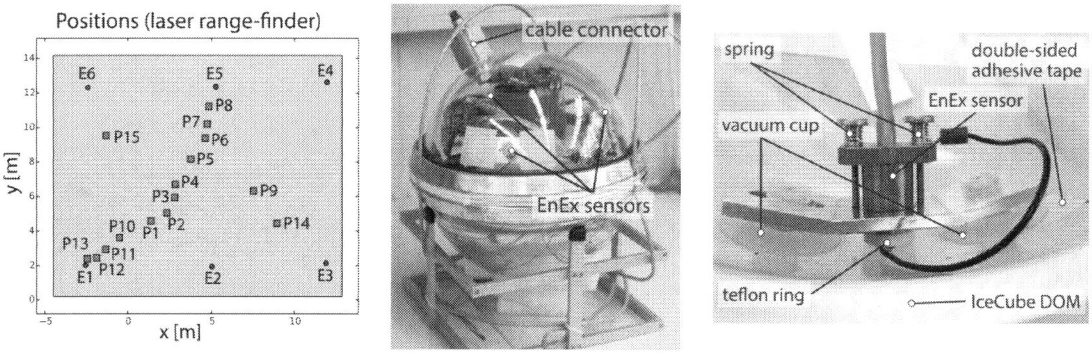

Figure 1. *left*: positions of the six acoustic emitters (E1 - E6, circles) and 15 measurement positions of the Acoustic DOM (P1 - P15, rectangles) in the swimming pool, measured with the laser range-finder (LRF); *center*: Acoustic DOM: IceCube DOM[1] with three EnEx sensors, read-out electronics and a pressure-tight cable connector; *right*: mounting of the acoustic sensor inside the IceCube DOM.

After measuring the positions of the six acoustic emitters with the LRF, the measurement procedure continues as follows: First, the reference position and heading of the Acoustic DOM at the respective measurement position is measured with the LRF. Then a sine-burst signal is sent consecutively from each emitter and data is taken synchronously with the Acoustic DOM for each emitter. After changing the position of the Acoustic DOM the above steps are repeated (see fig. 1). The position was reconstructed by determining the propagation time between the acoustic emitters and the acoustic sensors by detecting the start of the sine-burst as the point, where the signal exceeds 5σ of the preceding noise signal. After that, a least square fit was applied to the measured data for each channel of the sensors to reconstruct the position of the sensor.

4 Performance

The reconstructed positions are in good agreement with the reference positions measured with the LRF and the heading of the Acoustic DOM can be estimated (see fig. 2). However, there are some systematic errors that are not yet fully understood.

To quantify the performance of the system, the distance of the measured position from the reference position (fig. 2), both in full space and in the x-y-plane, were histogrammed. The accuracy in the x-y-plane (median: 6.31 cm) is better than the overall 3D accuracy (median: 26.48 cm) which results from smaller variation of the acoustic emitters' positions in z-direction compared to the x-y-plane, due

[1]kindly provided by Rolf Nahnhauer, DESY

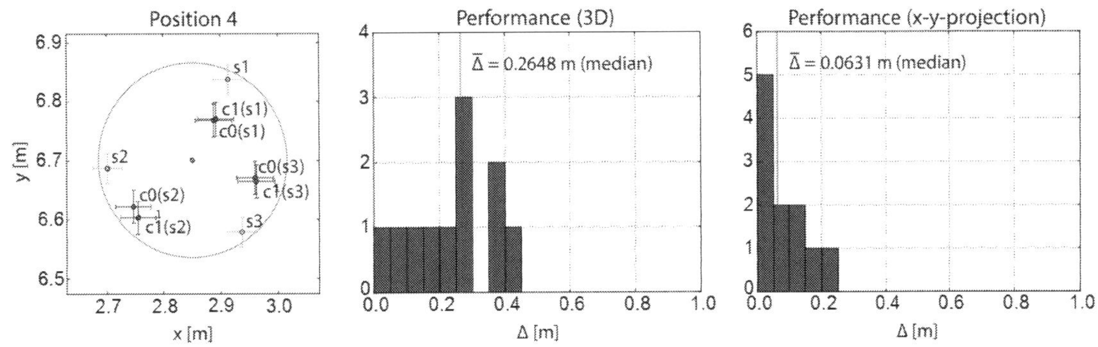

Figure 2. *left*: reconstructed positions of the acoustic sensors at Acoustic DOM pos. 4, cX(Y) labels channel X of sensor Y, s1 - s3 are the reference positions of the acoustic sensors deduced from the measurements with the LRF, the circle shows the diameter of the Acoustic DOM; *center, right*: Histogram of the distances of the acoustically measured positions from the true positions in full space and in the x-y-plane, respectively.

to the limited depth of the swimming pool. It is expected that a more sophisticated data selection for noise and outliers, as well as orientation dependent corrections would further improve the accuracy.

For use in ice, the media properties involve larger uncertainties which, however, do not vary with time. A similar performance is expected if the system provides sufficient redundancy for determining and correcting these uncertainties.

5 Conclusion

A concept of an acoustic positioning system for IceCube-Gen2 based on the EnEx positioning system has been evolved by developing a prototype sensor front end based on the EnEx sensor front end and a firmware module for IceCube-Gen2. The frequency ranges of the front end have been adapted to meet the requirements of IceCube-Gen2. Also the power consumption of the front end was reduced to a third of the power consumption of the EnEx sensor front end. The EnEx acoustic sensor has been integrated into an IceCube DOM and the acoustic positioning of an IceCube detector module has been demonstrated in water with a measured accuracy better than 7 cm on average.

The next steps to further evolving the concept of acoustic positioning for IceCube-Gen2 would be to improve the prototype front end by means of power consumption and actually integrate it into an IceCube-Gen2 DOM. Furthermore, a concept for design and distribution of acoustic emitters for IceCube-Gen2 needs to be developed.

References

[1] A. Achterberg et al., Astroparticle Physics **26**, 155 (2006)
[2] M.G. Aartsen et al. (2014), `arXiv:1412.5106`
[3] M. Aartsen et al., Nuclear Instruments and Methods in Physics Research Section A **711**, 73 (2013)
[4] R. Abbasi et al., Astroparticle Physics **34**, 382 (2011)
[5] G. Askariyan et al., Nuclear Instruments and Methods **164**, 267 (1979)
[6] J. Kowalski et al., Cold Regions Science and Technology **123**, 53 (2016)
[7] D. Eliseev et al., Annals of Glaciology **55**, 253 (2014)
[8] S. Wickmann, master's thesis, RWTH Aachen University (2016)

DOI: 10.1051/epjconf/201713506004

Acoustic Search for High Energy Neutrinos in Lake Baikal: Status and Perspectives

N.M. Budnev[3,★], *A.D.* Avrorin[1], *A.V.* Avrorin[1], *V.M.* Aynutdinov[1], *R.* Bannasch[7], *I.A.* Belolaptikov[2], *D.Yu.* Bogorodsky[3], *V.B.* Brudanin[2], *I.A.* Danilchenko[1], *G.V.* Domogatsky[1], *A.A.* Doroshenko[1], *A.N.* Dyachok[3], *Zh.-A.M.* Dzhilkibaev[1], *S.V.* Fialkovsky[5], *A.R.* Gafarov[3], *O.N.* Gaponenko[1], *K.V.* Golubkov[1], *T.I.* Gress[3], *Z.* Honz[2], *K.G.* Kebkal[7], *O.G.* Kebkal[7], *K.V.* Konischev[2], *A.V.* Korobchenko[2,3], *A.P.* Koshechkin[1], *F.K.* Koshel[1], *A.V.* Kozhin[4], *V.F.* Kulepov[5], *D.A.* Kuleshov[1], *V.I.* Ljashuk[1], *M.B.* Milenin[5], *R.R.* Mirgazov[3], *E.R.* Osipova[4], *A.I* Panfilov[1], *L.V.* Pankov[3], *A.A.* Perevalov[3], *E.N.* Pliskovsky[2], *E.V* Rjabov[3], *M.I.* Rozanov[6], *V.Yu.* Rubtzov[3], *B.A* Shaybonov[2], *A.A.* Sheifler[1], *M.D* Shelepov[1], *A.V.* Shkurihin[4], *A.A.* Smagina[2], *O.V.* Suvorova[1], *V.A* Tabolenko[3], *B.A.* Tarashansky[3], *S.A.* Yakovlev[7], *A.V* Zagorodnikov[3], and *V.L.* Zurbanov[3]

[1]*Institute for Nuclear Research, 60th October Anniversary pr. 7A, Moscow 117312, Russia*
[2]*Joint Institute for Nuclear Research, Dubna 141980, Russia*
[3]*Irkutsk State University, Irkutsk 664003, Russia*
[4]*Skobeltsyn Institute of Nuclear Physics MSU, Moscow 119991, Russia*
[5]*Nizhni Novgorod State Technical University, Nizhni Novgorod 603950, Russia*
[6]*St. Petersburg State Marine University, St. Petersburg 190008, Russia*
[7]*EvoLogics GmbH, Berlin*

Abstract. The status and perspectives of the feasibility study to detect high energy cosmic neutrinos acoustically in Lake Baikal is presented. The concept of on acoustic array as a part of the Baikal Gigaton Volume Neutrino Telescope GVD based on results of simulation and background measurements is described.

1 Introduction

Markov's suggestion about the creation of big optical detectors in natural reservoirs [1] had a fundamental significance for neutrino astrophysics development. In such a set-up natural water works as a target for neutrinos and as a medium for Cherenkov light. Last year, the IceCube collaboration detected first astrophysical neutrinos with PeV energies. In spite of big advances in the development of optical UHE neutrino detection they still have quite strong limitations. The main problem is the attenuation length in water and ice, even in purest media it can not be longer than tens of meters for Cherenkov light. That imposes considerable restrictions on the minimal density of the optical sensors in a detector. Since the first optical neutrino telescopes were put into operation it became clear that the volume of the detectors to search for UHE astrophysical neutrinos should be probably much more than one cubic kilometer. Thus, with optical sensors, the cost of such a system is very high. On the other hand in 1957 the Soviet physicist Askaryan showed that UHE particles may produce acoustic

★e-mail: nbudnev@api.isu.ru

pulses [2]. It was suggested to use this effect to search for UHE neutrino in the sea or in Lake Baikal. The considerable advantage of this approach is that acoustic signals with frequencies higher than 1 kHz are capable to propagate over very long distances with minimal attenuation due to absorption (attenuation length is two orders higher than for Cherenkov light), so the dominant attenuation factor is the geometrical divergence of the energy of the acoustic pulse. And it should be mentioned that the amplitude of pulses in the near-field zone decreases only as the square root of distance which is due to the cylindric form of the wave front produced by showers. So in principle an acoustic neutrino telescope may observe a bigger target volume with a lower number of detector units. Research for the feasibility of acoustic UHE neutrino registration in Lake Baikal was started in 2004. Since then substantial technical and methodological experience was gained [3], [4], [5]. It was revealed that the acoustic background in the Lake Baikal provides even better conditions for neutrino registration than the one in the Mediterranean Sea. In this paper we report the current state of research in the scope of the Baikal acoustic neutrino detection project.

2 Acoustic array design

For the development of the project of the neutrino acoustic detector we take into account results of our previous measurements of hydro-physical characteristics and acoustic background at the site of the Baikal neutrino experiment, as well as results of simulations of acoustic pulses produced by ultra-high energy neutrinos and the propagation of the signal in Baikal water [6], [7]. Cascades, generated by neutrino interactions in water, should produce bipolar acoustic transients. The expected bipolar pulse duration is of the order of several tens of microseconds. The amplitudes of such pulses are proportional to the thermal expansion coefficient $\alpha(T)$ [8], which is proportional to the difference $\Delta T(H)$ between the temperature of water $T(H)$ and temperature of its maximum density TMD $T_{MD}(H)$ at the current depth H. In Lake Baikal this difference $\Delta T(H)$ is close to zero at the depths of about 200-400 m and increases with depth, at 1400 m it is about $2C^o$ [9]. Thus, it is sensible to optimize the acoustic detector for "listening" of water layers at depths in the range 800 – 1400 m. The earth is not transparent for the ultra-high energy neutrinos, thus, at depths deeper than several hundreds meters the horizontal neutrinos dominate in total flux Fig. (1). Most of the acoustic signal energy is concentrated within a disk, the axis of which coincides with the axis of the cascade. The pulse amplitude varies very rapidly depending on angle from the median plane of the disk. The directions of the axes of cascades coincide with the direction of incoming neutrinos, in our case they are in the range of nadir angles from 80 to 180 degrees, but acoustic signals from that cascades will propagate mainly at the zenith angles 0-75 degrees.

For the search of such signals it is viable to distribute the Acoustic Sensor Modules (ASM) of the detector in the relatively thin layer with thickness of about 100 m at depths about 500-600 m. This distribution of ASMs provides the best signal-to-noise ratio (SNR) for the acoustic search of ultra-high energy neutrinos in Lake Baikal. On the one hand, the ASMs are in relatively close distances from the areas of the most effective generation of acoustic pulses. On the other hand, there is a very small number of the sources which mimic neutrino signals at large depths in Lake Baikal [6]. ASMs placed at depths larger than 500 m are able to detect the background 'neutrino-like' signals at zenith angles more than 110 degrees only when produced by near surface sources located at distances of more than 15 km or by signals reflected from the bottom. In both cases signals are significantly suppressed. As a result, acoustic signals from the most part of high energy cascades are distinguished from the background generated by surface sources. Thus, the distribution of the ASM at depths 500-600 m allows to significantly decrease the detection threshold of neutrino acoustic detection in the cold Baikal water and makes it comparable or even smaller than in relatively warm Mediterranean sea water with its higher thermal expansion coefficient"

 DOI: 10.1051/epjconf/201713506004

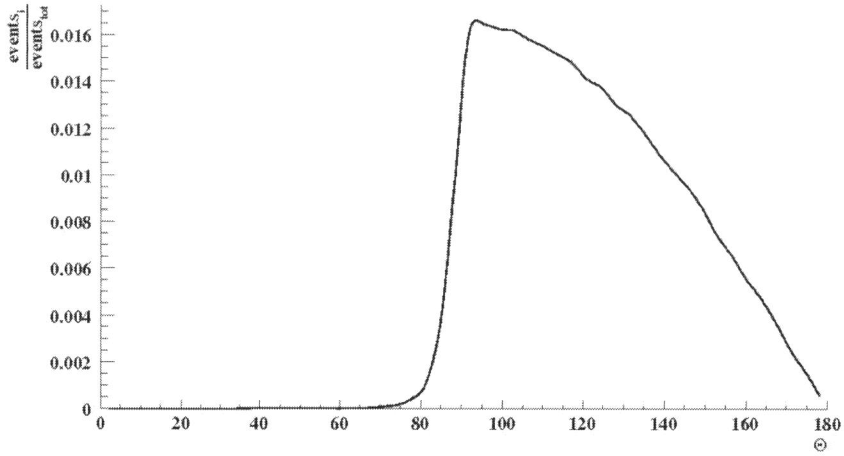

Figure 1. Nadir angle distribution of electron neutrino with energy $10^{17} - 10^{21}$ eV at depth 1 km.

The most realistic plan of the development of such an acoustic detector is to design it as an extension for the Gigaton Volume Detector (GVD) - kilometer-scale Baikal Cherenkov underwater neutrino telescope, which is currently under construction [10]. ASMs will be placed in the quasi-plane layer at the depths 500-600 m. We assume that on every string of GVD we will install up to 8 ASMs with a spacing of about 10 -15 m (Fig. (2)). In this case every acoustic pulse could be detected by several triplets of ASMs placed on the same string. The data about dimensional distribution of amplitudes is important for the suppression of short background signals which could appear as a result of interference of acoustic waves from distributed sources (rain, waves on the surface, etc.), and produced by some quasi-local sources (fish, boats, etc.) at shallow depths. The distribution of amplitudes and wave-factors in the first case will be quasi-random, in the second case we expect a quasi-spherical wave-front. In the both cases the time-dimensional distribution of pulses strongly differs from the pancake pulses from high energy cascades.

3 Baikal Acoustic Sensor Module

The main elementary unit of the Baikal acoustic neutrino detector - the Acoustic Sensor Module - is a device capable of detecting signals of predefined shape and to interpret them in terms of a plane wave, generated by a quasi- local source. To reconstruct the direction of the pulse's source each ASM will be equipped with an antenna with at least 4 hydrophones. The optimum distance between the hydrophones is defined by the condition that it must surely exceed several wave-lengths of the expected signal but, on the other hand, should not be too long in order to minimize the number of background pulses covered by the "time window" [6]. So, we arranged the hydrophones in the shape of a tetrahedron with equal distances of about 1.5 m between the hydrophones. All ASMs of every string are connected to the String Control Module (SCM), which provides acquisition and preprocessing of the data from the ASM, sending the data to the Shore Computer Center (SCC) of the Baikal Neutrino Observatory, operating with control input from SCC to ASM, synchronization and power supply for ASM.

The power supply and communication management within an acoustic string is organized as follows. The supply DC voltage of 300 V for the string is applied to the SCM where it is converted

 DOI: 10.1051/epjconf/201713506004

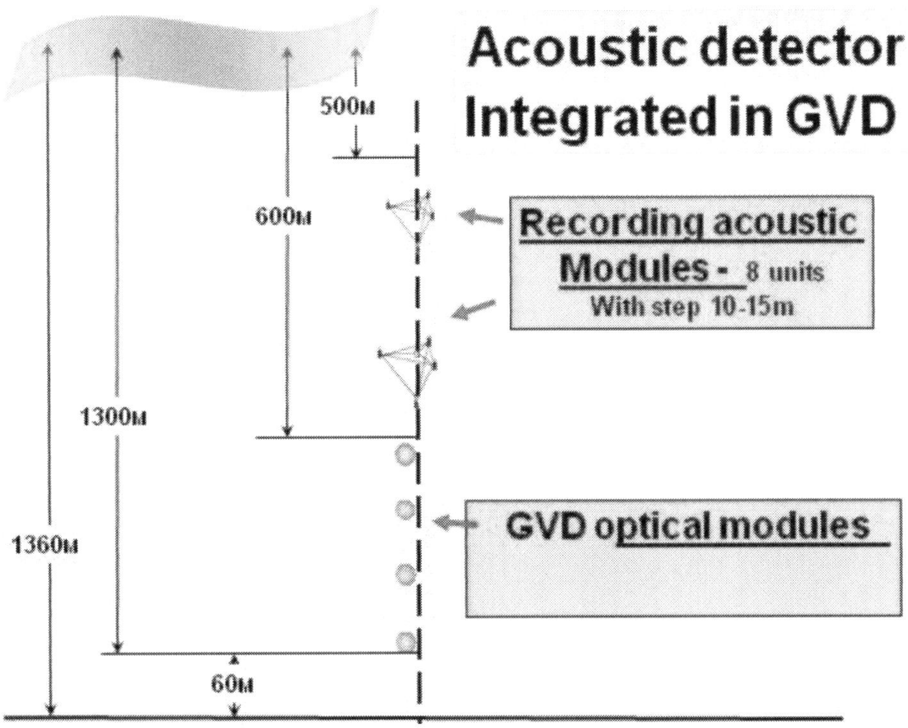

Figure 2. Acoustic Sensor Modules on a string of the Gigaton Volume Detector.

to the level of 48 V and distributed over the all ASM with power supply and synchronization (PSS) board, see Fig. (3). All ASM and power supply board controlling electronics is locally networked with star topology. Communication between acoustic strings and lakeside computer is realized with DSL modems. For electric power transmission to ASM and communication the same Ethernet cable is used.

The all data interchange is realized through a switch-board installed on the SCM. On the PSS board Ethernet traffic from the switch-board is redirected to special RJ-45 connectors manufactured by HALO company specialized on the support of the PoE technologies. In that connector the data flow is joined with the power supply and is transmitted through one net cable to the detection module. The detection module power supply chain is equipped with a resettable fuse designed for a current of 1.5 A. It will switch off the power supply channel of the ASM if a software protection does not switch off the power in case of a short circuit for whatever reason.

There are two parts on the PSS board: a control unit and a unit responsible for monitoring of the power parameters. The control unit performs the controlling of gates of solid state relays PVG612PBF of ASM power supply system. It's central element is the micro-controller STM32F437. The power monitoring unit is used for permanent checking of voltage and currents of acoustic string electronics. In case of identification of any malfunctions it informs the control unit in due time. The central element of the power monitoring unit is the STM32F303 micro-controller. The components belonging to different units are galvanically isolated from each other. As sensors for the hydro-acoustic antennas, piezoelectric hydrophones developed in the Far-East State University are used. They have sufficiently high and homogeneous sensitivity in a cone with opening angle of about 30 degrees in the frequency

 DOI: 10.1051/epjconf/201713506004

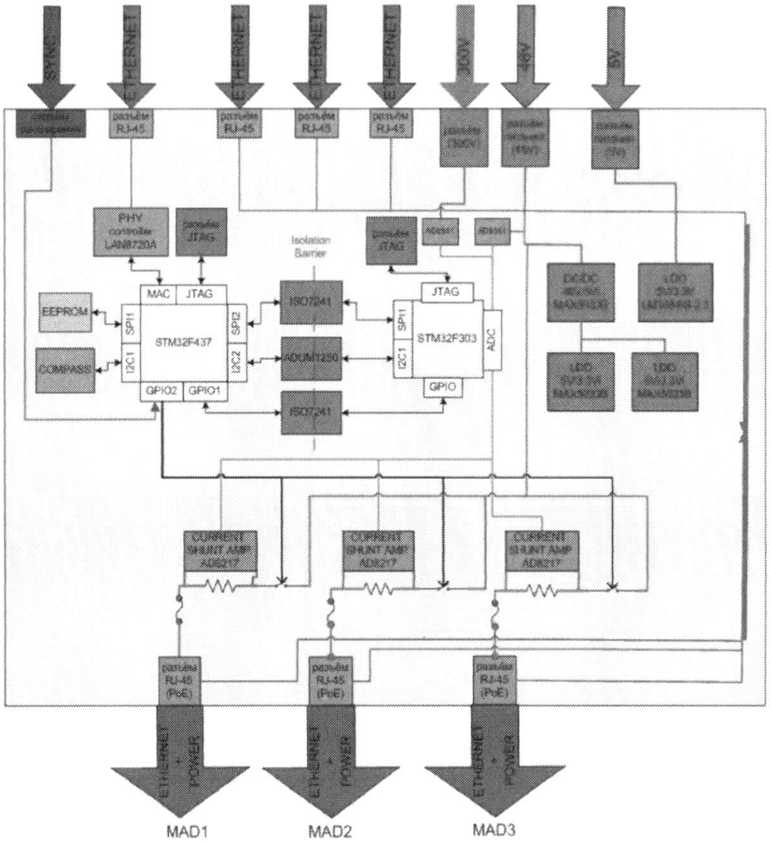

Figure 3. The functional scheme of the power supply and synchronization board.

range 5-30 kHz, from other directions the sensitivity is significantly lower. Such parameters of hydrophones allow to listen deep layers of the lake effectively and significantly suppress the influence of the background from the surface layer. Hydrophones have integrated pre-amplifiers with two amplifier stages. The first cascade is developed as charge amplifier, because the source of the signal has a very high output resistance. The second cascade transforms the asymmetric signal to a differential one, which comes via differential pair to the input of an analog chain of the ASM. Total amplitude amplification factor is 10. Test of the measurement chain in the lab gives the following results: input referred self-noise level of analogous channels is of less than 1 μV, bandwidth 53 kHz, ripple of amplitude-frequency characteristic in the frequency range of $3 - 38kHz$ is about $\pm 3dB$ hydrophone sensitivity is about 0.8 mV/Pa, beam width is 90 degrees. Total gain of analog channel is from 20 dB to 80 dB with possibility of setting with step of 3 dB.

The calibration of the acoustic detector was successfully performed in situ at Baikal neutrino observatory site with a specially developed emulator of neutrino acoustic signals.

4 A prototype acoustic string

To test the main technical and methodical solution we constructed a prototype acoustic string consisting of 3 Acoustic Sensor Modules and the String Control Module. This setup was arranged on

DOI: 10.1051/epjconf/201713506004

the engineering string of Baikal neutrino observatory. The data are in processing. An example of a background neutrino-like pulse detected by one of the ASM is shown in Fig. (4). Its zenith angle is 21 degrees.

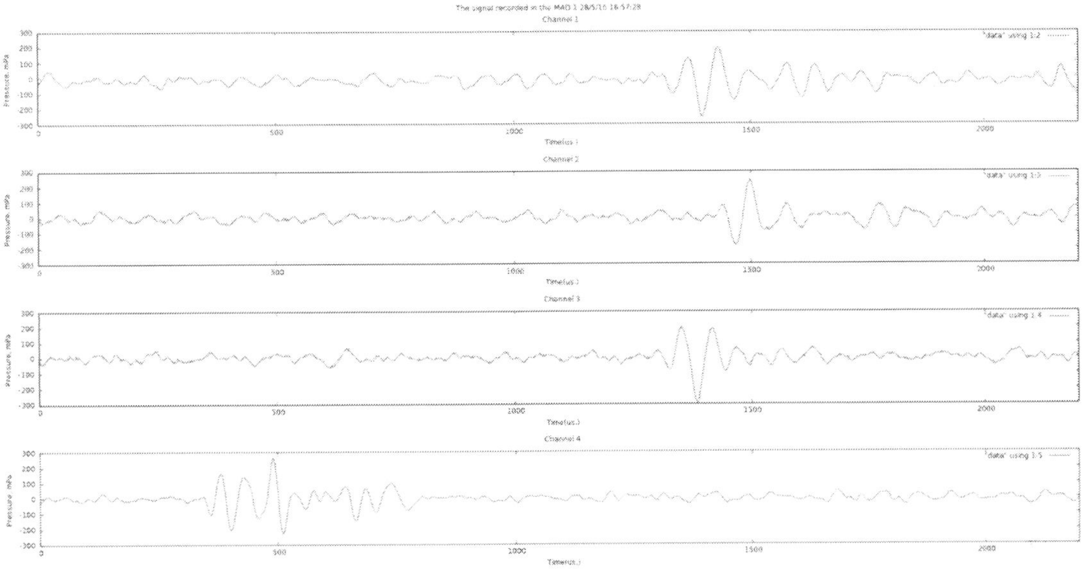

Figure 4. A sample of background neutrino-like pulse detected 28-05-2016.

5 Summary and outlook

Acoustic neutrino detection in Lake Baikal and in other natural reservoirs is far from being trivial due to high level of background noise. Comparison of results of acoustic pulse characteristic calculations and properties of noise in Lake Baikal shows that the energy threshold for neutrino detection in the lake should be of order $10^{19} - 10^{20}$ eV [6]. Taking into account that sources of background including "neutrino-like" short pulses are concentrated in a near-surface zone, we suggest an acoustic detector design that is based on the deployment of a grid of Acoustic Sensor Modules with rather compact antennas arranged at stings of the Gigaton Volume Detector at depths 500–600m at distances of 10-15 m. This configuration is optimized to identify interaction of ultra-high energy neutrino by listening from the top to signals from the deeper zones of Lake Baikal.

ACKNOWLEDGMENTS

This work was supported by the Russian Found for Basic Research (grant 14-02-00175)

References

[1] M.Markov, *Proceedings of the 1960 Annual International Conference on High Energy Physics*, (Rochester, 1960) 578.

[2] G.Askaryan, Atomnaya Energiya **3**, 152 (1957).

[3] V.Anutdinov, et al. (Baikal-Collaboration), Int. Journal of Modern Physics **A21**, 117 (2006).
[4] V.Anutdinov, et al. (Baikal-Collaboration), Acoustical Physics **52**, 581 (2006).
[5] V.Anutdinov, et al. (Baikal-Collaboration), Nucl. Inst. and Meth.A **604**, 130 (2009).
[6] V.Aynutdinov, et al (Baikal-Collaboration), Nucl. Inst. and Meth.A. **662**, S210 (2012).
[7] V.Aynutdinov, et al (Baikal-Collaboration), AIP Conf.Proc. **1535**, 176 (2013).
[8] G.Askaryan, B.Dolgoshein, A.Kalinovsky, N.Mokhov, Nucl. Instr. and Meth. **164**, 267 (1979).
[9] V.Aynutdinov et al. (Baikal-Collaboration) Nucl. Instr. and Meth.A, **598**, 282 (2009).
[10] A.Avrorin et al., (Baikal-Collaboration) EPJ Web Conf. **121**, 05003 (2016).

EPJ Web of Conferences **135**, 06005 (2017) DOI: 10.1051/epjconf/201713506005
ARENA 2016

Signal classification and event reconstruction for acoustic neutrino detection in sea water with KM3NeT

Dominik Kießling[1,*] for the ECAP Acoustic Detection Group

[1] *Friedrich-Alexander-Universität Erlangen-Nürnberg, Erlangen Centre for Astroparticle Physics, Erwin-Rommel-Str. 1, 91058 Erlangen, Germany*

Abstract. The research infrastructure KM3NeT will comprise a multi cubic kilometer neutrino telescope that is currently being constructed in the Mediterranean Sea. Modules with optical and acoustic sensors are used in the detector. While the main purpose of the acoustic sensors is the position calibration of the detection units, they can be used as instruments for studies on acoustic neutrino detection, too. In this article, methods for signal classification and event reconstruction for acoustic neutrino detectors will be presented, which were developed using Monte Carlo simulations. For the signal classification the disk–like emission pattern of the acoustic neutrino signal is used. This approach improves the suppression of transient background by several orders of magnitude. Additionally, an event reconstruction is developed based on the signal classification. An overview of these algorithms will be presented and the efficiency of the classification will be discussed. The quality of the event reconstruction will also be presented.

1 Introduction

The acoustic neutrino detection technique is a promising approach to create a neutrino telescope with several cubic kilometers effective volume in the energy range beyond 10^{18} eV. Measuring the cosmic neutrino flux in this energy range allows testing the flux predicted by the GZK-effect[1–3]. The acoustic signal generated by neutrinos is produced via the thermoacoustic effect, which was predicted by Askaryan [4]. The neutrino interacts with the sea water, generating a hadronic cascade. Superposition of the emitted elementary sound waves from each point of the shower generates a bipolar pulse. The emission strength is strongly peaked in the plane perpendicular to the direction of the incoming neutrino. This characteristic shape of the sound emission is often called "pancake". The properties of the sound emission pattern by neutrino signals can be used to classify the recorded events. As experience with the experimental test set up AMADEUS [5] has shown, using only the bipolar shape of the waveform to distinguish the background and the signal causes too many false positives [6]. The initial rate of events recorded by the AMADEUS project, where a reconstruction of the source position is possible, will be shown in Table 1. Several classifiers remove background events, but after all currently implemented steps are applied, 63 k events/year remain. The effective volume of AMADEUS is very small compared to the size needed for an actual acoustic neutrino detector. The GZK neutrino flux would have to be several orders of magnitude higher than the most optimistic

*e-mail: dominik.kiessling@fau.de

prediction to produce that many events. Therefore it is assumed that all remaining events are caused by background sources. However, further improvement of the background suppression would require a cut on the "pancake" emission pattern. The size of the AMADEUS set up unfortunately is not suited for this task. The future detector of the KM3NeT collaboration does not have this size limitation.

2 The KM3NeT neutrino telescope

The KM3NeT Collaboration is currently constructing a neutrino telescope of the same name in the Mediterranean Sea [7]. In this study, possibilities of acoustic neutrino detection with the device are investigated. The telescope will be composed of several detection units anchored at the sea bed, which are kept taut vertically by a buoy. Each detection unit of KM3NeT/ARCA has a length of about 700 hundred meters. There are 18 modules on each unit with spacings of 36 m, each housing photomultiplier tubes and one acoustic sensor. The detection units are about 90 m apart from each other. While the primary purpose of the acoustic sensors is the position calibration of the detector, it can also be used to study the feasibility of the acoustic neutrino detection. Since the response function has not been fully measured for the acoustic sensors used in the KM3NeT-modules, it is assumed that they have the same characteristics as the hydrophones in AMADEUS [5]. A building block for the final detector comprises 115 detection units. The signals from transient background sources and neutrino signals produce different signatures in such a large detector. The amount of data produced by approx. 2000 sensors, which have to be sampled at a frequency of at least 100 kHz for acoustic neutrino detection, is very large. It is not feasible to store all produced data, so the following signal classification will be using only the peak to peak amplitude and the arrival time of the sound.

3 Simulating the event samples

Both the background and the signals were simulated using the software SeaTray that was developed internally. The background consists of spherically emitting sound sources, that are placed randomly in a cylindrical volume around the detector. The initial amplitude is also chosen randomly in the range of signals recorded by AMADEUS. Another class of background signals are emissions from the acoustic positioning system. These differ from the first kind, since their source is always located close to the detector and their amplitude is fixed. Finally, the third part of the background are random coincidences between signals in multiple hydrophones. The relative occurrence for the samples was chosen as 75:20:5 respectively, which keeps the order of the actual rates and guarantees a statistical significant number of events for each class. The neutrino signals are generated by code adapted from the ACoRNe project [8]. The interaction vertex is chosen randomly in the same cylinder as for the background samples. The direction of the neutrino is randomly chosen from an uniform distribution on the upper hemisphere. This can be done because the earth is opaque for neutrinos with energies exceeding 10^{18} eV, which is roughly the energy threshold for acoustic detection. The shower energy is determined from neutrino energy by using an estimation for the Bjorken-y value, while the neutrinos are generated according to a E^{-1} power spectrum. The amplitude is calculated by a parametrization from the shower energy. The arrival time of the pressure pulse at the sensors is determined by raytracing the path from the source. The ration of background to signal events in the samples is varied from 1 to 3, while the typical size for a training or testing sample is 20 k events.

4 Signal classification

The signals are classified by using multivariate analysis. A set of features is calculated from the signature in the detector, which is independent of the exact layout of the detector. This ensures that

DOI: 10.1051/epjconf/201713506005

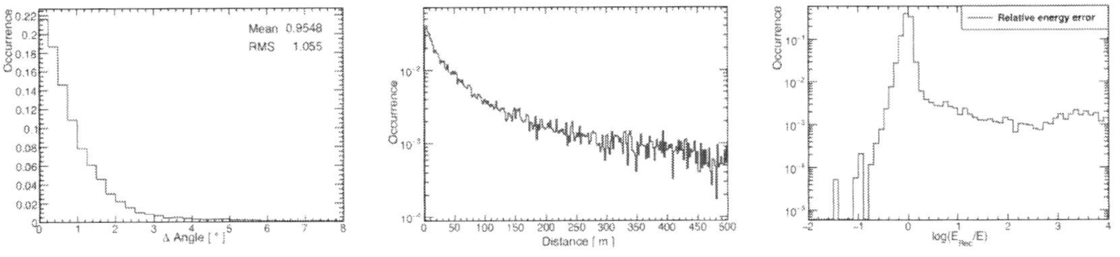

(a) Angular resolution (b) Vertex resolution (c) Energy resolution

Figure 1: Resolution of the event reconstruction for 20 k simulated neutrinos. Panel (c) shows the relative occurrence of the ratio of the reconstructed and the true shower energy. For 90% of the reconstructed events the error is smaller than 75% of the true energy.

the classification will still be functional if small changes to the detector occur, e.g. the failure of a detection unit. First, the position of the sound source is reconstructed. In order to calculate the other features, the positions of all sensors that triggered for that event are used to generate a 3-dimensional point cloud. The distances are normalized, so that the center of gravity of the point cloud is in the origin, and the variance of the distances from the origin is normalized to 1. A singular value decomposition is used to process the data. The singular values obtained this way are the first set of features. The singular vector corresponding to the smallest singular value is also the normal vector to the pancake plane. It represents the direction of the neutrino (if it is not a background event) with an accuracy of about 2.5°. The sound disk is reconstructed by using the latter vector as normal vector and the center of gravity of the point cloud in the detector as starting point. The distance between the reconstructed interaction vertex and this plane is another feature. Another value used is the sum of the squared distances of the sensors from the reconstructed plane. Since the amplitude for neutrino signals should decrease with the distance from the plane, the correlation coefficient between those values is also used as feature. Lastly, a "likelyhood" L of the signature is calculated by summing up $L = \sum_i \left(\frac{\alpha_i}{\sigma}\right)^2$, where α is the angle between the pancake plane and the direction in which the sound is emitted to reach the sensor with index i. The value $\sigma = 3.2°$ represents the width of the pancake, taking into account the error of the vertex and direction reconstruction.
This feature vector is then analyzed by using the machine learning tools from the OpenCV [9] library. Boosted decision trees, random decision trees and a simple decision tree are used here. The algorithmns were trained on an individual sample of 20 k events each, minimizing the cross entropy. The clssification is carried out as a majority vote. In total, this removes 99.99% of background events (tested on 120 k events), while 98% of neutrinos remain in the sample, see Table 1.

5 Energy reconstruction

Now only the energy of the neutrino is missing for the complete event reconstruction. It should be noted that there is no way to determine the Bjorken-y, so even a perfectly reconstructed shower energy would only impose a lower limit for the neutrino energy. In the previous section, the parameters of the event have been reconstructed in a step by step manner. The energy however is extremely dependent on the direction and vertex reconstruction, so a simple calculation using the previously reconstructed values will often produce wrong results due to small errors in the input values. One possible way to circumvent this is using a combined fit for all parameters. A toy Monte Carlo algorithm is used

Table 1: Remaining events after each step of the background reduction. The first three steps were already implemented for AMADEUS.

Background reduction method	Events per year
Source position reconstruction possible	10000000
Waveform classified as bipolar pulse	315000
Temporal and spatial clusters removed	63000
Pancake–signature detected	7

to create the arrival times and amplitudes at the sensors for a given set of vertex position, shower direction and shower energy. The MINUIT package from ROOT is used to minimize the following function f by varying the starting values, where the sum runs over:

$$f = \sum_i \left[\frac{\Delta t_i^2}{\sigma_t^2} + \left(H(A_i^{rec} - A_i) \frac{A_i - A_i^{rec}}{\sigma_A} + H(A_i - A_i^{rec}) \frac{A_i (A_i - A_i^{rec})}{\sigma_A A_i^{rec}} \right)^2 \right] \tag{1}$$

where Δt_i is the difference between the measured and reconstructed arrival time, $H(x)$ is the Heaviside step function, A_i the (reconstructed) amplitude at sensor i and the σ are the respective expected deviations. The resulting errors for this combined fit are much smaller than for the step by step reconstruction. The angular resolution for a KM3NeT Building Block is 1° and the distribution of the errors is shown in Fig. (1a). The average vertex error is 250 m, see Fig. (1b). The energy resolution for the shower energy is 30%, as shown in Fig. (1c).

6 Conclusions

The rate of transient background sources for acoustic neutrino detection can be reduced by 4 orders of magnitude when the sound emission pattern is taken into account. With the rates shown in Table 1, this would reduce the background events from 63 k events/year to 7 events/year. A further reduction of the background rate is still necessary, as the the expected rate of neutrinos is extremely small. This can be achieved by using quality cuts on the reconstruction of the combined fit of vertex position, shower direction and energy. Preliminary results show a possible further reduction of the background by 80%, while also removing about 20% of the neutrinos.

References

[1] K. Greisen, Phys. Rev. Lett. **16**, 748 (1966)
[2] G.T. Zatsepin, V.A. Kuzmin, JETP Lett. **4**, 78 (1966)
[3] V.S. Berezinsky, G.T. Zatsepin, Phys. Lett. B **28**, 423 (1969)
[4] G.A. Askariyan et al., Nucl. Instr. and Meth. **164**, 267 (1979)
[5] J. Aguilar et al., Nucl. Instrum. Methods Phys. Res., Sect. A **626-627**, 128 (2011)
[6] M. Neff, Dissertation, Friedrich-Alexander-Universität Erlangen-Nürnberg (2012)
[7] S. Adrián-Martínez et al. (KM3NeT collab.), J. Phys. G: Nucl. Part. Phys. **43**, 130 pages (2016)
[8] S. Bevan et al., Astrop. Phys. **28**, 366 (2007)
[9] *Open source computer vision library* (2014), `http://opencv.org/`

A large fiber sensor network for an acoustic neutrino telescope

Ernst-Jan Buis[1,*], *Ed* Doppenberg[1], *Robert* Lahmann[2], and *Peter* Toet[1]

[1] *TNO Technical Sciences, P.O. Box 155, 2600 AD Delft, The Netherlands*
[2] *Erlangen Centre for Astroparticle Physics, University of Erlangen, Erwin-Rommel-Str. 1, 91058, Erlangen, Germany*

Abstract. The scientific prospects of detecting neutrinos with an energy close or even higher than the GKZ cut-off energy has been discussed extensively in literature. It is clear that due to their expected low flux, the detection of these ultra-high energy neutrinos ($E_\nu > 10^{18}$ eV) requires a telescope larger than 100 km^3. Acoustic detection may provide a way to observe these ultra-high energy cosmic neutrinos, as sound that they induce in the deep sea when neutrinos lose their energy travels undisturbed for many kilometers. To realize a large scale acoustic neutrino telescope, dedicated technology must be developed that allows for a deep sea sensor network. Fiber optic hydrophone technology provides a promising means to establish a large scale sensor network [1] with the proper sensitivity to detect the small signals from the neutrino interactions.

1 Introduction

The measurements of neutrinos with energies higher than what has been detected so far may provide input in one of the long standing issues in astrophysics, which is the mystery of cosmic rays and the mechanism of their acceleration to the highest energies measured [2]. To detect the neutrinos that originate from the GKZ cut-off, a neutrino telescope is required with a detection volume of at least 100km^3. To establish such as large detection volume one could employ the technique of acoustic detection of neutrinos [3–5]. In this paper we motivate the use of optical sensor networks as a building block of an acoustic neutrino telescope and provide an overview of the research and development of the fiber hydrophone technology at TNO.

2 Acoustics neutrino telescope concept

As the absorption length of sound in seawater is as large as ~1 km (at an acoustic frequency of 25kHz), the acoustic detection of high energy cosmic rays provides a means to create a truly large neutrino telescope. An example of an acoustic signal is shown Fig. 1 (a), where a typical bipolar pulse is plotted that has an amplitude at the order of a mPa. The dominant frequency components of this signal are found between 1 and 40kHz.

To discriminate the acoustic signal from background noise, a telescope is required that is equipped with many (>1000) hydrophones. It means that a cost-effective technology needs to developed that

*e-mail: ernst-jan.buis@tno.nl

can be deployed on industrial scale, but that provides the proper sensitivity to detect the mPa signals at the same time. Optical fiber hydrophone technology might provide a solution here as the hydrophones can produced and deployed on a cost effective way. Moreover, the technology is based on (passive) optical fibers, so that (in contrast to conventional piezo-hydrophones) no electronics (or electrical power) is needed at the sensitive parts of the system. Finally, since the signals from the hydrophones may travel through optical fibers unattenuated for many kilometers, the technology is suited for a large sensor network. A concept of an acoustic neutrino telescope is depicted in Fig. 1 (b), where a 3D grid of hydrophone strings is deployed a depth of several km, similar to the concept of KM3NeT.

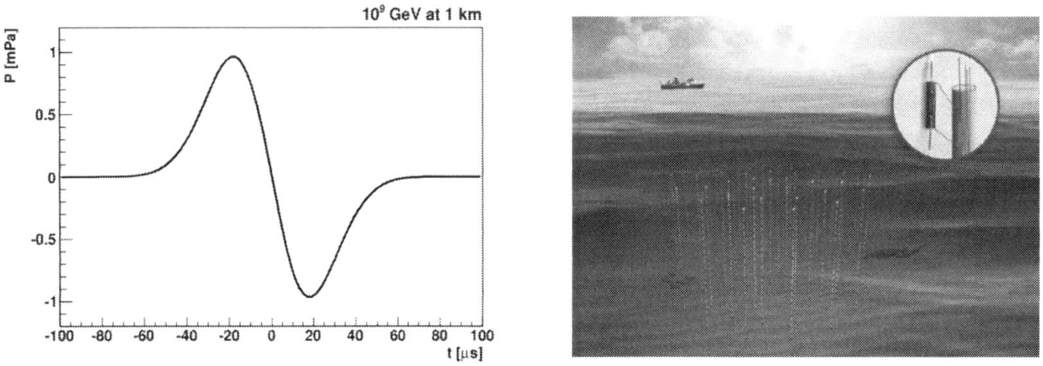

(a) (b)

Figure 1. (a) Typical acoustic signal from a energetic neutrino induced particle shower. Adapted from [6]. In (b) an underwater neutrino telescope is schematically depicted.

3 Optical fiber hydrophone technology

The optical fiber hydrophone that has been developed at TNO is explained in detail in [7, 8] and shown in Fig. 2. A hydrophone consists of a mechanical transducer that converts a pressure wave in to strain in the optical fiber. At the position of the transducer a fiber laser structure is locally inserted in the fiber, which schematically depicted in figure Fig. 2 (b). A fiber laser consists of two fiber Bragg gratings (FBG) and a gain medium that includes a high concentration of Erbium ions. The cavity that is defined by the two FBGs at either side is pumped using an external pump laser that can be located at a large distance. The emitted laser light is highly coherent as it is defined by the cavity and the reflectivity of the FBGs In Fig. 2 (b) the reflectivity spectrum of the FBGs is shown. The strain that is induced on the cavity by the transducer will result in a wavelength shift of the emitted laser light. Finally, this shift is a measure for the acoustic signal.

4 Interrogation

Using the optical hydrophones as discussed above a sensor network can be created by implementing many transducers on a single fiber as shown in Fig. 3. On a single fiber about 10 hydrophones can be housed [8]. A dedicated interrogator has been developed that is able to detect minute wavelength shifts. The interrogator is based on a 3x3 interferometer, which is used to determine the wavelength shift of the light that is coupled into the interrogator. The output of the 3x3 coupler consists of three

DOI: 10.1051/epjconf/201713506006

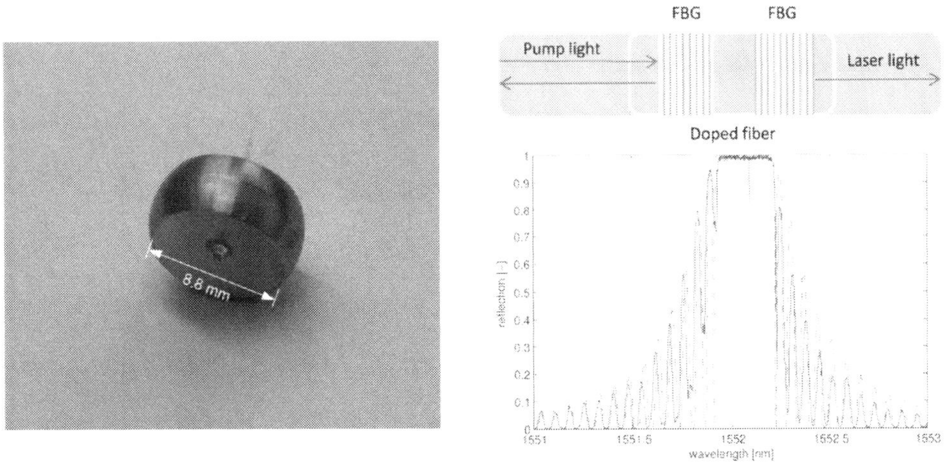

(a) (b)

Figure 2. Optical hydrophone technology: in (a) a transducer is shown attached to a optical fiber. The transducer is designed to have a strain sensitivity of $\eta = 70\ n\varepsilon/mPa$. In (b) the fiber laser cavity is shown that is located in the optical fiber at the position of the transducer. Also shown is the reflectivity spectrum of the FBGs.

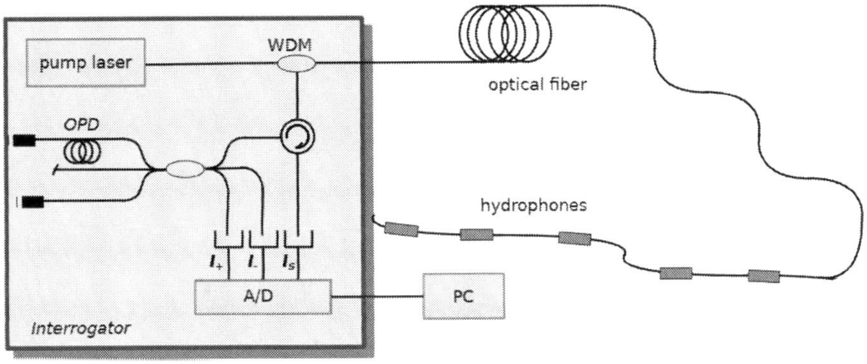

Figure 3. Concept of the optical fiber sensor network: An optical fiber with several fiber lasers and sensors are read out using an interrogator. Once deployed in the deep-sea, several of these strings make up a neutrino telescopes.

light intensities (I_x) as indicated in Fig. 3. Note that the outputs of the three arms are identical, but each is shifted in phase over $\frac{2}{3}\pi$ as given in the following expression.

$$\begin{aligned} I_S(OPD) &= I_0(1 + V\cos(\tfrac{2\pi}{\lambda}OPD)) \\ I_+(OPD) &= I_0(1 + V\cos(\tfrac{2\pi}{\lambda}OPD + \tfrac{2\pi}{3})) \\ I_-(OPD) &= I_0(1 + V\cos(\tfrac{2\pi}{\lambda}OPD - \tfrac{2\pi}{3})) \end{aligned} \tag{1}$$

where I_0 is the normalized amplitude of the signal and V is the visibility. Using the phase shift and the intensity of each arm, the optical path length (OPD) is given by the following equation:

$$OPD = \frac{\lambda}{2\pi}\arctan\left(\sqrt{3}\frac{I_+ - I_-}{2I_s - I_+ - I_-}\right) \tag{2}$$

DOI: 10.1051/epjconf/201713506006

Because the optical path difference (OPD) is known, the wavelength shift of each hydrophone in the string can be accurately determined when the three output intensities are measured. In Fig. 4 the performance of the interrogator is plotted as a function of the (acoustic) frequency. One can see from the plot that the strain noise is at the order of $10^{-13}\varepsilon/\sqrt{Hz}$, while the strain sensitivity of the transducer is around $\eta = 70\ n\varepsilon/mPa$. As an example, a measurement of a synthetic pulse (after filtering) is shown Fig. 4 (b), in which the typical bipolar structure is clearly visible.

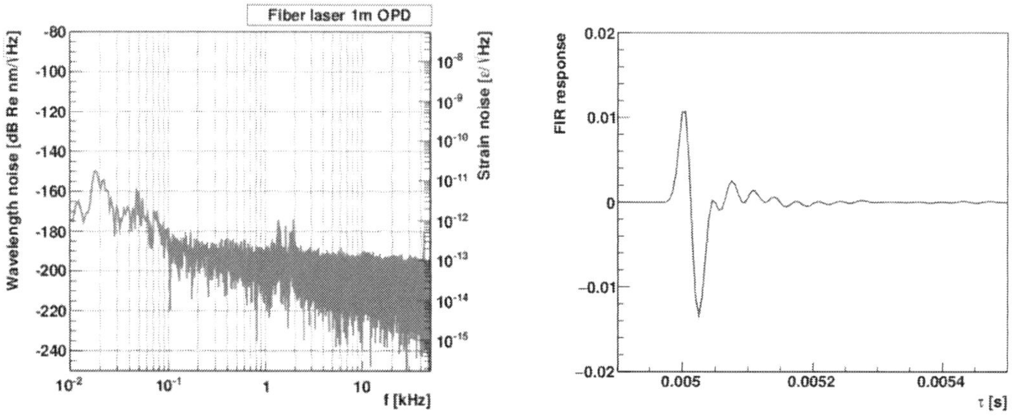

(a) (b)

Figure 4. (a) The noise performance of the interrogator. In (b) an example of a measured pulse is shown after filtering. While a large part of the mechanical resonance was filtered, the graph still shows some ringing.

5 Conclusions

This conference has shown that the perspectives for both radio and acoustic detection of high energetic neutrino are bright as new technologies to detect these particles become available. It is of high importance that both detection methods will be pursued for many reasons, such as cross calibrations and background removal. The technology for the acoustic detection of neutrinos as discussed in this paper has the potential to be implemented on large scale with the proper sensitivity and is hence suitable for a large volume neutrino telescope. Next steps in the development are targeted on the multiplexing of the output of a large number of hydrophones.

References

[1] E.J. Buis et al., Journal of Instrumentation **9**, C03051 (2014)

[2] E. Waxman, *Neutrino Astronomy: What did we learn, Where should we aim?*, (2016), these proceedings

[3] G.A. Askaryan, Zemlia i Vselennaia **1**, 13 (1979)

[4] J.G. Learned, Phys. Rev. D **19**, 3293 (1979)

[5] R. Nahnhauer, Nucl. Instr. Meth. A **692**, 58 (2012)

[6] R. Lahmann, Ph.D. thesis, Friedrich-Alexander Universität Erlangen (2011)

[7] E.J. Buis et al., *Design studies for a neutrino telescope based on optical fiber hydrophones*, (2015), in proceedings of the "34th International Cosmic Ray Conference"

[8] V.M. Baas et al., *Fiber laser design and measurements for fiber optical hydrophones*, (2015), in proceedings of the "34th International Cosmic Ray Conference"

EnEx-RANGE - Robust autonomous Acoustic Navigation in Glacial icE

Dirk Heinen[1,*], *Dmitry* Eliseev[1], *Christoph* Henke[2], *Sabina* Jeschke[2], *Peter* Linder[1], *Sebastian* Reuter[2], *Sebastian* Schönitz[2], *Franziska* Scholz[1], *Lars Steffen* Weinstock[1], *Stefan* Wickmann[1], *Christopher* Wiebusch[1], and *Simon* Zierke[1]

[1] *III. Physikalisches Institut, RWTH Aachen University, Germany*
[2] *IMA/ZLW & IfU, RWTH Aachen University, Germany*

Abstract. Within the Enceladus Explorer Initiative of the DLR Space Administration navigation technologies for a future space mission are in development. Those technologies are the basis for the search for extraterrestrial life on the Saturn moon Enceladus. An autonomous melting probe, the EnEx probe, aims to extract a liquid sample from a water reservoir below the icy crust.
A first EnEx probe was developed and demonstrated in a terrestrial scenario at the Blood-falls, Taylor Glacier, Antarctica in November 2014. To enable navigation in glacier ice two acoustic systems were integrated into the probe in addition to conventional navigation technologies. The first acoustic system determines the position of the probe during the run based on propagation times of acoustic signals from emitters at reference positions at the glacier surface to receivers in the probe. The second system provides information about the forefield of the probe. It is based on sonographic principles with phased array technology integrated in the probe's melting head. Information about obstacles or sampling regions in the probe's forefield can be acquired. The development of both systems is now continued in the project EnEx-RANGE. The emitters of the localization system are replaced by a network of intelligent acoustic enabled melting probes. These localize each other by means of acoustic signals and create the reference system for the EnEx probe.
This presentation includes the discussion of the intelligent acoustic network, the acoustic navigation systems of the EnEx probe and results of terrestrial tests.

1 Enceladus Explorer – EnEx

The measurements by the CASSINI mission to the Saturn moon Enceladus discovered a global ocean below the icy crust [1]. Hence, it is a possible candidate for extraterrestrial life within our solar system. The mission identified active cryovolcanoes at Enceladus' south pole region. This region is a point of interest for a future mission to Enceladus.
The Enceladus Explorer Initiative (EnEx) managed by the DLR Space Administration aims for an Enceladus mission. Within this initiative navigation base technologies are developed. The idea is to develop a probe, which is able to navigate through ice, find a liquid water reservoir, take a sample and

*e-mail: heinen@physik.rwth-aachen.de

analyze it. Several projects within the initiative focus on different navigation techniques.
A first EnEx probe for a terrestrial test scenario was developed in the years 2012 – 2015 and successfully demonstrated in 2014 [2]. For the demonstration a clean sample was retrieved from a water filled crack at a depth in about 20 m at the Blood Falls, Taylor Glacier, Antarctica.
For this scenario the carrier probe, the EnEx-IceMole, was equipped with navigation, decontamination and sampling systems. The EnEx-IceMole is able to drive curved trajectories through ice. To determine the position and orientation of the probe an acoustic positioning system, an inertial measurement unit and a magnetometer were integrated into the EnEx probe. To identify objects, e.g. stones or liquid water, in front of the probe the forefield was scanned with ultrasonic phased arrays.
The EnEx probe was tested in glaciers in the Alps and Antarctica in advance of the final demonstration at the Blood Falls.

2 EnEx-RANGE

The EnEx-RANGE project is an interdisciplinary project between physics and engineering within the EnEx Initiative. The abbreviation RANGE stands for Robust autonomous Acoustic Navigation in Glacial icE.
The main subject of the project is the development of an autonomous acoustic sensor network (figure 1). The nodes of the network are Autonomous Pinger Units (APUs), which are described in section 3. In addition to the network the acoustic navigation systems for the forefield exploration and the positioning of the first EnEx probe will be enhanced. The new systems are going to be validated and demonstrated in collaboration with the other EnEx Initative projects in the Alps in 2017 and 2018.

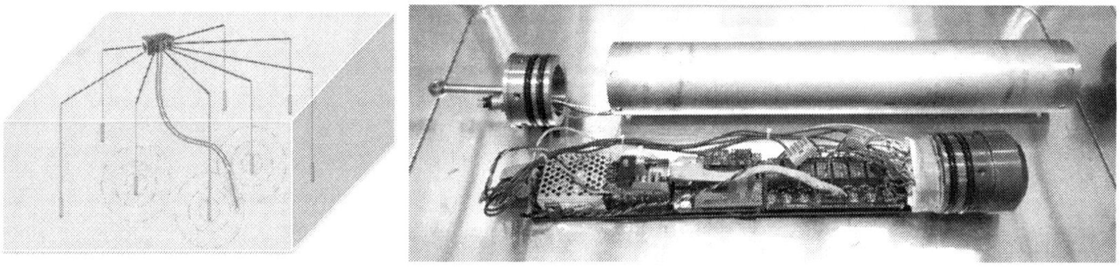

Figure 1. The scheme on the left side shows an artistic impression of the APU network (typically 50 to 100 m distance between APUs). The EnEx probe navigates through the network and determines its position in reference to the network. The picture on the right shows the first APU without acoustic instrumentation. This APU was successfully tested in the Alps in 2015 and 2016.

3 APU Network

The Autonomous Pinger Units (APU) are melting probes, equipped with local intelligence, an acoustic pinger, acoustic receivers, house keeping sensors and a melting control system. The APUs are able to communicate via acoustic signals and determine the distances between each APU by the propagation of these signals. This information is exchanged by the APUs to determine the relative position of each APU within the network. The local intelligence optimizes the APU network in size and positioning error. After the APU network is set up, it provides the reference for the EnEx probe to navigate trough.

The APU carrier system is mainly a watertight, pressure housing with an outer diameter of 8 cm and a length about 60 cm (figure 1). The melting head with heating elements allows the APU to melt vertically into ice at a speed of about 3 m/h. A stainless steel tube connects the melting head and the back plate. A cable, for power and communication, and a rope, to extract the probe after the scenario, are attached to the back plate. The APUs' control system is system on a chip (SoC) based. The SoC manages the heater control, the data acquisition, the signal generation and provides the computing power for the autonomy. For house keeping a MARG [1] sensor, a pressure sensor and several temperature sensors are integrated into the carrier system. A first APU without the acoustic instrumentation was tested in test campaigns in summer 2015 and 2016 on glaciers in the Alps.

The acoustic pinger design aims for long-range communications between the APUs. To achieve such a long range the pinger will be optimized to a high output power with respect to the limitation in size within the APU carrier system. For a good coupling between the pinger and ice, the acoustic transducer is pressed against the melting head. The acoustic transducer consists of a stack of lead zirconate titanate (PZT) rings. The APUs' SoC controls an arbitrary function generator for signal generation. These signals are amplified by a transistor based, high power amplifier and fed to the transducers. A transducer integrated into an APU was first tested in the 2016 test campaign.

The design of the acoustic receivers is matched to the frequency range of the pingers' signal frequency. Three acoustic transducers, lead zirconate titanate (PZT) disks, are glued on the inner sides of the APU tube. Each transducer is coupled to front end electronics, which filter, amplify and digitize the signals. For first in ice measurements an APU instrumented with the acoustic receivers was tested in the 2016 test campaign. To test the frequency response of the APU receivers coupled to glacial ice, signals were emitted by a reference transducer and received by a demonstrator APU over a short distance of about 1 m. Furthermore the communication between the APU pinger and the APU receivers was successfully demonstrated (figure 2).

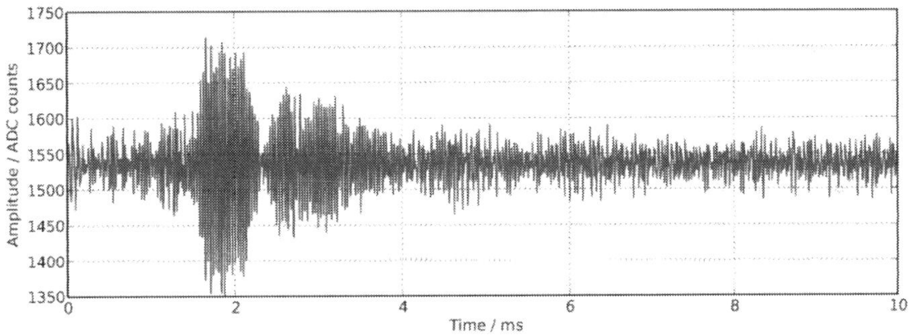

Figure 2. The plot shows a waveform measured during a test campaign on the Langenferner glacier in the Alps in 2016. The acoustic signal was emitted by an APUs' pinger and measured by another APUs' receiver.

The network is designed to be able to configure itself autonomously and to compensate system failures of individual APUs. Each single APU is controlled by a software agent. This agent is realized by a rule-based expert system. The expert system evaluates available information of the environment and creates an internal world model. This internal world model is shared with the other APUs. Based on the shared world model each APU decides its further actions. The APU positions are iteratively determined to define the network topology. To study different scenarios depending on various constraints (e.g. fixed starting positions of APUs, air filled cracks, inhomogeneities in ice quality, size of APU network) simulation modules for all APU components and a physical model of the ice are

[1]MARG, acronym for Magnetic, Angular Rate, and Gravity

 DOI: 10.1051/epjconf/201713506007

available. The first test of an APU network in ice will be in 2017 in the Alps. For this test in total seven fully instrumented APUs will be available and form the network.

4 Acoustic Positioning

Similar to the APU, an APU SoC and four acoustic receivers will be integrated into the EnEx probe. The APUs will be able to share their internal world model with the information about their positions. By recording and analyzing the acoustic signals from each APU, the EnEx probe will determine its position in reference to the network.

5 Acoustic Forefield Exploration

This system provides information about the forefield of the EnEx probe. It is based on sonographic principles with transducer arrays integrated in the probe's melting head. By modulation of the phase of the acoustic signal generation at each transducer the acoustic beam can be steered and focused. Any objects, e.g. stones, crevasses or liquid water, in front of the probe reflect this signal. The reflections are acquired at the transducers and visualized for the operator in a B-scan. After the identification of an object, the operator is able to act, e.g. steer the probe around a stone.

6 Conclusion and Outlook

The APU network is a robust approach for navigation within a glacier. The APUs' hardware is in development and a first network with seven APUs will be demonstrated in a test campaign in 2017. The EnEx probe, with the acoustic navigation systems, was successfully demonstrated in a terrestrial scenario and will take part in the up coming test campaigns.
Furthermore, synergies exist between the EnEx activities and research projects within the ARENA community. At the moment we plan to support the ARIANNA project [4] with the development of an APU like probe for deploying antennas at a depth of about 100 m into the Ross Ice Shelf. Furthermore, we developed an acoustic sensor based on the EnEx receiver concept for the IceCube-Gen2 neutrino detector for module positioning and acoustic neutrino detection [5].

Acknowledgment

The Enceladus Explorer Initiative is managed by the DLR Space Administration. The EnEx-RANGE project is funded by the German Federal Ministry of Economics and Energy (BMWi) by resolution of the German Federal Parliament under the funding code 50NA1501.

References

[1] D. A. Patthoff and S. A. Kattenhorn, Geophys. Res. Lett. **38**, L18201 (2011)
[2] J. Kowalski et al., Cold Regions Science and Technology **123**, 53-70 (2016)
[3] D. Eliseev et al., Annals of Glaciology **55(68)**, 253-259 (2014)
[4] A. Nelles et al., EPJ WoC, in these proceedings, ARENA 2016
[5] S. Wickmann et al., EPJ WoC, in these proceedings, ARENA 2016

EPJ Web of Conferences **135**, 06008 (2017) DOI: 10.1051/epjconf/201713506008
ARENA 2016

Acoustics in water: synergies with marine biology

Mike van der Schaar[1,*], *Robert* Lahmann[2], *Kay* Graf[2], *Ludwig* Houégnigan[1], *Alba* Solsona[1], *Steffen* De Vreese[1], *Antonio* M. Sánchez[1], and *Michel* André[1]

[1] *Laboratory of Applied Bioacoustics, UPC, Barcelona, Spain.*
[2] *Erlangen Centre for Astroparticle Physics (ECAP), Friedrich-Alexander-Universität, Erlangen, Germany.*

Abstract. This paper presents some of the bioacoustics related analysis that was performed on the ANTARES data, focussing on the year 2014. The data was processed for sperm whale, dolphin and shipping presence and grouped by hour of the day. It seemed that dolphins were more socially active during the day and foraging during the night. Sperm whales were mostly foraging during the day, but they may have been moving to other areas during the night. The most intense shipping noise came from a ferry that passed the platform twice a day. Although beaked whales were expected to be present in the area, so far their biosonar signal has not been conclusively found.

1 Introduction

The principal aim of the ANTARES collaboration is the operation of a Cherenkov Neutrino telescope [1]. In addition to the optical sensors required for this purpose, an array of 36 hydrophones was installed with the aim of investigating acoustic neutrino detection techniques. This acoustic data stream has been active since 2008 and is managed under AMADEUS [2]. Coincidentally, ANTARES is installed in the Pelagos marine sanctuary, home to many marine mammals species, both whales and dolphins, all of which are acoustically active. The fact that ANTARES is cabled (providing real-time access to data) makes it a unique observation platform for the local marine fauna. The Laboratory of Applied Bioacoustics (LAB) has been analysing the acoustic data, relayed through AMADEUS, since 2010 with a focus especially on sperm whales and dolphins. The ability to study trends in animal presence and a possible relationship with anthropogenic activities for such a long and continuous time frame is unique and is impossible as a purely biological initiative; the costs for installing such a platform just to measure sound levels and to monitor the environment are too high. Opening up platforms such as ANTARES, and for example the NEPTUNE Ocean Observatory (part of Ocean Networks Canada [3]) which offers open access to data from both acoustic and environmental sensors, to marine researchers will not only help to improve marine research but may also help to obtain funding by widening the scope of applications.

Presented here are some of the signal processing techniques that have been used for the acoustic event detection and classification algorithms on data recorded during 2014 at ANTARES. A few examples of both biological and anthropogenic sonar are shown, and daily cetacean activity is evaluated.

*e-mail: mike.vanderschaar@upc.edu

In the evaluation of cetacean presence or the measurement of shipping noise it has to be taken into account that the hydrophones are installed deep under water. The sound speed profile of the Mediterranean Sea does not have a common sound channel at depth and is mostly refracting upwards [4]. This means that most surface sources, such as dolphins and ships, may have to be on top of the array for the sound to be able to reach the hydrophones. Deep diving animals, such as beaked whales [5] and especially sperm whales [6] who can regularly dive below 1 km depth, will have a better horizontal detection range particularly when they vocalize downwards.

2 Bioacoustics at ANTARES

As part of the ESONET project [7] a server was installed at the ANTARES data centre to analyse the incoming data. For detection a single channel was used, depending on which one was available. The data was run through software developed by the LAB for bioacoustics data analysis called Listening to the Deep Ocean Environment (LIDO [8]). LIDO takes acoustic data in segments (normally aournd 16 seconds in length) and processes these segments as a whole. It includes modules for sound level measurements, acoustic event detection and classification, localization, spectrogram creation, acoustic data compression and other functions.

The LIDO software uses two broad acoustic categories. First are impulsive signals which are broadband in frequency and well defined in time [9]. Examples of impulsive sounds are biosonar, noise from shipping, echo sounders, etc. The second type of signals are named short tonal [10]. These signals are well defined in frequency and can have a duration from half a second to half a minute or longer. They may be frequency modulated. A signal that remains constant over the duration of an entire segment (e.g. power line interference) is removed by the detector (hence "short tonal"). Examples of these events are dolphin whistles, whale calls, fish, and e.g. some ships produce tonal sounds. Many baleen whales produce low frequency sweeps that depending on the time-frequency resolution used for the analysis can be detected by both impulsive and tonal detectors. However, at ANTARES baleen whales cannot be detected due to the hardware configuration where low frequency sound is filtered.

Normally, multiple versions of these detectors are used to operate in specific frequency bands. In some cases, the detector can be specific enough to primarily pick up events from a particular source. When a detector is not sufficient to identify a source, classifiers are used. The main classifier used in LIDO is a Gaussian Mixture Model (GMM [11]). A separate GMM is fit on the feature space of each competing class, with features composed from the time-frequency domains. When a new acoustic event is detected it is passed through each GMM model, and the model with the highest likelihood decides the class. Under this approach a single event can have a high likelihood for multiple models. In that case the software might not take an immediate decision on the class, indicating that one of the high scoring classes may be present. Over time it is likely that the real class will maintain high scores while the other false classes will score lower. Alternatively, all models may have a low score on a new event. When this happens frequently in a short time period these events are flagged and recorded for manual analysis. They can be used to retrain a class or to introduce a new class. At ANTARES classifiers are used for impulsive signals to distinguish between sperm whale and dolphin biosonar, and shipping impulses.

After classification spectrograms and compressed audio streams are created for real-time playback in the software interface. Especially the spectrograms are useful for post-analysis and obtaining

 DOI: 10.1051/epjconf/201713506008

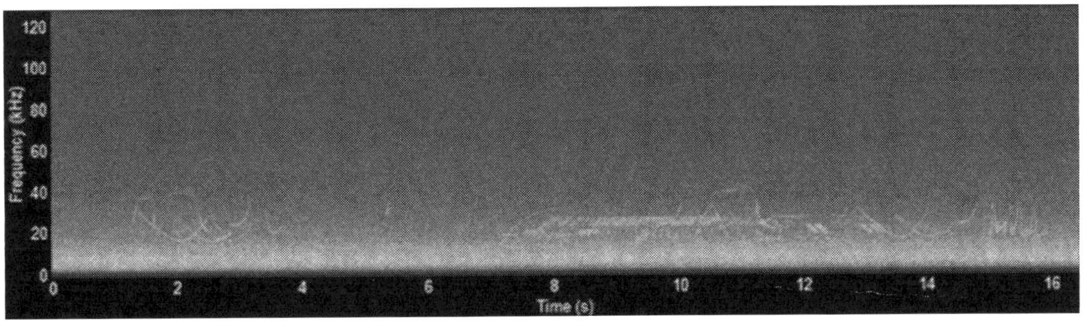

(a)

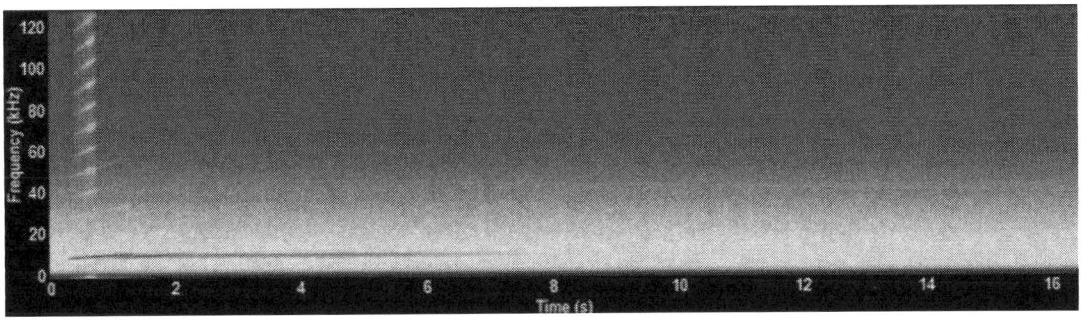

(b)

Figure 1: The top image (a) shows typical dolphin whistles and between 8 and 12 seconds a modulated sound slightly above 20 kHz that is not from a dolphin. The bottom image (b) shows a sonar signal.

a quick overview of the quality of the automatic detection or classification. Sometime multiple spectrograms are created to focus on low and high frequencies separately.

2.1 Short tonal events

Short tonal events detected between 2.5 - 20 kHz at ANTARES were almost always produced by dolphins. Occasionally other signals were encountered such as shown in Fig. (1). On the top image Fig. (1a) there are common dolphin sounds mixed with a frequency modulated tone that continues for a few seconds; the tone might be related to an acoustic release. On the bottom image Fig. (1b) there is signal from what is believed to be a sonar exercise. Fig. (2) shows the short tonal detections throughout the year and for each hour of the day. All histogram frequency axes in this report are provided as percentages representing the percentage of the time that the activity was measured. The monthly histogram Fig. (2a) indicates that dolphins were present all year round, with increased activity at the end of the summer and in fall. The hourly image Fig. (2b) suggests that the dolphins were mostly socially active during daylight hours. These findings are currently being studied in more detail.

2.2 Ultra sonic cetacean clicks

Ultra sonic cetacean clicks (UCC) are defined as impulses detected between 20 - 46 kHz and classified as UCC. Signals at these frequencies are produced mostly by dolphins and beaked whales. Sperm whale sonar and occasionally impulsive shipping sound is detected in this range as well. For this

DOI: 10.1051/epjconf/201713506008

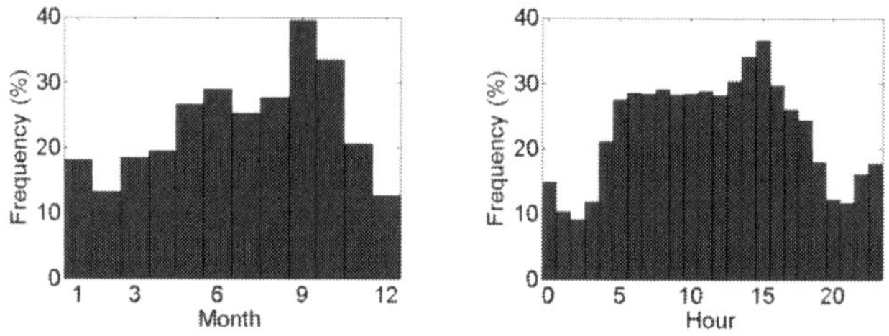

Figure 2: The histogram on the left (a) shows short tonal activity for each month during 2014 with peak detections at the end of the summer. The histogram on the right (b) shows the short tonal activity for each hour of the day with a higher intensity during daylight hours

reason separate classifiers for these classes were defined. Fig. (3a) shows an example of dolphin biosonar, with impulses around 60 kHz while there is another anthropogenic sonar in the image just below 20 kHz. The histogram Fig. (4a) shows the output of the dolphin classifier for each hour during the day. In comparison with the short tonals in Fig. (2b) it is interesting that they peak at different hours, which can be related to the function of the two different signals: biosonar for feeding and whistles for social behaviour.

2.3 Sperm whale clicks

Sperm whale presence was evaluated using a classifier as their biosonar signal often overlaps in frequency with shipping generated impulsive noise. An example of a click train from a single animal is shown in the spectrogram in Fig. (3b). The impulses have energy up to 40 kHz, but this changes depending on the orientation between the animal and the hydrophone. Surface recordings often record off-axis clicks which have less energy at high frequencies [12]. The histogram Fig. (4b) shows the hourly output of the classifier during 2014 indicating more activity during the day. It may be that the whales forage elsewhere during the night. The presence of sperm whales at ANTARES is studied in more detail in a forthcoming paper.

2.4 Shipping activity

The contribution of shipping noise at ANTARES was not found to be very high, except that twice a day a ferry to north Africa would pass by completely saturating the hydrophone acquisition. Due to overlap of other impulsive signals in the shipping band a shipping impulse classifier was trained. Fig. (3c) shows an example of shipping noise. It is clear that low frequency biological sounds are quickly lost in the broadband shipping noise. Higher frequency dolphin biosonar could still be detected in this example. The histogram Fig. (4c) shows the output of the shipping classifier with two peaks in the morning and evening when the ferry passes the platform. Otherwise, most shipping occurred during the day, but for a small percentage of the time.

DOI: 10.1051/epjconf/201713506008

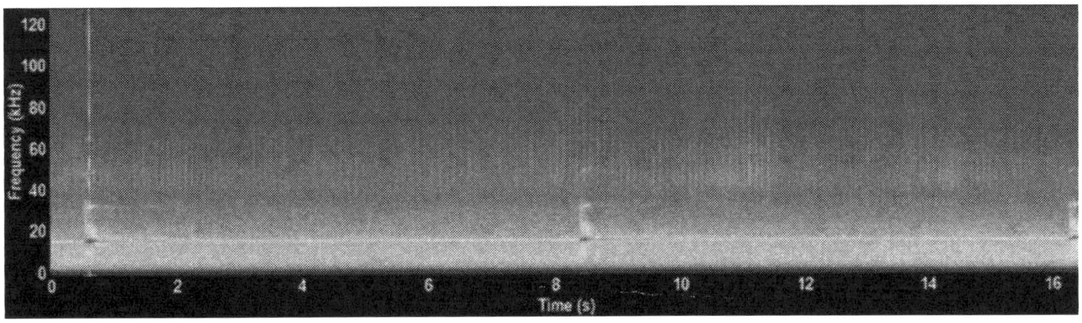

(a)

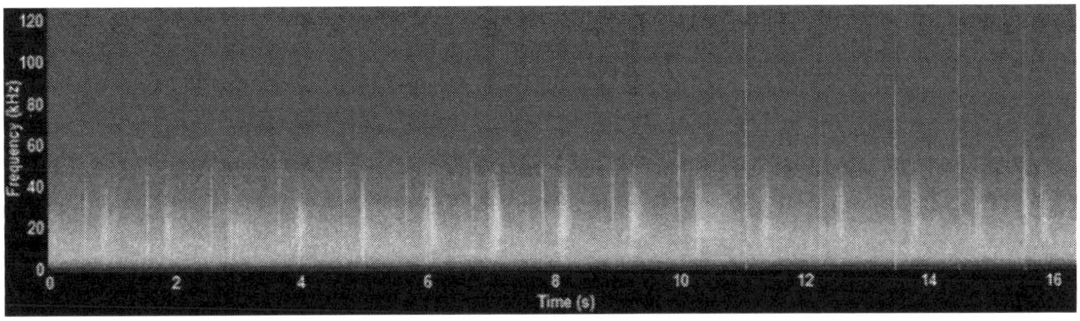

(b)

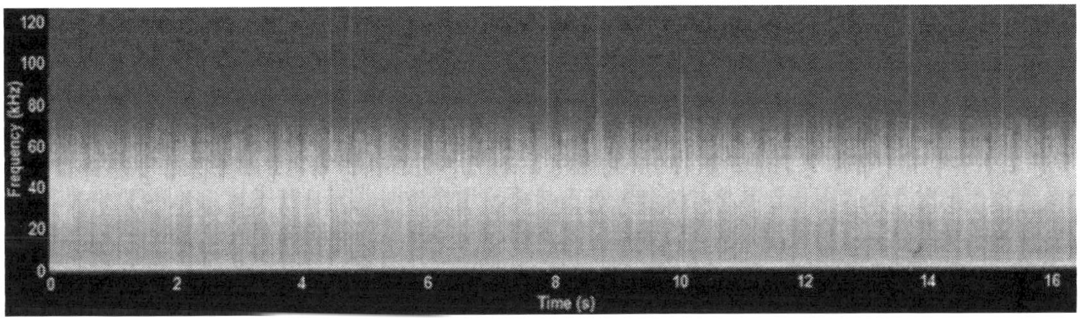

(c)

Figure 3: The top spectrogram (a) is an example of typical high frequency dolphin sonar, visible as the rapid impulses around 60 kHz. There is an unknown sonar signal visible around 20 kHz. The middle spectrogram (b) shows an example of sperm whale biosonar clicks, with slow pulses up to 40 kHz. The bottom spectrogram (c) shows an example of a ship passing nearby, with both very strong broadband noise and impulsive noise.

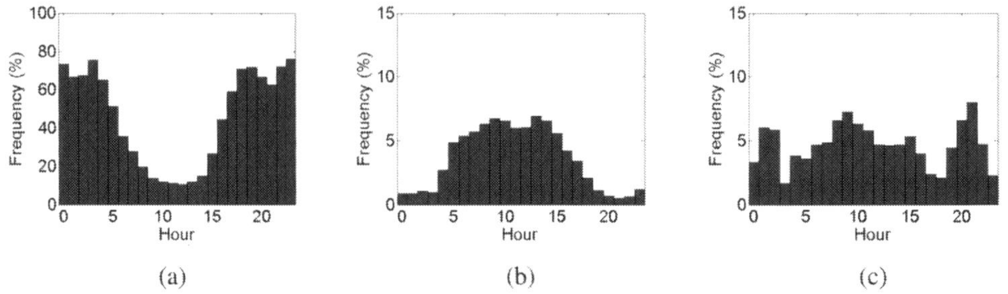

Figure 4: All histograms show classifier output for each hour of the day during 2014. From left to right the classifiers are: high frequency biosonar (a), sperm whale (b), and shipping (c). Biosonar was especially detected during night, while sperm whales appear to be predominately active during the day. Shipping was mostly present during the day, with a ferry passing by each evening and early morning.

3 Discussion

While this paper presented an overview of data analysis of only one year, the strength of platforms such as ANTARES is that data is available during the entire year for many years. This allows establishing baseline levels of cetacean and anthropogenic activities and estimating the common background noise level. Over time, these baseline levels can be tracked and deviations noted. There is especially interest in the increase of background noise levels due to shipping and establishing such a relationship requires many years of data. Even though the position of the hydrophones at ANTARES is not ideal for measuring surface based sources, these deep water platforms can still provide a large amount of unique data. In that sense, it would be advantageous to take into account design requirements of other fields when deploying new monitoring platforms (e.g. KM3). For example, for acoustics it would be preferred not to use hardware high-pass filters and to store or transfer the raw data. This would allow evaluation of the Marine Strategy Framework Directive [13] acoustic criteria.

The results shown here were based on a combination of detectors and classifiers and are considered to be preliminary. In order to verify the monthly or daily patterns such as shown for dolphins between foraging and social behaviour or for sperm whales in their foraging behaviour, a better evaluation of the reliability of the classifiers has to be made. This to ensure that e.g. masking by shipping traffic or anthropogenic impulses misclassified as biosonar are not the reason for reduced or increased detections respectively during a certain time period. This requires manual analysis of the data and analysis results which is currently being done. A more fine-tuned result is expected to be published in 2017.

Acknowledgement

The main author (MvdS) wishes to thank the organisers of the ARENA 2016 workshop for the invitation to give this presentation. He also wishes to thank his ANTARES colleagues for the fruitful interdisciplinary collaboration.

References

[1] J. Aguilar, ANTARES Collaboration, Nuclear Instruments and Methods in Physics Research A **570**, 107 (2007)
[2] J. Aguilar, ANTARES Collaboration, Nuclear Instruments and Methods in Physics Research A **626**, 128 (2011)
[3] M. Heesemann, T. Insua, M. Scherwath, S. Juniper, K. Moran, Oceanography **27**, 151 (2014)
[4] S. Salon, A. Crise, P. Picco, E. de Marinis, O. Gasparini, Annales Geophysicae **21**, 833 (2003)
[5] P. Tyack, M. Johnson, N. Aguilar Soto, A. Sturlese, P. Madsen, The Journal of Experimental Biology **209**, 4238 (2006)
[6] S. Watwood, P. Miller, M. Johnson, P. Madsen, P. Tyack, Journal of Animal Ecology **75**, 814 (2006)
[7] R. Person, L. Beranzoli, C. Berndt, J. Danobitia, M. Diepenbroecke, P. Favali, M. Gillooly, V. Lykousis, J. Miranda, J. Mienert et al., *ESONET: An European Sea Observatory Initiative* (OCEANS 2008 - MTS/IEEE Kobe Techno-Ocean, 2008), pp. 1–6
[8] M. André, M. van der Schaar, S. Zaugg, L. Houégnigan, A. Sánchez, J. Castell, Journal of the Marine Biological Association **87**, 59 (2007)
[9] S. Zaugg, M. van der Schaar, L. Houégnigan, C. Gervaise, A. Sánchez, M. André, Applied Acoustics **71**, 1011 (2010)
[10] S. Zaugg, M. van der Schaar, L. Houégnigan, M. André, Applied Acoustics **73**, 281 (2012)
[11] C. Bishop, *Neural Networks for Pattern Recognition* (Oxford university press, 1995)
[12] B. Møhl, M. Wahlberg, P. Madsen, A. Heerfordt, A. Lund, Journal of the Acoustical Society of America **114**, 1143 (2003)
[13] European parliament and the council of the European Union, Official Journal of the European Union **L164**, 19 (2008)

AUTHOR INDEX

AUTHOR INDEX